Titles in This Series

21 **D. Frank Hsu, Arnold L. Rosenberg, and Dominique Sotteau, Editors,** Interconnection Networks and Mapping and Scheduling Parallel Computations

20 **William Cook, László Lovász, and Paul Seymour, Editors,** Combinatorial Optimization

19 **Ingemar J. Cox, Pierre Hansen, and Bela Julesz, Editors,** Partitioning Data Sets

18 **Guy E. Blelloch, K. Mani Chandy, and Suresh Jagannathan, Editors,** Specification of Parallel Algorithms

17 **Eric Sven Ristad, Editor,** Language Computations

16 **Panos M. Pardalos and Henry Wolkowicz, Editors,** Quadratic Assignment and Related Problems

15 **Nathaniel Dean and Gregory E. Shannon, Editors,** Computational Support for Discrete Mathematics

14 **Robert Calderbank, G. David Forney, Jr., and Nader Moayeri, Editors,** Coding and Quantization: DIMACS/IEEE Workshop

13 **Jin-Yi Cai, Editor,** Advances in Computational Complexity Theory

12 **David S. Johnson and Catherine C. McGeoch, Editors,** Network Flows and Matching: First DIMACS Implementation Challenge

11 **Larry Finkelstein and William M. Kantor, Editors,** Groups and Computation

10 **Joel Friedman, Editor,** Expanding Graphs

9 **William T. Trotter, Editor,** Planar Graphs

8 **Simon Gindikin, Editor,** Mathematical Methods of Analysis of Biopolymer Sequences

7 **Lyle A. McGeoch and Daniel D. Sleator, Editors,** On-Line Algorithms

6 **Jacob E. Goodman, Richard Pollack, and William Steiger, Editors,** Discrete and Computational Geometry: Papers from the DIMACS Special Year

5 **Frank Hwang, Clyde Monma, and Fred Roberts, Editors,** Reliability of Computer and Communication Networks

4 **Peter Gritzmann and Bernd Sturmfels, Editors,** Applied Geometry and Discrete Mathematics, The Victor Klee Festschrift

3 **E. M. Clarke and R. P. Kurshan, Editors,** Computer-Aided Verification '90

2 **Joan Feigenbaum and Michael Merritt, Editors,** Distributed Computing and Cryptography

1 **William Cook and Paul D. Seymour, Editors,** Polyhedral Combinatorics

DIMACS

Series in Discrete Mathematics
and Theoretical Computer Science

Volume 21

Interconnection Networks and Mapping and Scheduling Parallel Computations

DIMACS Workshop
February 7–9, 1994

D. Frank Hsu
Arnold L. Rosenberg
Dominique Sotteau
Editors

NSF Science and Technology Center
in Discrete Mathematics and Theoretical Computer Science
A consortium of Rutgers University, Princeton University,
AT&T Bell Labs, Bellcore

American Mathematical Society

This DIMACS volume on Interconnection Networks and Mapping and Scheduling Parallel Computations contains refereed versions of papers from a workshop held at Rutgers University from February 7–9, 1994.

1991 *Mathematics Subject Classification.* Primary 68–06, 94–06;
Secondary 68Q10, 78–08, 94C99.

Library of Congress Cataloging-in-Publication Data

Interconnection networks and mapping and scheduling parallel computations : DIMACS workshop, February 7–9, 1994 / D. Frank Hsu, Arnold L. Rosenberg, Dominique Sotteau, editors.
p. cm. — (DIMACS series in discrete mathematics and theoretical computer science, ISSN 1052-1798; v. 21)
"NSF Science and Technology Center in Discrete Mathematics and Theoretical Computer Science, a consortium of Rutgers University, Princeton University, AT&T Bell Labs, Bellcore."
Includes bibliographical references.
ISBN 0-8218-0238-0
1. Parallel processing (Electronic computers)—Congresses. 2. Computer networks—Congresses. I. Hsu, D. Frank (Derbiau Frank), 1948– . II. Rosenberg, Arnold L., 1941– . III. Sotteau, Dominique, 1950– . IV. NSF Science and Technology Center in Discrete Mathematics and Theoretical Computer Science. V. DIMACS (Group) VI. Series.
QA76.58.I494 1995
004′.35—dc20

95-14029
CIP

The paper used in this book is acid-free and falls within the guidelines
established to ensure permanence and durability.
Printed on recycled paper.

This volume was prepared by the authors using AMS-TEX and AMS-LATEX,
the American Mathematical Society's TEX macro systems.

10 9 8 7 6 5 4 3 2 1   00 99 98 97 96 95

Contents

Foreword

This DIMACS volume on Interconnection Networks and Mapping and Scheduling Parallel Computations contains refereed versions of papers from a workshop held at Rutgers in February of 1994. The Workshop was one of eight workshops sponsored by DIMACS as part of a Special Year in Massively Parallel Computing.

We wish to thank in particular the Special Year organizers, Tom Leighton and Bruce Maggs; and the Workshop organizers, Frank Hsu, Arnold Rosenberg, and Dominique Sotteau, for their considerable efforts to bring the workshops and this volume to fruition.

Andras Hajnal
Director, DIMACS

Andrew Yao
Co-Director, DIMACS

Stephen R. Mahaney
Associate Director, DIMACS

Preface

The interconnection network is one of the most basic components of a massively parallel computer system. Each system consists of hundreds or thousands of processors, interconnected to work cooperatively on computations. One of the central problems in parallel computing is the task of mapping a collection of processes onto the processors and routing network of a parallel machine. Once processors are mapped onto the physical processors, it is critical to schedule computation within and communication among processors so that the necessary inputs for a process would be available where and when the process is scheduled to be computed.

In the context of the 1993–94 DIMACS Special Year on Massively Parallel Computation, funded jointly by the National Science Foundation and the New Jersey Commission for Science and Technology, a three-day workshop on Interconnection Networks and Mapping and Scheduling Parallel Computations was held on February 7–9, 1994. The Workshop focused on the interconnection networks of parallel architectures of today and of the near future, examining the task of mapping and scheduling parallel computations on these parallel machines. Subject areas covered included network topologies, network properties, message routing, network embeddings, network emulation, mappings, and efficient scheduling.

The Workshop was held at DIMACS at Rutgers University, Piscataway, New Jersey. DIMACS is the National Science Foundation science and technology center for discrete mathematics and computer science. It is a consortium of Rutgers and Princeton Universities, AT&T Bell Laboratories, and Bellcore.

There were 49 presentations in the 12 scheduled sessions of the Workshop. Over 115 scientists, mathematicians, and engineers from inside and outside DIMACS, and from 14 countries attended the Workshop. Participants included researchers from universities and research laboratories, as well as practitioners who are involved with the design, implementation, and application of massively parallel computer systems.

All speakers were invited to submit their complete papers following talks given at the Workshop. All manuscripts have been refereed according to the standard of a high-quality journal. This volume contains 21 accepted papers. All of the papers are in final form. A DIMACS technical report, which contains titles and abstracts for the talks and the program and schedule of the Workshop, is available upon request.

We would like to thank the DIMACS Executive Committee and the Special Year Organizing Committee for sponsoring the Workshop. Diane Souvaine, Pat Toci, and Maryann Holtsclaw all provided great help. Tom Leighton and Bruce Maggs were of enormous help in the process of organizing the Workshop. We also want to thank Christine Thivierge, Donna Harmon, Shirley Hill, and Jennifer Sharp at the AMS for helping to prepare this volume.

D. Frank Hsu, New York, U.S.A.
hsu@murray.fordham.edu

Arnold L. Rosenberg, Amherst, MA, U.S.A.
rsnbrg@cs.umass.edu

Dominique Sotteau, Orsay, France
sotteau@lri.lri.fr

January 1995

DIMACS Series in Discrete Mathematics
and Theoretical Computer Science
Volume **21**, 1995

RANKING ALGORITHMS FOR HAMILTONIAN PATHS IN HYPERCUBIC NETWORKS

FRED S. ANNEXSTEIN

ABSTRACT. Given a labeled set that is linearly ordered, a ranking algorithm returns the rank-position of an element in the linear order when input with the label of that element. In this paper we provide ranking (and unranking) algorithms for certain classes of graphs where the linear order on the vertex-set of a graph is determined by a Hamiltonian path. The classes of graphs we consider include the Hypercube, the De Bruijn, and the Butterfly, the so-called hypercubic networks. These graphs are widely recognized as important interconnection topologies for parallel computations. Our ranking and unranking algorithms can be applied to yield efficient implementations of certain network emulations using SIMD-style parallel algorithms for translating node labels.

1. Introduction

One unavoidable step in the implementation of parallel algorithms on interconnection networks is the generation of software that "reconfigures" the physical architecture, given the algorithm's specified logical interconnection topology. Reconfigurations must be programmed efficiently or they may dominate the costs of a parallel computation.

In this paper we focus on the problem of reconfigurations involving the use of long embedded paths (in particular, Hamiltonian paths) in hypercubic networks. The existence of Hamiltonian paths in the Hypercube, the De Bruijn, and the Butterfly is well known. However, the existence of such paths does not imply that they may be used effectively in parallel computations on networks. Indeed, embedding algorithms for Hamiltonian paths are sometimes specified sequentially and thus run in time linear (or worse) in the size of the networks. This may not be useful in practice, especially when implementing sub-linear time algorithms.

Paths can be embedded efficiently in parallel networks by developing fast *ranking* and *unranking* algorithms. The existence of such algorithms constitutes an efficient parallel implementation of a graph embedding. In a previous paper [1] the author provided a general framework for such embeddings. The basic problem in providing parallel implementations of graph embeddings is the following: given a logical graph structure G, and an algorithm for embedding G into a physical network H, we seek an efficient method for translating between the labels of G-nodes and H-nodes that respects the embedding.

1991 *Mathematics Subject Classification.* 68Q22, 68Q25, 68R10, 05C38.
Key words and phrases. Hypercubic networks, De Bruijn graphs, parallel architectures.
The author was supported in part by NSF Grant CCR–93–09470

Specifically, the problem we consider in this paper is as follows. Given an n-node labeled graph, find a Hamiltonian path along with associated ranking and unranking algorithms. A ranking algorithm takes as input the label of a node and returns the rank r $(0 \leq r \leq n-1)$ of that node in the Hamiltonian path. An unranking algorithm takes as input an integer r $(0 \leq r \leq n-1)$ and returns the label of the rth-ranked node in the Hamiltonian path. The ranking and unranking algorithms we describe all run in (sequential) time polynomial in the length of the node labels. When such algorithms are applied to embed paths in a network, they run as SIMD-style programs in parallel-time polylogarithmic in the size of the network. This time bound is exponentially faster than implementations that rely on an explicit listing of all the nodes in the path.

2. The Hypercube

In this section we review a well-known recursive construction for a Hamiltonian path in a Hypercube. This construction leads to very natural and efficient ranking and unranking algorithms.

The *order-n Hypercube* has 2^n nodes labeled by distinct n-bit strings. Pairs of nodes are adjacent if their labels differ in precisely one bit position. Each edge is associated with one of n dimensions: An edge of dimension $0 \leq i \leq n-1$ connects a node $w = a_{n-1} \cdots a_0$ to the node $w^i = a_{n-1} \cdots \bar{a}_i \cdots a_0$, which denotes the string w with the ith bit changed.

One Hamiltonian path in a Hypercube is known as the *(binary reflected) Gray code*. This sequence of nodes (bit-string labels) is named after Frank Gray who used it in the design of an A/D conversion circuit [5]. However, this sequence was recognized in the last century as a solution method to an old puzzle known as the *Chinese Rings*; see [2, 3] for discussions of this connection.

The Gray code sequence can be recursively described using a method we call *recursive reversing*. Recursive reversing is a powerful technique that has been used to generate "Gray-like" lists of various combinatorial objects [8, 9]. In this work, we use the technique to generate lists of labels that have simple and efficient ranking algorithms. Roughly speaking, recursive reversing is a method in which we build a list of order-$(n+1)$ via the concatenation of the list of order-n followed by the list of order-n in reverse order. Sometimes it is required to make small modifications in the lists to make the constructions work properly.

The Gray code sequence is defined via recursive reversing in just this way. Let $\mathcal{G}_n$ be the list of all 2^n n-bit node-labels in the binary reflected Gray code of order-n. Then $\mathcal{G}_{n+1}$ is defined as $\mathcal{G}_n$ with a zero prefixing each string, followed by the reverse of $\mathcal{G}_n$ with a one prefixing each string. For example, see Figure 1 for the construction of $\mathcal{G}_4$ from $\mathcal{G}_3$. The following proposition gives a simple ranking algorithm for Gray codes.

Proposition 2.1. *(Ranking Algorithm) Given a node-label $w = w_{n-1} \cdots w_0$, the Rank(w) in the list $\mathcal{G}_n$ can be expressed in binary as $r_{n-1} \cdots r_0$, where $r_\ell = w_{n-1} + w_{n-2} + \cdots + w_\ell \pmod 2$, for each $0 \leq \ell \leq n-1$. In other words, the rank number (in binary) is an xor-scan of the node-label.*

Proof. The proof follows by induction on n. The proposition clearly holds for the case $n = 1$. Now suppose the proposition holds for $\mathcal{G}_n$. Let $w = w_n \cdots w_0$. First,

Rank	$\mathcal{G}_3$	*Rank*	$\mathcal{G}_4$	*Rank*	$\mathcal{G}_4$
0	000	0	0000	15	1000
1	001	1	0001	14	1001
2	011	2	0011	13	1011
3	010	3	0010	12	1010
4	110	4	0110	11	1110
5	111	5	0111	10	1111
6	101	6	0101	9	1101
7	100	7	0100	8	1100

FIGURE 1. Illustration of the recursive construction of Gray codes.

suppose $w_n = 0$. The formula above states that $Rank(w) = Rank(w_{n-1} \cdots w_0)$. Hence, using the inductive hypothesis, for each $0 \le \ell \le n-1$, $r_\ell = 0 + w_{n-1} + \cdots + w_\ell \ (mod\ 2)$, and so the statement of the proposition is satisfied for the case $w_n = 0$. Now suppose $w_n = 1$. Then it follows from the construction that $Rank(w) = 2^{n+1} - 1 - r$, where $r = Rank(w_{n-1} \cdots w_0)$. This equation simply corresponds to the bit-complement of the binary representation of r (with a one as prefix). Hence, using the inductive hypothesis, for each $0 \le \ell \le n$, $r_\ell = 1 + w_{n-1} + \cdots + w_\ell \ (mod\ 2)$, and so the statement of the proposition is satisfied. □

Proposition 2.2. *(Unranking Algorithm) Suppose the node-labeled* $w = w_{n-1} \cdots w_0$ *has rank* r *in the list* $\mathcal{G}_n$*, where in binary* $r = r_{n-1} \cdots r_0$*. Then, for each* $0 \le \ell \le n-1$, $w_\ell = r_{\ell+1} + r_\ell \ (mod\ 2)$*. (Note we define* $r_n = 0$*.)*

Proof. Follows immediately from the previous proposition by cancellation. □

The time complexity of both the ranking and unranking algorithms is easily seen to be proportional to the length of the labels $O(n)$.

3. The De Bruijn Graph

In this section we derive ranking and unranking algorithms for a Hamiltonian path on the De Bruijn graph which is constructed in manner similar to the Gray code construction of Section 2—using the technique of recursive reversing.

The 2^n-node *De Bruijn graph* $\mathcal{D}_n$ is the digraph whose node-set is Z_2^n. The arcs of $\mathcal{D}_n$ are labeled by either π_S or π_E as follows: Given $a \in Z_2$ and $w \in Z_2^{n-1}$, there is an arc $aw \to wa$ labeled by π_S, and an arc $aw \to w\bar{a}$ labeled by π_E. Our interest is in the undirected version of $\mathcal{D}_n$, hence, we will make use of the oppositional arcs $\pi_{S^{-1}}$ and $\pi_{E^{-1}}$. We call the arcs labeled by π_S, π_E the *positive-directed* arcs, and the arcs labeled by $\pi_{S^{-1}}, \pi_{E^{-1}}$ the *negative-directed* arcs.

The particular method we use to construct Hamiltonian paths in De Bruijn graphs is reminiscent of an inductive method of generating *De Bruijn sequences* found in [7].

Consider the set of all n-bit strings B_n, and let the mapping $L : B_n \to B_{n-1}$ be defined by $L(w_{n-1} \cdots w_0) = (w_{n-1} + w_{n-2}) \cdots (w_2 + w_1)(w_1 + w_0)$. It is easily verified that L is a 2-to-1 onto mapping that induces a graph homomorphism of De Bruijn graphs $\mathcal{D}_n \to \mathcal{D}_{n-1}$. Except for the self-loops, each arc in $\mathcal{D}_{n-1}$ is the image under L of a pair of node disjoint arcs in $\mathcal{D}_n$. Hence, by induction it follows that any simple path in $\mathcal{D}_{n-1}$ is the image under L of a pair of node disjoint paths. Furthermore, for each i, the ith node in one of the paths has a node-label that is the bit-complement of the ith node in the other path. This follows since such bit-complementary pairs have the same image under L. We now use these facts about the mapping L to describe an inductive construction of a Hamiltonian path P_n in $\mathcal{D}_n$.

For the inductive hypothesis, let P_{n-1} be a Hamiltonian path in $\mathcal{D}_{n-1}$ that begins with the node labeled all zeros 00...0, and terminates with the node labeled all ones 11...1. Then P_{n-1} is the image under L of a pair of node disjoint paths P^0 and P^1 in $\mathcal{D}_n$ such that P^0 begins at the node labeled 00...0, and P^1 begins at node labeled 11...1. Each of P^0 and P^1 terminates at one of the two nodes labeled with the "alternating" bit-vectors ($10101\ldots$ or $01010\ldots$). Let $P_n = P^0 + (P^1)^{rev}$, that is, P_n is the concatenated sequence of node-labels that begins with the path P^0 and is followed by the path P^1 in the reverse order. Note that the sub-paths P^0 and $(P^1)^{rev}$ are adjacent since the two nodes labeled with the alternating bit-vectors are adjacent. Hence, P_n defines a Hamiltonian path in $\mathcal{D}_n$.

A ranking algorithm for the Hamiltonian path P_n can be described as follows. Given $w = w_{n-1}\cdots w_0$, recursively compute $Rank(L(w)) = r$. Then, it follows from the construction that

$$Rank(w) = \begin{cases} r & \text{if } w \in P^0 \\ 2^n - 1 - r & \text{if } w \in P^1 \end{cases}$$

Hence, to determine the rank, we are left with the problem of overcoming the uncertainty of whether $w \in P^0$ or $w \in P^1$. We show that we can solve this problem by keeping track of the amount of cyclic shifting (positive or negative directed) of the node-label taking place at each step in the Hamiltonian path. Such shifting is "preserved" under L, that is to say positive-directed arcs are mapped by L to positive-directed arcs, and negative-directed arcs are mapped by L to negative-directed arcs.

Let $ShiftAmt(n,i)$ denote the sum-total amount of cyclic shifting performed to reach the rank i node in P_n starting from the node $00\ldots0$. A recursive formula for $ShiftAmt(n,i)$ can be defined as follows.

Proposition 3.1. *$ShiftAmt(0,0) = 0$, and for $n \geq 1$, $0 \leq i < 2^n$*

$$ShiftAmt(n,i) = \begin{cases} ShiftAmt(n-1,i) & \text{if } i < 2^{n-1} \\ 1 + ShiftAmt(n-1, 2^n - 1 - i) & \text{if } i \geq 2^{n-1} \end{cases}$$

Proof. The validity of the recursive formula for $ShiftAmt(n,i)$ follows by induction on n. We show that for all valid i, the correctness of the formula for $ShiftAmt(n-1,i)$ implies the correctness of the formula for $ShiftAmt(n,i)$ and $ShiftAmt(n,2^n-1-i)$.

In the construction of the path P_n, the cyclic direction of each step in the path's first half P^0 is the same as the corresponding step in the path P_{n-1}, since L preserves the direction of the arcs. Hence, the sum-total shifting after $i < 2^{n-1}$ steps $ShiftAmt(n,i)$ is equal to the value $ShiftAmt(n-1,i)$.

In the construction of the path P_n, the cyclic direction of each step in the path's second half $(P^1)^{rev}$ is the reverse of the corresponding step in the path P_{n-1}. Hence, this portion of the path simply "undoes" the shifting accomplished by P^0. Thus, the sum-total shifting after $i \geq 2^{n-1}$ steps $ShiftAmt(n,i)$ is equal to the value

$ShiftAmt(n-1, 2^n-1-i)$ plus the amount of cyclic shifting used to bridge from P^0 to P^1, which can be chosen to be either $+1$ or -1; we choose it to be $+1$. The proof follows by induction. □

From the formula above it is apparent (by induction) that for each $n \geq 0$ and all $0 \leq i \leq 2^n - 1$ that $0 \leq ShiftAmt(n,i) \leq n$. Hence, during the first half of the Hamiltonian path (i.e., the list of node-labels in P^0) the original least significant bit (a zero) can never be changed. This is due to the fact that the cyclic shifting never moves it past the most significant bit position, nor past the least significant bit position, and it is therefore left unadulterated. Hence, if the node-label $w = w_{n-1} \cdots w_0$ is part of P^0 then the $ShiftAmt(n,r)$-bit of w is zero, where $r = Rank(L(w))$; likewise, if w is part of P^1 then the $ShiftAmt(n,r)$-bit of w is one, since w is the complementary string to a string in P^0 whose $ShiftAmt(n,r)$-bit is zero. This is the key observation sufficient to eliminate the uncertainty of whether a particular node w lies in the P^0 or the P^1 portion of the Hamiltonian path P_n. We can now refine our ranking algorithm for the Hamiltonian path P_n as follows.

Proposition 3.2. *(Ranking Algorithm) Given $w = w_{n-1} \cdots w_0$, we compute the rank of w as follows. First, recursively compute $Rank(L(w)) = r$. Second, compute $ShiftAmt(n,r) = k$ (via the recursive formula of Proposition 3.1). Then,*

$$Rank(w) = \begin{cases} r & \text{if } w_k = 0 \\ 2^n - 1 - r & \text{if } w_k = 1 \end{cases}$$

We can invert this procedure to obtain an unranking algorithm as follows. Let an integer $0 \leq r \leq 2^n - 1$ represent the rank of a node in the path P_n. If $r < 2^{n-1}$, then recursively compute the node-label of the rank-r node in P_{n-1}; otherwise, if $r \geq 2^{n-1}$, recursively compute the node-label of the rank-$(2^n - 1 - r)$ node in P_{n-1}—in either case, call this $(n-1)$-bit node-label w. Next, determine the two n-bit strings w^0 and w^1, such that $L(w^0) = L(w^1) = w$. If $r < 2^{n-1}$ and the $ShiftAmt(n,r)$-bit of w^0 is 0, or if $2^{n-1} \leq r \leq 2^n - 1$ and the $ShiftAmt(n, 2^n - 1 - r)$-bit of w^0 is 1, then w^0 is the label of the rank-r node in P_n. Otherwise, w^1 is the label of the rank-r node in P_n.

Both the ranking and unranking algorithms have a time complexity bounded by the recurrence relation $T(n) = T(n-1) + O(n)$. This follows since, for any r, the time to compute $ShiftAmt(n,r)$ is $O(n)$. In fact, the reader should note that the function $ShiftAmt()$ can be efficiently computed in a manner quite similar to the xor-scan method used in Proposition 2.1 for the ranking algorithm for the Gray code. It follows that $T(n) = O(n^2)$.

4. The Butterfly

In this section we derive ranking and unranking algorithms for a Hamiltonian path of the Butterfly. Here too, the construction of the path is recursive, however, no reversing is required.

The *order-n Butterfly graph* $\mathcal{B}(n)$ has node-set $V_{\mathcal{B}} = Z_n \times Z_2^n$. The subset $V^{(\ell)} = \{\ell\} \times Z_2^n$ of $V_{\mathcal{B}}$ $(0 \leq \ell < n)$ is called the *level-ℓ* of $\mathcal{B}(n)$; the string $w \in Z_2^n$ is the *column* string of each node in $Z_n \times \{w\}$. The edges of $\mathcal{B}(n)$ form *butterflies* (i.e., copies of the complete bipartite graph $K_{2,2}$) between consecutive levels of vertices, with wraparound in the sense that level 0 is identified with level n. The edges of the Butterfly connect each node $v = \langle \ell;\ a_{n-1} \cdots a_\ell \cdots a_0 \rangle$ on level ℓ of $\mathcal{B}(n)$ $(0 \le \ell \le n-1)$ via a *straight-edge* with the node $\langle \ell + 1(mod\ n);\ a_{n-1} \cdots a_\ell \cdots a_0 \rangle$, and v is connected via a *cross-edge* with the node $\langle \ell + 1(mod\ n);\ a_{n-1} \cdots \bar{a}_\ell \cdots a_0 \rangle$. We call these edges the level-ℓ edges of $\mathcal{B}(n)$.

There is a well-known inductive construction for a Hamiltonian path in a Butterfly graph (see [6]). We review this construction now. For the inductive hypothesis, assume that $\mathcal{B}(n)$ has a Hamiltonian path P_n that begins at $\langle 0;\ 0\ 0\ \ldots\ 0 \rangle$ and ends at $\langle n-1;\ 1\ 0\ \ldots\ 0 \rangle$; in fact, finishing with the cross-edge forms a Hamiltonian cycle. Now consider $\mathcal{B}(n+1)$, which we can view as a pair of order-n Butterflies (a left one and a right one) in which level-n nodes and edges have been added. Using the inductive hypothesis, we start with the sequence of edges defined by P_n, and then splice in level-$(n-1)$ straight edges to form a path in $\mathcal{B}(n+1)$ that begins at $\langle 0;\ 0\ 0\ \ldots\ 0 \rangle$ and ends at $\langle n;\ 0\ \ldots\ 0 \rangle$ that visits every node in the left sub-Butterfly. A cross-edge joins $\langle n;\ 0\ \ldots\ 0 \rangle$ in the left sub-Butterfly to $\langle 0;\ 1\ 0\ \ldots\ 0 \rangle$ in the right sub-Butterfly. Now, following the same sequence of edges just defined (for the left side), our path then visits the entire right sub-Butterfly, and terminates at the node $\langle n;\ 1\ 0\ \ldots\ 0 \rangle$; thus completing the Hamiltonian path P_{n+1}. A base case is easily verified, and the proof of Hamiltonicity follows by induction. The next proposition follows as a simple consequence of this construction.

Proposition 4.1. *In the path P_n, a cross-edge is used when leaving from a level-i node and enters into a level-$i+1$ $(mod\ n)$ node if and only if $i = 0$ or the rightmost $i > 0$ bits of the column string are all zeros.* □

We provide a ranking algorithm for P_n by first showing how to rank all nodes on level 0. Note that in the inductive construction when the path P_n enters a new, previously unvisited column, it is always the column immediately following the set of visited columns; hence, columns are visited in the same order as the integer values of their column strings. Also, once a new column is entered, P_n always visits the node in level-0 before leaving that column. Successive level-0 nodes are thus always reached after every n steps. Therefore, we have the following ranking function for the level-0 nodes:

$$Rank(\langle 0; a_{n-1}a_{n-2}\ldots a_0 \rangle) = n \cdot Int(a_{n-1}a_{n-2}\ldots a_0),$$

where the function $Int()$ returns the integer value of binary number.

To compute the rank of the remaining nodes it is a simple matter to determine the column-string of the level-0 node that is the nearest in the path P_n. This can be accomplished by repeated application of Proposition 4.1. To rank the node, simply add the rank of the nearest level-0 predecessor to the level number of the node in question, or equivalently, we can subtract ($n-$*level number*) from the rank of the nearest level-0 successor. For example, to rank the node $w = \langle 2;\ 1\ 1\ 0 \rangle$ in the path P_3 in the graph $\mathcal{B}(3)$, we first determine (by applying Proposition 4.1) that the predecessor of w is $v = \langle 1;\ 1\ 0\ 0 \rangle$, and the predecessor of w is $u = \langle 0;\ 1\ 0\ 1 \rangle$. Now $Rank(u) = 3 \cdot Int(101) = 15$. Adding the level number we have that the $Rank(w) = 15 + 2 = 17$. Equivalently, we find (by applying Proposition 4.1) that the nearest level-0 successor to w is $x = \langle 0;\ 1\ 1\ 0 \rangle$; hence, subtracting $(3-2) = 1$ from the $Rank(x) = 3 \cdot Int(110)$, we also get 17.

The algorithm to unrank is equally easy. Given an integer k, we want to determine the label of the node of rank k in P_n. First, compute the binary representation of $k \, div \, n$, for this is the column string of the level-0 node that is the nearest predecessor of the node we want. Now by repeatedly applying Proposition 4.1 we arrive at the node-label associated with rank k which lies in level $k \, mod \, n$ of the Butterfly.

In determining the time complexity for the ranking algorithm on P_n, we note that it takes time $O(n)$ to traverse the column-string to determine the path back to the nearest level-0 predecessor, or the path forward to the nearest level-0 successor. Assuming multiplication of an n-bit number by a $\log n$-bit number can be computed in $O(n)$ time, it follows that the time complexity for the ranking algorithm for P_n on the Butterfly is $O(n)$. The time complexity for the unranking is seen to be $O(n)$, assuming that is time sufficient to do integer division (of an n-bit number by a $\log n$-bit number) plus a traversal of the bit-string to facilitate application of Proposition 4.1.

5. Conclusion

We have described efficient ranking and unranking algorithms for Hamiltonian paths on certain classes of hypercubic networks, including the Hypercube, the De Bruijn, and the Butterfly networks. Using similar techniques to those described in the paper, one can find ranking and unranking algorithms for other hypercubic networks, e.g. the Cube-Connected-Cycles. Recently, it has been shown that there is a Hamiltonian path in the shuffle-exchange graph [4]. It is an interesting open question whether an efficient ranking algorithm can be derived for this class of graphs. Another open problem is to find efficient ranking and unranking algorithms for De Bruijn sequences, i.e., Hamiltonian circuits in De Bruijn digraphs.

References

1. F. Annexstein, *Parallel implementations of graph embeddings*, in Parallel Architectures and Their Efficient Use, Lecture Notes in Computer Science 678, Springer Verlag, 1993, 207–217.
2. W. W. Rouse Ball and H. S.M. Coexter, Mathematical Recreations and Essays, University of Toronto Press, 1972.
3. E. R. Berlekamp, J. H. Conway, and R. K. Guy, *Winning Ways* (vol. 2), Academic Press, 1982.
4. R. Feldmann and P. Mysliwietz, *The shuffle exchange network has a Hamiltonian path*, Proc. of 17th Math Foundations of Computer Science (MFCS '92), Lecture Notes in Computer Science 629, Springer Verlag, 1992, 246–254.
5. F. Gray, *Pulse code communication*, 1953, U.S. Patent Number 2,632,058.
6. F. T. Leighton, *Introduction to Parallel Algorithms and Architectures*, Morgan Kaufmann, San Mateo, 1992.
7. A. Lempel, *On a homomorphism of the De Bruijn graph and its applications to the design of feedback shift registers*, IEEE Trans. Comput. C-19, (1970), 1204–1209.
8. F. Ruskey, *Simple combinatorial gray codes constructed by reversing sublists*, in Algorithms and Computation, Lecture Notes in Computer Science 762, Springer Verlag, 1993.
9. H. S. Wilf, *Combinatorial algorithms: an update*, CBMS-55, SIAM, 1989.

Department of Electrical and Computer Engineering and Computer Science, University of Cincinnati, Cincinnati, Ohio 45221

E-mail address: fred.annexstein@uc.edu

DIMACS Series in Discrete Mathematics
and Theoretical Computer Science
Volume **21**, 1995

Dense Bus Networks of Diameter 2

J.-C. BERMOND, J. BOND AND S. DJELLOUL

ABSTRACT. In this paper we consider networks in which processors share a communication medium called a bus. These networks are called bus networks and are modeled by hypergraphs. The vertices of the hypergraph represent the processors of the network and edges represent buses. We give some techniques used to design dense bus interconnection networks of given maximum degree Δ, diameter D and maximum bus size r. In particular we construct a new family of bus interconnection networks of diameter 2 having a large number of processors.

1. Introduction

A bus interconnection network is a collection of processing elements (processors) and communication elements (buses). The processors produce and/or consume messages and the buses provide communication channels to exchange messages among the processors. Every bus provides a communication link between two or more processors.

For practical reasons, a processor may only be connected to a limited number of buses (this number is known as the processor degree) and a bus may only connect a limited number of processors (this number is known as the bus size). Therefore, messages may have to be relayed by a number of intermediate processors before arriving at their destinations, and thus the message transmission time becomes a function of the distance (measured in terms of the number of buses traversed by a message) between processors. The maximum distance over all pairs of processors is the network diameter.

1991 *Mathematics Subject Classification.* Primary 68M10; Secondary 05B30, 68R05, 94C30.

The three authors where supported by the French GDR-PRC C^3.

The second author was supported by the French Israeli Cooperation.

The third author was supported by the Algerian government.

For given upper bounds on the processor degree Δ, bus size r, and network diameter D, the construction of bus networks with maximal number of processors is an important problem in the design of interconnection networks. We refer to the survey [**10**] for a state of art on this problem. Other design parameters such as network reliability, symmetry properties, ease of message routing, balanced message traffic throughout the network, implementation issues (algorithms and architecture) should also be taken into consideration.

Our aim is to give some new techniques to design dense bus connection networks of given maximum degree Δ, diameter D and maximum bus size r. In particular we construct a new family of bus connection networks of diameter 2 having a large number of processors.

This paper is organized as follows : In section 2 we give the necessary definitions and notation both from hypergraph theory and from design theory. We also state the (Δ, D, r)-hypergraph problem and give a Moore like upper bound. In section 3 we recall some earlier results obtained in the case of diameter 2. In section 4 we give our main theorem and deduce as corollary some lower bounds and construction of dense bus connection networks of diameter 2.

2. Definitions and notation

2.1. (Δ, D, r)-hypergraph problem. An *(undirected) hypergraph* H is a pair $H = V(H), E(H)$ where $V(H)$ is a non-empty set of elements, called *vertices*, and $E(H)$ is a finite set of subsets of $V(H)$ called *edges*. The number of vertices in the hypergraph is $n(H) = |V(H)|$ and the number of edges is $m(H) = |E(H)|$ where the vertical bars denote the cardinality of the set. The *degree* of a vertex v is the number of edges containing it and is denoted by $d_H(v)$. The *maximum degree* over all of the vertices in H is denoted by $\Delta(H)$. The *size* of an edge $E \in E(H)$ is its cardinality, and is denoted by $|E|$. The *rank* of H is the size of its largest edge, and is denoted by $r(H)$. A *path* in H from vertex u to vertex v is an alternating sequence of vertices and edges $u = v_0, E_1, v_1, \cdots, E_k, v_k = v$ such that $v_{i-1}, v_i \subseteq E_i$ for all $1 \leq i \leq k$. The *length* of a path is the number of edges in it. The *distance* between two vertices u and v is the length of a shortest path between them. The *diameter* of H is the maximum of the distances over all pairs of vertices, and is denoted by $D(H)$.

We call a hypergraph with maximum degree Δ, diameter D, and rank r, a (Δ, D, r)-*hypergraph*. The problem on bus networks we considered in the introduction is known as the (Δ, D, r)-*hypergraph problem* and consists of finding (Δ, D, r)-hypergraphs with the maximum number of vertices or finding large (Δ, D, r)-hypergraphs. We will denote by $n(\Delta, D, r)$ the maximum number of vertices of any (Δ, D, r)-hypergraph.

In the case $r = 2$ (graph case), this problem has been extensively studied and is known as the (Δ, D)-graph problem (see for example [**8**], [**9**]), and the maximum number of vertices in any (Δ, D)-graph is denoted by $n(\Delta, D)$.

Note that parts of this problem have been studied in other contexts with different notation. For example d or r is used for maximum degree, k or d is used for diameter, and b or k is used for rank. (In the notation of Design Theory r and k are used for maximum degree and rank, respectively.) We follow the notation of Hypergraph Theory [**3**].

Finally, let us mention that the drawing of hypergraphs can be very complex and therefore it is useful to represent a hypergraph H with a bipartite graph,

$$R(H) = (V_1(R) \cup V_2(R), E(R))$$

called the *bipartite representation graph.* Every vertex v_i in $V(H)$ is represented by a vertex v_i in $V_1(R)$ and every edge E_j in $E(H)$ is represented by a vertex e_j in $V_2(R)$. We draw an edge between $v_i \in V_1(R)$ and $e_j \in V_2(R)$ if and only if $v_i \in E_j$ in H.

If H is a (Δ, D, r)-hypergraph and $R(H)$ is its bipartite representation graph, then the maximum degrees in $V_1(R)$ and in $V_2(R)$ are Δ and r, respectively. The distance between two vertices of $V_1(R)$ is at most $2D$, but the diameter of $R(H)$ can be $2D$, $2D+1$ or $2D+2$ as the vertices of $V_1(R)$ and $V_2(R)$ do not play the same role. So, the (Δ, D, r)-hypergraph problem is partly related but different from the $(\Delta_1, \Delta_2; D')$-bipartite graph problem, i.e. finding large bipartite graphs with maximum vertex degrees Δ_1, Δ_2 and diameter D' (for details of this problem see [**13**]).

2.2. Duality tools. The *dual* of a hypergraph $H = (V(H), E(H))$ is the hypergraph $H^* = (V(H^*), E(H^*))$ where the vertices of H^* correspond to the edges of H, and the edges of H^* correspond to the vertices of H. A vertex e_j^* is a member of an edge V_i^* in H^* if and only if the corresponding vertex v_i is a member of E_j in H.

Bermond, Bond and Peyrat [**4**] observed the following relationship between a hypergraph and its dual :

PROPOSITION 2.1. [*J.-C. Bermond, J. Bond, C. Peyrat*] *If H is a (Δ, D, r)-hypergraph then its dual hypergraph H^* is a (r, D^*, Δ)-hypergraph where*

$$D - 1 \leq D^* \leq D + 1.$$

Note that, if G is a graph of maximum degree Δ and diameter D then its dual is a $(2, D^*, \Delta)$-hypergraph. Furthermore,

PROPOSITION 2.2. [*J.-C. Bermond, J. Bond, C. Peyrat*] *If G is a bipartite (Δ, D)-graph then its dual hypergraph H^* is a $(2, D^*, \Delta)$-hypergraph where*

$$D^* \leq D.$$

In what follows we will call the diameter of the dual of H the *line diameter* of H. In particular if a hypergraph H has line diameter 2 it means that for any pair of edges E and F either $E \cap F \neq \emptyset$ or there exists an edge I such that $E \cap I \neq \emptyset$ and $I \cap F \neq \emptyset$.

2.3. Design Theory. We will need some concepts of design theory (for more details and results see [**17**] and [**18**]). In particular, we need the concept of transversal design.

A *transversal design* $T[\Delta, \lambda; k]$ is a triple $(X, \mathcal{G}, \mathcal{P})$ where X is a finite set of elements, $\mathcal{G}$ a partition of X into Δ classes G_i, $i = 1, 2, \ldots, \Delta$, called *groups* or *classes* each containing exactly k elements and $\mathcal{P}$ a family of subsets, called *blocks* of X such that :

(i) For any block P of $\mathcal{P}$ and any group G_i $|P \cap G_i| = 1$.
(ii) Each pair $\{x, y\}$ of elements of X where x and y belong to distinct groups is contained in exactly λ blocks of $\mathcal{P}$.

A transversal design $T[\Delta, \lambda; k]$ has exactly Δk elements. Each block contains exactly Δ elements. Easy countings show that each element belongs to exactly $k\lambda$ blocks of $\mathcal{P}$ and that the number of blocks is $k^2\lambda$.

The case $\lambda = 1$ can be viewed as a partition of the edges of a complete Δ-partite graph into cliques of size Δ. It is known that there exists a $T[\Delta, 1; k]$ if and only if there exist $\Delta - 2$ mutually orthogonal latin squares of order k (see [**17**]). In particular $T[3, 1; k]$ transversal designs always exist and $T[4, 1; k]$ transversal designs exist if and only if $k \neq 2, 6$. For a general Δ, a $T[\Delta, \lambda; k]$ exists if k is large enough by a result of Wilson [**22**].

2.4. Moore-like bounds. As for graphs, we can determine an upper bound on the maximum number of vertices of a (Δ, D, r)-hypergraph . Indeed, by computing the number of vertices at distance $0, 1, \ldots, D$ from a given vertex, we find at most one vertex at distance 0, $\Delta(r-1)$ vertices at distance 1, $\Delta(\Delta - 1)(r-1)^2$ vertices at distance 2, $\Delta(\Delta-1)^2(r-1)^3$ vertices at distance 3, ..., $\Delta(\Delta-1)^{D-1}(r-1)^D$ vertices at distance D. Therefore, we obtain the Moore-like bound given by :

PROPOSITION 2.3. *The maximum number $n(\Delta, D, r)$ of vertices of a (Δ, D, r)-hypergraph satisfies :*

$$n(\Delta, D, r) \leq 1 + \Delta(r-1)\Sigma_{i=0}^{i=D-1}(\Delta-1)^i(r-1)^i.$$

The topologies whose number of vertices achieves the Moore bound are called *Moore geometries.* Several authors have tried to know for what values of Δ, D and r Moore geometries exist. Now, it is established that the Moore bound may be achieved only in case $r = 2$ or D=1 or D=2 with $r > 3$ (see [**5**] for references).

In case $D = 1$, the Moore bound is $1 + \Delta(r-1)$. The following result appears in [**6**] :

PROPOSITION 2.4. [*J.-C. Bermond, J. Bond, J.-F. Saclé*] *The maximum number $n(\Delta, 1, r)$ of vertices of a $(\Delta, 1, r)$-hypergraph is equal to the Moore bound if and only if there exists a $(v, r, 1)$-BIBD with $v = 1 + \Delta(r-1)$.*

The problem of existence of $(v, r, 1)$-BIBD is itself a large domain of research and the results are complete only for $r \leq 6$.

3. Earlier results

Many results have been obtained on the construction of "good" (Δ, D, r) hypergraphs (see the survey [**10**]) in particular in the case $D = 1$ or $\Delta = 2$. Let us recall what is known in the case of diameter 2. The Moore bound is $(\Delta^2 - \Delta)r^2 - (2\Delta^2 - 3\Delta)r + (\Delta - 1)^2$. The case $D = 2$ and $\Delta = 2$ was studied in [**4**] and will be recalled in paragraph 4.

For $D = 2$ and $\Delta \geq 3$ one construction is based on the following propositions :

PROPOSITION 3.1. [*J. Bond*] *For any* $\Delta \geq 2$, *any* $D \geq 2$ *and any* $r \geq 2$,

$$n(\Delta, D, r) \geq rn(\Delta - 1, D - 1, r).$$

PROOF. Take r copies of a $(\Delta - 1, D - 1, r)$ hypergraph and join the r vertices having the same label in each copy with a hyperedge. Therefore we obtain a (Δ, D, r) hypergraph [**12**]. □

PROPOSITION 3.2. [*J.-C. Bermond, J. Bond, J.-F. Saclé*] *If a projective plane of order* q *exists then, for* $\Delta = q + 1$ *and* $r \geq \Delta$, *there exists a* $(\Delta, 1, r)$-*hypergraph with* $\Delta r - (\Delta - 1)\lceil \frac{r}{\Delta} \rceil$ *vertices and then, in this case,* $n(\Delta, 1, r) \geq \Delta r - (\Delta - 1)\lceil \frac{r}{\Delta} \rceil$.

PROOF. see [**6**]. □

PROPOSITION 3.3. *For any* $\Delta \geq 3$ *such that* $\Delta - 1 = q + 1$ *where* q *is a prime power and for any* $r \geq \Delta - 1$,

$$n(\Delta, 2, r) \geq (\Delta - 1)r^2 - (\Delta - 2)r\lceil \frac{r}{\Delta - 1} \rceil.$$

PROOF. According to proposition 3.1, $n(\Delta, 2, r) \geq rn(\Delta - 1, 1, r)$. On the other hand, there exists a projective plane of order q and therefore a $(q + 1, 1, r)$ hypergraph for any prime power q. Then, using proposition 3.2, we have, for any $r \geq \Delta - 1$, $n(\Delta - 1, 1, r) \geq (\Delta - 1)r - (\Delta - 2)\lceil \frac{r}{\Delta - 1} \rceil$. □

REMARK 3.1. *For any* $\Delta \geq 3$ *such that* $\Delta - 1 = q + 1$ *where* q *is a prime power,*

$$n(\Delta, 2, r) \geq f(\Delta) \text{ with } f(\Delta) = \frac{\Delta^2 - 3\Delta + 3}{\Delta - 1}r^2 + O(r).$$

We give here for $\Delta = 3, 4$ the lower bounds $f(\Delta)$ on $n(\Delta, 2, r)$, deduced from proposition 3.3 :

Δ	$f(\Delta)$	Moore bound
3	$\frac{3}{2}r^2 + O(r)$	$6r^2 - 9r + 4$
4	$\frac{7}{3}r^2 + O(r)$	$12r^2 - 20r + 9$

Table 1.

When $D = 2$ and $\Delta \geq 16$, Δ and r even, a better construction is obtained by considering the de Bruijn bus networks which have $\frac{\Delta^2}{16}r^2$ vertices [**7**].

REMARK 3.2. *In the particular case where r is a multiple of Δ, the lower bounds of table 1 will be improved by the technique of construction we shall propose in section* 4.

4. A new technique to design dense bus connection networks of diameter 2

4.1. Case $\Delta = 2$. The technique presented below is inspired from the construction presented in [**4**] for the case $D = 2$, $\Delta = 2$. We repeat this construction :

Step 1 : Consider the graph $G_0 = C_5$, the cycle of length 5. Note that C_5 has line-diameter 2.

Step 2 : Replace each vertex x_i of C_5 by a set X_i of k vertices and each edge $x_i x_j$ of C_5 by the edges of the complete bipartite graph constructed on $X_i \cup X_j$. We therefore obtain a graph G regular of degree $2k$, on $5k$ vertices and having $5k^2$ edges. One can check that G has still line-diameter 2.

Step 3 : Take the dual G^* of G. G^* is a $(2,2,r) - hypergraph$ with $r = 2k$ and $N = 5k^2 = \frac{5r^2}{4}$ vertices.

So we have shown that $n(2,2,r) \geq \frac{5r^2}{4}$ when r is even. In fact Kleitman [**19**] (see also [**21**] and [**14**]) has proved that $n(2,2,r) \leq \frac{5r^2}{4}$ so

THEOREM 4.1. *If r is even, then $n(2,2,r) = \frac{5r^2}{4}$.*

If r is odd, $r = 2k+1$ we have to modify slightly the construction by replacing two adjacent vertices x_1 and x_2 of the C_5 each by a set of $k+1 = \lceil \frac{r}{2} \rceil$ vertices and the 3 other vertices by a set of $k = \lfloor \frac{r}{2} \rfloor$ vertices. The graph G obtained has maximum degree r and has $\frac{5r^2-2r+1}{4}$ edges. By taking its dual we obtain $n(2,2,r) \geq \frac{5r^2-2r+1}{4}$. In [**21**] and [**14**] they have also proved that there is in fact equality. So

THEOREM 4.2. *If r is odd, then $n(2,2,r) = \frac{5r^2-2r+1}{4}$.*

Note that the value given in [**4**] for r odd is incorrect.

4.2. General case : Main theorem. For $D = 2$ and $\Delta \geq 2$ we generalize the construction given before :

Step 1 : We construct a hypergraph H_0, with Δ-uniform rank and with line-diameter 2 (in the case $\Delta = 2$, H_0 was choosen as C_5). Let d be the maximum degree, n the number of vertices and m the number of edges of H_0.

Step 2 : Replace each vertex x_i of H_0 by a set X_i of k vertices and each edge $E = (x_1, x_2, \ldots, x_\Delta)$ of H_0 by the edges (blocks) of a transversal design $T[\Delta, 1, k]$ constructed on the set $X = \cup X_i$, with classes(groups) X_i, $(i = 1, 2, ..., \Delta)$ corresponding to the vertices x_i of E. If such a transversal design exists it follows that the hypergraph H obtained from H_0 has nk vertices, mk^2 edges and maximum degree dk. Figure 1 shows an example with $\Delta = 4$, $k = 3$ and H_0 reduced to an edge $\{a, b, c, d\}$.

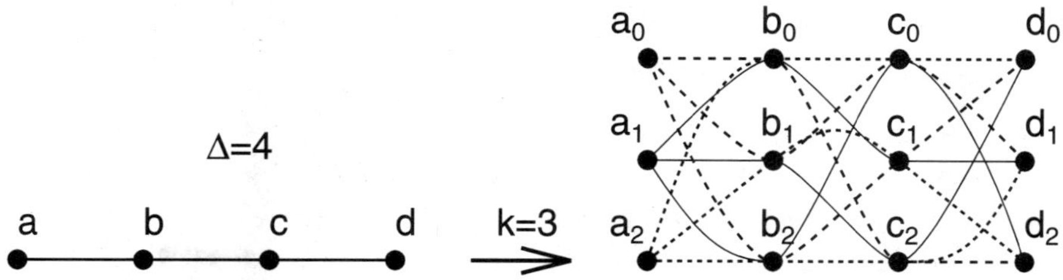

FIGURE 1. H_0 H_1

Furthermore we will prove that

LEMMA 4.1. *If there exists a transversal design $T[\Delta, 1, k]$ then the hypergraph H constructed from H_0 as described above has line-diameter 2.*

(Note that for $\Delta = 2$ we use a $T[2, 1, k]$ transversal design which is nothing else than a bipartite complete graph).

Step 3 : Take the dual H^* of H. By step 2 and the lemma H^* is a $(\Delta, 2, r)$-hypergraph with $r = dk$, $N = mk^2 = \frac{mr^2}{d^2}$ vertices. So we have

THEOREM 4.3. *If there exists a Δ-uniform hypergraph of line-diameter 2 with maximum degree d and m edges and if there exists a transversal design $T[\Delta, 1, k]$ then $n(\Delta, 2, dk) \geq mk^2$.*

PROOF. *(of lemma 4.1).* Let P and Q be two distinct edges of H. P (resp. Q) corresponds to some block of a transversal design $T[\Delta, 1, k]$associated to an edge E (resp. F) of H_0.

Case 1 : $E = F = (x_1, x_2, \ldots, x_\Delta)$. Let p be a vertex of P which does not belong to Q; p belongs to some set X_i. Let q be a vertex of Q not in X_i. By the definition of a tranversal design there exists an edge R containing p and q. R intersects both P and Q and so the distance between P and Q is at most 2.

Case 2 : $E \neq F$, $E \cap F \neq \emptyset$. Let $x_i \in E \cap F$ and let $p = P \cap X_i$ (by definition $P \cap X_i$ is not empty) and $q = Q \cap X_i$. If $p = q$, P and Q are at distance 1. If $p \neq q$, let p' be an element of P different from p and let R be the edge of the transversal design associated to E containing p' and q. R intersects both P and Q and so the distance between P and Q is at most 2.

Case 3 : $E \cap F = \emptyset$. As H_0 is of line-diameter 2 there exist an edge E' intersecting both E end F. Let $x_i \in E \cap E'$ and $x_j \in F \cap E'$. Let $p = P \cap X_i$ and $q = Q \cap X_j$. By definition there exists in the transversal design associated to E' an edge R containing both p and q. So R intersects both P and Q and again the distance between P and Q is at most 2 (in fact exactly 2 in that case). □

REMARK 4.1. *This theorem can easily be generalized for a line-diameter D giving a lower bound for $n(\Delta, D, dk)$. Unfortunately this bound is of order mk^2 and therefore cannot be good asymptotically.*

4.3. Results. Note that to apply theorem 4.3 we need the existence of a transversal design $T[\Delta, 1, k]$ (see some known results in 2.3) and the construction of a Δ-uniform hypergraph H_0 of line-diameter 2. We can have many possible choices. For a given degree d of H_0 we will try to maximize m. For a given r we will choose the hypergraph H_0 optimizing $\frac{m}{d^2}$, with d divisor of r. (We can also in that case use a solution for a value near for r like we did for r odd in the case $\Delta = 2$.)

So all the difficulty consists in finding optimal or good hypergraphs H_0. The line-diameter of H_0 correspond to half the maximum distance between vertices belonging to the class $V_2(R)$ associated to the buses. As we want a line-diameter 2, we have to consider bipartite graphs of diameter 4, 5 or 6. In the literature (see for example [**13**]) has been considered the case of regular graphs (in that case $\Delta = d$). Using these results for diameter 5 we have a good family of hypergraphs.

Let us denote by $b'(\Delta, 5)$ the maximum number of vertices of a biregular bipartite graph with degree Δ and diameter 5, then we have a family of hypergraphs H_0 with $m = \frac{b'(\Delta,5)}{2}$ and $d = \Delta$. So we obtain :

THEOREM 4.4. *Let $\Delta \geq 3$ and r be a multiple of Δ, $r = \Delta k$, then if there exists a transversal design $T[\Delta, 1, k]$*

$$n(\Delta, 2, r) \geq \frac{b'(\Delta,5) r^2}{2\Delta^2}.$$

For example, in the case $\Delta = 3$, we have $b'(3, 5) \geq 56$. So as there exists a $T[\Delta, 1, k]$ for any k, $n(3, 2, r) \geq \frac{28}{9} r^2$ for r multiple of 3.

In the case $\Delta = 4$, we have $b'(4, 5) \geq 144$. So as there exists a $T[\Delta, 1, k]$ for any $k \neq 2, 6$, $n(4, 2, r) \geq \frac{9}{2} r^2$ for r multiple of 4, $r \neq 8, 24$.

Asymptotically we have a family better than those obtained before. In the case $\Delta = 3$, recall that $n(3, 2, r)$ was of order $\frac{3}{2} r^2$ and for $\Delta = 4$, $n(4, 2, r)$ was of order $\frac{7}{3} r^2$.

If one is particularly interested in some values of r not multiple of Δ, we can either use a d different from Δ (but in that case we have to construct "good" (d, Δ, D)-bipartite biregular graphs) or modifiy the construction by starting from a solution with a slightly different r.

5. Conclusion

The technique we have proposed in this paper provides a new family of dense bus connection networks of parameters $\Delta \geq 3$, diameter 2 and edge size r. Improvement in the case where r is not a multiple of Δ could be obtained by finding new (d, Δ, D)-bipartite biregular graphs. Using these results one can also improve, via compound techniques, some of the known values for $D \geq 3$. Unfortunately we are still far from the Moore bound and so new ideas have to be proposed in that case.

References

1. C.T. Benson. Minimal graphs of girth 8 and 12. *Canadian Journal of Mathematics*, 18, pp. 1091–1094, 1966.
2. C. Berge. *Graphes et Hypergraphes.* Dunod, 1970.
3. C. Berge. *Hypergraphs.* North-Holland, 1989.
4. J.-C. Bermond, J. Bond, and C. Peyrat Interconnection networks with each node on two buses. In *Proceedings Coll. Int. Algorithmes et Architectures Parallèles, Marseille*, pp. 155–167. North-Holland, 1986.
5. J.-C. Bermond, J. Bond, M. Paoli, and C. Peyrat. Graphs and interconnection networks: Diameter and vulnerability. *Survey in Combinatorics, Invited Papers for the 9th British Combinatorial Conference London Math-Soc. Lect. Note Ser.*, 82, pp. 1–30, 1983.
6. J.-C. Bermond, J. Bond, and J.-F. Saclé. Large hypergraphs of diameter 1. In *Graph Theory and Combinatorics, Proc. Coll. Cambridge*, pp. 19–28. Academic press, 1983.
7. J.-C. Bermond, R.W. Dawes, and F.O. Ergincan. *De Bruijn and Kautz bus networks.* Submitted to Networks.
8. J.-C. Bermond, C. Delorme, and J-J. Quisquater. Strategies for interconnection networks : Some methods from graph theory. *Journal of Parallel and Distributed Computing*, 3, pp. 433–449, 1986.
9. J.-C. Bermond, C. Delorme, and J.-J. Quisquater. Table of large (Δ, D)-graphs. *Discrete Applied Mathematics*, 37/38, pp. 575–577, 1992.
10. J.-C. Bermond and F.O Ergincan. Bus interconnection networks. To appear in Discrete Applied Mathematics.
11. N. Biggs. *Algebraic graph theory.* Cambridge University Press, 1974.
12. J. Bond. *Construction de grands réseaux d'interconnexion.* PhD thesis, Université de Paris-Sud, France, 1984.
13. J. Bond and C. Delorme. New large bipartite graphs with given degree and diameter. *Ars Combinatoria*, 25-C, pp. 123–132, 1988.
14. F.R.K. Chung, A. Gyarfas, Z. Tuza and W.T. Trotter. The maximum number of edges in $2K_2$-free graphs of bounded degree *Discrete Mathematics*, 81, pp. 129-135, 1990.
15. C. Delorme. Large bipartite graphs with given degree and diameter. *Journal of Graph Theory*, 8, pp. 325–334, 1985.
16. K.W. Doty. Dense bus connection networks. In *Proceedings of the 1983 International Conference on Parallel Processing*, pp. 158–160, 1983.
17. M.Jr. Hall. *Combinatorial Theory.* Wiley-Interscience Series in Discrete Mathematics, 1986.
18. H. Hanani. Balanced incomplete block designs and related designs. *Discrete Mathematics*, 11, pp. 255–369, 1975.
19. D.J. Kleitman. Private communication. 1983.
20. M.D. Mickunas. Using projective geometry to design bus connection networks. In *Proc. of the workshop on Interconnexion Networks for Parallel and Distributed Processing, West Lafayette*, pp. 47–55, 1980.
21. M. Paoli, G.W. Peck, W.T. Trotter and D.B. West. The maximum number of edges in regular K_2-free graphs preprint, 1986.

22. R.M. Wilson. An existence theory for pairwise balanced designs, III : Proof of the existence conjectures. *Journal of Combinatorial Theory, ser. A*, 18, pp. 71–79, 1975.

I3S, CNRS, U.R.A. 1376, UNIVERSITÉ DE NICE SOPHIA-ANTIPOLIS, 930 ROUTE DES COLLES, B.P. 145, 06903 SOPHIA-ANTIPOLIS CEDEX, FRANCE
E-mail address: bermond@unice.fr

I3S, CNRS, U.R.A. 1376, UNIVERSITÉ DE NICE SOPHIA-ANTIPOLIS, 930 ROUTE DES COLLES, B.P. 145, 06903 SOPHIA-ANTIPOLIS CEDEX, FRANCE
E-mail address: bond@unice.fr

LRI, UA 410 CNRS, UNIVERSITÉ DE PARIS-SUD, 91405 ORSAY CEDEX, FRANCE
E-mail address: djelloul@lri.fr

DIMACS Series in Discrete Mathematics
and Theoretical Computer Science
Volume **21**, 1995

ON BROADCASTING SCHEMES IN RESTRICTED OPTICAL PASSIVE STAR SYSTEMS

P. BERTHOMÉ AND A. FERREIRA

ABSTRACT. In this paper, we address some fundamental communication operations in parallel Optical Passive Star systems with restricted communication capabilities. Optimal algorithms are proposed for broadcasting a message from one processor to all the others, even when the number of available wavelengths is limited. We also study multi-broadcasting and gossiping schemes.

1. INTRODUCTION

For massively parallel architectures, the hardware complexity of the interconnection network is much higher than that of the processing units, employing most of the hardware involved. In many optical communication networks, Optical Passive Star technology using wavelength division multiplexing access (WDMA) offers optical multiple access channels that allow significant reduction in the complexity of the specialized routers connected to the processors, along with a substantial reduction in the intermediary latencies for one-to-many communications [LZGK92].

WDMA allows us to share the enormous bandwidth of optical fibers among several messages, through the creation of multiple channels by the use of different wavelengths. This can be implemented by tunable feedback laser diodes for the transmitters, and wavelength tunable filters for the receivers [TD93]. In this paper, we use the processors capability of freely tuning the wavelength they want to access at a given moment, in order to design efficient global communication schemes for Optical Passive Star (OPS) networks. Such schemes are commonly used in all areas of parallel computing, ranging from neural networks to linear algebra.

One important feature of OPS systems is the number of available wavelengths. This being high means that the system is expensive to build and to maintain, and that the tuning time ρ would be large. Thus, we study broadcasting also under the assumption of bounded wavelengths in the system.

Our paper is organized as follows. The Optical Passive Star model is precisely described in the next section. Then, in Section 3 we consider the broadcasting (one-to-all) problem, where a message has to be sent from an initiator to all other processors. We also examine the multiple broadcasting problem in Section 4 and

---

1991 *Mathematics Subject Classification.* Primary 68M10, 68Q22, 68Q25, 78A99; Secondary 78A99.

This work has been partially supported by ANM and PRS of the French CNRS, and by the French-Israeli project on Optical Models for Parallel Computing.

the gossiping (all-to-all) problem in a simple case in Section 4.1. We close the paper with concluding remarks and ways for further research.

2. Optical Passive Star networks

In this section, we give a brief overview of existing Optical Passive Star networks, followed by a description of the general model and of the particular one used in this paper.

2.1. Existing Optical Passive Star. The concept of Optical Passive Star has been used to design networks that are mostly experimental. In [BDP+92] and [TD93] the authors show a table where most of the characteristics of some existing implementations are listed. Interested readers can find another survey on existing OPS systems in [Muk92].

The ALOHA based system proposed by Dowd [Dow91] is designed for a multiprocessor and uses only one Optical Passive Star where, for each processor, both the transmitter and the receiver are tunable.

The STAR-TRACK system [LGA90] proposes some devices where only the receivers are tunable. The RAINBOW [DGL+90] system is a prototype working under the same conditions. There is no direct implementation of the *opposite* system, i.e., in which each processor has only one tunable transmitter. However, the FOX system provides two Optical Stars of this type [ACG+88].

There are numerous variations on the theme. The papers mentioned above show several possible combinations of different Optical Passive Stars. One further example is the HYPASS system [AGKV88] that proposes for each processor two transmitters and two receivers, one of each category.

2.2. The general Optical Passive Star model. The OPS model can be seen as a bipartite graph $G = (T, R, E)$, where each processor of the network P_i is decomposed into two parts, a transmitter T_i and a receiver R_i. The set E of the edges symbolizes the optical connections between those two sets. There exists an edge labeled λ between T_i and R_j, if there exists a physical optical mean connecting T_i and R_j that can use the wavelength λ. In other words, the transmitter T_i can choose the wavelength λ to transmit a message and the receiver R_j can read it using the same wavelength λ. To perform a communication in the general case, transmitters and receivers have first to choose the same wavelength for the given communication pattern (they are said to *tune* to that wavelength). As we shall see later, this tuning operation can be seen as a parameter of the system. There exists one major condition for a message to be correctly received by a given receiver: two transmitters cannot transmit on the same wavelength at the same time, otherwise this would produce either an error message, or no message at all.

2.3. Our model in this paper. In this paper, we propose to work on a sub-model of the completely tunable Optical Passive Star model. We assume that we have n processors, each with one optical tunable transmitter and one optical tunable receiver, the tuning time being ρ. Furthermore, let k be the *life-span* of an optical message (number of receivers that can access the message before it fades away because of the attenuation caused by the receivers accesses, as defined in [DFT94]).

In this model, we assume that transmitters and receivers are always tuned to the same wavelength. This strong assumption may be obtained by using one acousto-optic device per processor, used by both transmitter and receiver. Furthermore, we assume that the de-tuning time is negligible, i.e., when a receiver tunes to another wavelength, the one it was tuned to becomes available to the whole system allowing any other processor to receive a message on this wavelength.

This model may be easily connected to the postal model for communication in current parallel computers' interconnection networks. This other model assumes that you may send one or k different messages from one node at each step, and each node may receive k messages, the elapsed time between the moment where the message is sent and the moment where it reaches its destination being a parameter of the machine (see [BNK92]). We remark that the postal model is potentially more powerful than the OPS model, since one node can send/receive distinct messages in the same step.

3. Broadcasting on the Optical Passive Star

In this section, we discuss some broadcasting algorithms for Optical Passive Star systems, under the restrictions described above. The main difference among the following algorithms resides in the number of wavelengths available in the system. We first study the trivial case where we have only one wavelength. Although this is not a very realistic case, it would be useful afterwards. Then, we examine the opposite case, where there are as many available wavelengths as needed. In this case, some connections with the postal model can be shown. Finally, we address the case where the number of available wavelengths is limited. The resulting algorithm is a combination of the previous ones.

We consider that we have n processors in the broadcasting operation, namely $P_0, \dots P_{n-1}$. Each processor P_i is divided into one transmitter T_i and one receiver R_i. The number of available wavelengths is Λ. Without loss of generality, we assume that the initiator of the broadcasting operation is P_0.

3.1. One available wavelength. In this case, all the transmitters and receivers are always tuned to the same wavelength, so the tuning time ρ is not a real parameter of this sub-problem. We need $\lceil (n-1)/k \rceil$ steps to complete the broadcasting operation, using the trivial algorithm: T_0 keeps on transmitting, and receiver R_i receives the message at step $\lceil (i-1)/k \rceil$.

3.2. Unbounded number of wavelengths. We first show a naive approach not using the whole possibilities of the OPS and then propose an algorithm that achieves optimality.

3.2.1. *Naive approach.* This strategy is a simple succession of the following sequence: tune to the designated wavelength and communicate. At a given time step, the number of processors that have been informed with this algorithm follows an exponential law. The number of steps is:

$$T_{naive}(n) = (\rho + 1) \left\lceil \frac{\log n}{\log(k+1)} \right\rceil \tag{1}$$

This simple strategy does not use the complete possibilities of the Optical Passive Star. We can improve this strategy using the following fact. When the first round is completed, the initiator of the broadcasting is not forced to change its sending wavelength. Then, while the first k elements, that have received the message, prepare themselves to tune to another wavelength, the initiator can send its message provided some receivers are already tuned to the same wavelength. This strategy is described next.

3.2.2. *Optimality.* In order to achieve optimality, we shall use a more complicated strategy. When a message is received by processor P_i from processor P_j, while processor P_i tunes to its sending wavelength, P_j can send its message ρ times and inform other processors. This is a kind of overlapping between communication and computation. Such a strategy has an expansion equation given by a generalized Fibonacci suite. The number of processors than can be informed with the above strategy at step t, i.e., $F(t)$, follows the equation:

$$F(t) = \begin{cases} 1 & \text{for } 0 \leq t \leq \rho \\ F(t-1) + kF(t-(\rho+1)) & \text{for } t > \rho \end{cases} \tag{2}$$

Let $f(n)$ be the inverse function of F, defined by:

$$f(n) = \min\{t \mid F(t) \geq n\} \tag{3}$$

These two functions are a kind of generalization of the functions defined by Bar-Noy and Kipnis for the postal model in [BNK92] that have also been considered in [BNK93]. The algorithm we obtain is very similar, with time complexity $f(n)$. This strategy is optimal since the expansion of informed processors is maximal. Then,

$$T^{\infty}_{bcast}(n) = f(n). \tag{4}$$

Note that the number of wavelengths used during this algorithm can be expressed as:

$$\Lambda^{\infty}(n) = F(T^{\infty}_{bcast}(n) - (\rho+1)) \tag{5}$$

3.3. Bounded number of wavelengths. When the number of wavelengths is bounded by a given number Λ, then we have to combine the two previous algorithms. First, we broadcast the message using the algorithm above, until the number of processors reached by the message is Λ. Then, we divide the remaining $(n-\Lambda)$ processors into Λ subsets of equal size, and once the transmitters and receivers have tuned to the correct wavelength, they perform the naive algorithm. We must add to each subset an initiator in order to perform locally the naive algorithm. Thus, the size of each subset is bounded by $\lceil \frac{n}{\Lambda} \rceil$. Consequently, the time complexity for this strategy is:

$$T^{\Lambda}_{bcast}(n) \leq T^{\infty}_{bcast}(\Lambda) + \rho + \left\lceil \frac{n}{k\Lambda} \right\rceil \tag{6}$$

Notice that the previous equation has been obtained considering two optimal algorithms in order to complete each part of the algorithm. However, this does not give an optimal solution since between the two phases some processors remain inactive.

3.3.1. *Optimality.* The optimal broadcasting strategy when the number of wavelengths is bounded is given by the following theorem.

Theorem 1. *Given an Optical Passive Star with n processors and life-span k, having tunable ends with tuning time ρ, the broadcasting of a message using only Λ wavelengths in this system can be optimally achieved in time*

$$T_{bcast}^{\Lambda}(n) = T_{bcast}^{\infty}(\Lambda) + \rho + \left\lceil \frac{n + \Lambda - F(T_{bcast}^{\infty}(\Lambda) + \rho)}{k\Lambda} \right\rceil .$$

Proof: Note that the processors that should receive a message during the second step can be organized correctly in order to be ready when the first phase is finished. Therefore, this theorem follows by simply observing that the actual number of informed processors at time $T_{bcast}^{\infty}(\Lambda) + \rho$ is $F(T_{bcast}^{\infty}(\Lambda) + \rho)$, and not only Λ, as quietly supposed in equation 6.

Optimality stems from the fact that now all informed processors work all the time at maximal possible rate. □

4. Multiple broadcasting

In this section, we address the problem of multiple broadcasting in the OPS. A processor has m messages $M_1, \ldots M_m$ to broadcast to the other processors.

In this case, the differences with the postal model are more significant. In the postal model, when a message is sent at a given step, another message may be sent elsewhere at the next step, the messages being received after a certain delay. In the Optical Passive Star model, the system imposes different restrictions. First, transmitters and/or receivers have to tune to the correct wavelength, taking ρ steps time, and then the data transmission can be achieved. The sending power is very limited by this fact. There is no difference regarding the maximal number of processors that can be informed in a given time, but in the way these processors are informed.

A simple lower bound can be given by the following equation.

$$T_{mbcast}(n, m) \geq m - 1 + T_{bcast}^{\infty}(n). \tag{7}$$

This bound could be matched if messages were totally pipelined in the model. Unfortunately, the OPS cannot implement this kind of technique. On the other hand, this bound is clearly achieved when $k = n$.

A first approximation of the time needed to solve this problem is to consider the set of messages as a large message of size m and broadcast it using the naive algorithm. Then we have:

$$T_{bcast}^{\infty}(n, m) \leq (\rho + m) \left\lceil \frac{\log n}{\log(k+1)} \right\rceil \tag{8}$$

4.1. A simple gossiping in a simple case. In this section, we give a simple gossiping (all-to-all) algorithm that will be useful in the following. We just consider the case where only up to $k+1$ processors gossip their respective informations.

The algorithm *clique-gossiping* used here is as follows. All the processors tune to the same wavelength, and send their messages in a round robin fashion. The time complexity of such a scheme is:

$$T_g = \rho + k + 1.$$

This result can be extended to larger cases, where there are $(k+1)^d$ processors. We divide the processors into sets of size $k+1$ and proceed within each subset. The *general clique-gossiping* can be described as follows.

(1) Choose one wavelength λ per subset.
(2) For $i := 1$ to $k+1$
    (a) The i-th processor in the subset sends its data on wavelength λ.
    (b) The other processors receive the data on wavelength λ.
(3) The new message to gossip is the concatenation of all the collected messages during the previous steps.

Time analysis The size of the data to gossip will grow considerably using this algorithm, so we need to take it into account in our analysis. Let L be the size of the message at a given step of the global algorithm. Step 1 requires ρ steps to tune to the correct wavelength, $(k+1) \times L$ time units are necessary to complete Step 2, and Step 3 does not require any time, since it can be performed automatically at the reception of the different messages. The new size of the messages is $(k+1) \times L$.

Note that the processors can be logically organized in a $(k+1)$-based generalized hypercube as follows. If d is the dimension on the generalized hypercube, i.e., $n = (k+1)^d$, we say that two processors are on the same *clique* on *dimension* r if and only if the decomposition of their indices in base $(k+1)$ only differs in the r-th digit, $1 \leq r \leq d$. The final algorithm follows the dimensions of this hypercube and performs for each one the simple clique-gossiping algorithm described above. Let L_i be the size of the message to be sent at the i-th step. Clearly, we have $L_i = (k+1)^i$. Then, the global time complexity is:

$$T_G((k+1)^d, k) = \sum_{i=0}^{d-1} (L_i \times (k+1) + \rho), \tag{9}$$

leading to:

$$T_G((k+1)^d, k) = d\rho + \frac{k+1}{k}(n-1) \tag{10}$$

The case where $n \neq (k+1)^d$ has to be treated differently. It is an open question to design an efficient gossiping operation in such a general case.

4.2. Tree decomposition. We provide in this section another algorithm for multiple broadcasting based on a tree decomposition of the processors. We first show how to deal with a simple decomposition and then provide an improved algorithm.

Assume that $k > 1$. The processors are logically organized in one wide tree where the root, the initiator of the multiple broadcasting, has $k(k+1)$ children, and each such child is the root of a balanced k-ary tree. The exact number of nodes that are in depth i is either $k(k+1)$ if $i = 1$, or k times the number of nodes at level $(i-1)$. If the processors cannot be organized in such a logical structure, it is not hard to adapt the following results from the smallest tree having more than n nodes. Let h be the height of the tree defined as before. For example, if $h = 1$, the number of processors is between 1 and $1 + k(k+1)$.

Property 1. *Given a tree composed as above and having n nodes, its height verifies:*

$$h \leq \left\lceil \frac{\log n}{\log k} \right\rceil - 1.$$

The idea of our algorithm is to perform a kind of pipelining of the messages. There are two different algorithms for this multi-broadcasting, whether we consider the initiator processor or any other. The initiator processor alternates a phase sending in a round robin fashion $(k+1)$ messages, and a phase waiting the other processors to be ready to communicate in the next distribution of messages. An internal processor in the tree works all the time of the computation. When it has received a message from its parent, it first sends it to its children, if any, then it participates into a gossiping operation in order to mix the messages arrived at its level within the tree.

Such an algorithm needs the definition of subsets (or cliques) of size $(k+1)$ that will perform the clique-gossiping. In the global tree, the root has $k(k+1)$ children. These children can be organized as $k+1$ groups G_i of size k, $1 \leq i \leq k+1$. Within each group G_i the nodes can be ordered from 1 to k. All the children of the other nodes can also be organized in the same way, with the nodes of each of the k groups being numbered from 1 to k. The cliques we first consider are composed by the children of the root, more precisely each clique is formed by the nodes having the same numbering within the sets G_i, $1 \leq i \leq k+1$. The cliques in the other levels are defined recursively in such a way that they are formed by internal nodes having the same order number, and whose parents also form a valid clique.

The algorithm can be detailed as follows. We assume that the number of messages is such that $m = \alpha(k+1)$, $\alpha \in \mathbb{N}$.

*Algorithm **A1***
Initiator:

(1) Tune to the sending wavelength.
(2) For $i := 0$ to $\alpha - 1$ do
    (a) For $j := 1$ to $(k+1)$ do
        Transmit message $M_{i\alpha+j}$ to the group of nodes G_j.
    (b) Do nothing during $\beta = 3\rho + k + 2$ time steps.

All the other nodes:

(1) Tune to the receiving wavelength.
(2) For $i := 0$ to $\alpha - 1$ do
    (a) Wait for a message to arrive.
    (b) Send the message to the children, if any.
    (c) Wait for the completion of the preceding step by the whole clique of processors this node belongs to.

(d) Perform a simple clique-gossiping operation within its clique.
(e) Tune back to its parent's wavelength.

Note that, for a leaf node, Step 2b cannot be omitted, since the remaining messages will arrive considering this delay. However, it can be omitted when the last bunch of messages arrives, saving $\rho+1$ time steps. A way to implement Step 2c is to tune to the wavelength of the simple gossiping, assuming that this wavelength is used only in this algorithm to perform this kind of operation, and then the following step will start when the last processor in the clique sends its message, being the first in that operation. Knowing the position of each node within its clique, it is not hard to know how long any node has to wait before the beginning of the gossiping step.

The correctness of the algorithm is easily obtained by studying the time needed for the initiator and internal nodes to complete one phase, and be ready for the following one. After having sent $k+1$ messages, the initiator has to wait for $(\rho+1)$ steps for its children to send their data to their direct children, $(\rho+k+1)$ for the gossiping operation to complete, and another ρ steps in order to tune back to the initiator's wavelength. There is no conflict among internal nodes since this schedule is also respected.

The time complexity of this algorithm is given by the moment where the last message starts going into the pipe plus the time needed for this message to get out of the pipe. The last message is sent at time $(k+1+\beta)(\alpha-1)+(k+1)$. Then, the time needed for the last message to reach the leaves of the tree is obviously $(h-1)(\rho+1)$ and the time required by the last gossiping operation is $(\rho+1+k)$. Thus, the global time complexity is:

$$(11)\quad T_{A1}(n,m) \leq (k+1+\beta)(\alpha-1)+(k+1)+(h-1)(\rho+1)+(\rho+k+1)$$

Proposition 1. *The multi-broadcasting of m messages on an Optical Passive Star having n nodes and life-span k can be performed in time:*

$$T_{A1}(n,m) \leq (\rho+1)\left(3\left\lceil\frac{m}{k+1}\right\rceil - 3 + \left\lceil\frac{\log n}{\log k}\right\rceil\right) + 2k\left\lceil\frac{m}{k+1}\right\rceil - 2k+1.$$

In order to give some comparison with the postal model, we can give an approximation of the broadcasting time in our model. The cost of a broadcasting operation is between $(\rho+1)\left\lceil\frac{\log n}{\log k}\right\rceil$ and $(\rho+1)\left\lceil\frac{\log n}{\log(k+1)}\right\rceil$, given by the algorithm in Section 3.2.1. These two numbers are usually not very far from each other. Thus, $(\rho+1)\left\lceil\frac{\log n}{\log k}\right\rceil$ provides an upper approximation of the broadcast complexity. Thus, when m becomes large, the time complexity of the multi-broadcast operation is approximatively $2m+T_{bcast}^{\infty}(n)$. This latter complexity reaches the one obtained in the postal model using the algorithm PARTITION [BNK92].

4.3. Improvements on the tree decomposition. Note that the initiator processor is idle during a long time ($\beta=3\rho+k+2$) compared to the time it is really sending messages. One possibility to fill this waste of time is by dividing the processors into $\gamma=\left\lceil\frac{\beta}{k+1}\right\rceil$ sets of equal size and performing the previous algorithm on the different sets also in a round robin fashion. The general algorithm is changed only for the initiator.

However, using this strategy, the height of the new tree has decreased only by an additive constant factor. Moreover, there is some further delay implied by the new pipeline. Consequently, the time complexity of such an algorithm is not considerably better than the previous one.

The second improvement we propose tries to decrease the latency β of the initiator of the multi-broadcasting. In this case, we assume that the life-span k is greater than the tuning time ρ. Then, we can perform some interleaving of the first two phases for an internal processor.

Figure 1 shows how algorithm $A1$ works. Notice that we can perform the gossiping operation earlier, when the first bunch of messages has been received. Thus, only the first $k - \rho$ sub-trees had the possibility to inform a part of their elements. After having performed the gossiping operation, the last ρ sub-trees send the information to their children and tune back to the initiator's wavelength, while the others have only to tune back to the initiator's wavelength. The improvement is shown in Figure 2 where the reception times are depicted. Using this technique, β can be decreased from $3\rho + k + 2$ to $2\rho + k + 1$.

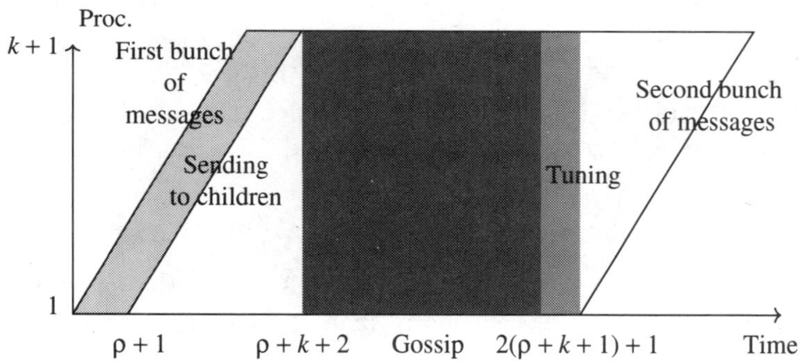

FIGURE 1. Evolution of the algorithm $A1$ of multi-broadcasting.

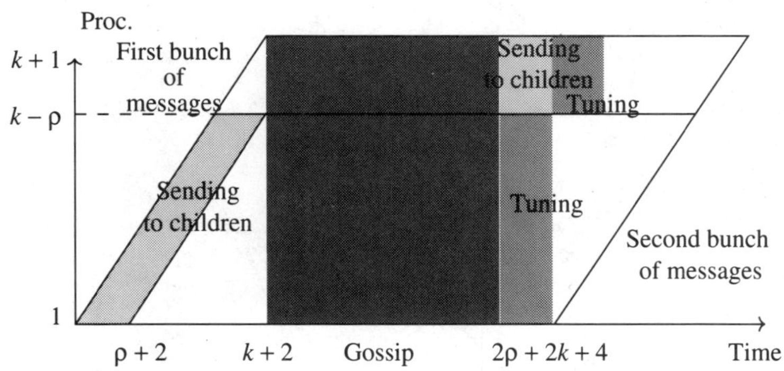

FIGURE 2. Proposed improvement: the clique-gossiping begins earlier.

Then, using Equation 11, we obtain the following time complexity.

Proposition 2. *The multi-broadcasting of m messages on an Optical Passive Star having n nodes and life-span k can be performed in time:*

$$T_{A3}(n,m) \leq 2m + 2\rho \left\lceil \frac{m}{k+1} \right\rceil - \rho + (\rho+1) \left\lceil \frac{\log n}{\log k} \right\rceil .$$

5. Conclusion

In this paper, we have proposed new algorithms for broadcasting and multi-broadcasting in Optical Passive Star networks with restricted communication capabilities. In this model, it is assumed that for any message sent by a given transmitter, at most k receivers may try to obtain the message. This kind of model is interesting in the case of very large networks, since the value k would be rather small compared to the number of processors. Our algorithms could be applied quite efficiently, provided the generalized Fibonacci function (See Equation 2) can be easily computed or given in tables.

We also addressed one of the main issues in the design of OPS systems, which is the limitation on the number of available wavelengths. Also in this case, we gave an optimal solution for the broadcasting problem on the OPS. On the other hand, much has to be done in this case concerning multi-broadcasting and gossiping.

We remark that this work can be included in a more general classification of the OPS systems. Let $\Lambda\text{-}\mathcal{S}_X^Y$ be the model of Optical Passive Star using Λ wavelengths, having the transmitters in state X ($X \in \{\text{tunable, fixed}\}$), and the receivers in state Y (also in $\{\text{tunable, fixed}\}$) . Then, in [BDS94], it is proposed a simple hierarchy of those models stating that:

$$\forall \Lambda \geq n \; \Lambda\text{-}\mathcal{S}_T^T \sim \Lambda\text{-}\mathcal{S}_F^T \succ \Lambda\text{-}\mathcal{S}_T^F \succ \Lambda\text{-}\mathcal{S}_F^F,$$

where $\sim$ means the equivalence between models and $\succ$ the fact that a model can solve a problem strictly more efficiently than another.

References

[ACG+88] E. Arthurs, J.M. Cooper, M.S. Goodman, H. Kobrinski, M. Tur, and M.P. Vecchi, *Multiwavelength Optical Crossconnect for Parallel Processing Computers*, Electronnic Letters **24** (1988), 119–120.

[AGKV88] E. Arthurs, M.S. Goodman, H. Kobrinski, and M.P. Vecchi, *HYPASS: An Optoelectronic Hybrid Packet-Switching System*, IEEE Journal on Selected Areas in Comm. **6** (1988), 1500–1510.

[BDP+92] S. Bhattacharya, D.H.C. Du, A. Pavan, S.R Tong, R. Vetter, and K. Williams, *A Network Architecture for Distributed High Performance Heterogeneous Computing*, Tech. Report TR 92-49, University of Minnesota - Department of Computer Science, August 1992.

[BDS94] P. Berthomé, T. Duboux, and A. Schuster, *Simulation Problems and Classification of the Optical Passive Star Model*, manuscript, 1994.

[BNK92] A. Bar-Noy and S. Kipnis, *Designing Broadcasting Algorithms in the Postal Model for Message-Passing Systems*, SPAA, 1992.

[BNK93] A. Bar-Noy and S. Kipnis, *Broadcasting Multiple Messages in Simultaneous Send/Receive Systems*, Symposium on Parallel and Distributed Processing, 1993, pp. 344–347.

[DFT94] F. Desprez, A. Ferreira, and B. Tourancheau, *Efficient Communication Operations on Passive Optical Star Networks*, Proceedings of the First International Workshop on Massively Parallel Processing Using Optical Interconnections, IEEE Press, apr 1994, pp. 52–58.

[DGL+90] N.R. Dono, P.E. Green, K. Liu, R. Ramaswami, and F. F.-K. Tong, *A Wavelength Division Multiple Access Network for Computer Communication*, IEEE Journal on Selected Areas in Communications **8** (1990), no. 6, 983–994.

[Dow91] P.W. Dowd, *Random Access Protocols for High-Speed Interprocessor Communication Based on an Optical Passive Star Topology*, Journal of Ligthwave Technology **9** (1991), no. 6, 799–808.

[LGA90] T. Lee, M. Goodman, and E. Arthurs, *A Broadband Optical Multicast Switch*, ISS'90, 1990.

[LZGK92] P. Lalwaney, L. Zenou, A. Ganz, and I. Koren, *Optical Interconnects for Multiprocessors Cost Performance Trade-Offs*, FRONTIERS'92, October 1992, pp. 278–285.

[Muk92] B. Mukherjee, *WDM-Based Local Lightwave Networks Part I: Single-Hop Systems*, IEEE Networks (1992), 12–27.

[TD93] S.R. Tong and D.H.C. Du, *Design Principles for Multi-Hop Wavelength and Time Division Multiplexed Optical Passive Star Networks* , GlobeCom'93, 1993.

CNRS - LABORATOIRE LIP, CNRS-URA 1398, ENS LYON, 46, ALLÉE D'ITALIE, 69364 LYON CEDEX 07, FRANCE

E-mail address: [berthome,ferreira]@lip.ens-lyon.fr

DIMACS Series in Discrete Mathematics
and Theoretical Computer Science
Volume **21**, 1995

Restricted Routing and Wide Diameter of the Cycle Prefix Network

W. Y. C. CHEN, V. FABER AND E. KNILL

ABSTRACT. The cycle prefix network is a Cayley coset digraph based on sequences over an alphabet, which has been proposed as a vertex symmetric communication network. This network has been shown to have many remarkable communication properties such as a large number of vertices given degree and diameter, simple shortest path routing, Hamiltonicity, optimal connectivity, and others. The considerations for designing symmetric and directed interconnection networks are well justified in practice. Among the properties of a good network, efficient routing is probably one of the most important. In this paper, we further study routing algorithms for the cycle prefix network. We confirm an observation first made based on computer experiment regarding the diameter change when certain links are removed in the original network, and we completely determine the wide diameter of the network. The wide diameter of a network is now perceived to be even more important than the diameter. We show by construction that the wide diameter of the cycle prefix network is very close to the ordinary diameter. This means that routing in parallel in this network costs little extra time compared to the ordinary single path routing.

1. Introduction

The cycle prefix network is a vertex symmetric, directed graph which has recently been proposed for use as a communication network [**5**]. It has been shown that the cycle prefix network has many remarkable communication properties such as a large number of vertices given degree and diameter, simple shortest path routing, Hamiltonicity, optimal connectivity, and others [**3, 5, 8**]. These considerations for designing symmetric and directed networks are well justified in practice. In the search for highly efficient network models and in the study

---

1991 *Mathematics Subject Classification.* Primary 68M10, 90B12; Secondary 68R10, 05C12, 05C38.

This work was performed under the auspices of the U. S. Department of Energy. We thank D. F. Hsu for help discussions, B. Q. Li and the referee for valuable comments.

This paper is in final form and no version of it will be submitted for publication elsewhere.

of communication algorithms, Cayley graph techniques have been used successfully in discovering new models and in analyzing network efficiencies. Another interesting idea in network design which has been used for the construction of many networks is to represent nodes as certain sequences over an alphabet with links represented by some operations on those sequences. The hypercube, de Bruijn, Kautz, the star and the pancake networks can all be constructed in this fashion. In the case of the cycle prefix network, both the idea of Cayley graphs (Cayley coset digraphs, to be precise) and that of sequences over an alphabet can be used as the underlying representations, and each has its own advantage. The former idea was the point of departure for the discovery of the cycle prefix network, motivated by the fundamental theorem of Sabidussi [**9**] which claims that all vertex symmetric digraphs are Cayley coset digraphs. The sequence representation of a cycle prefix network is more useful for studying its properties and for implementing it in practice. For this reason, we shall utilize the sequence representation of the cycle prefix digraph throughout the paper.

In the performance evaluation of a network, the diameter and routing are among the most critical concerns. In this paper we shall further study these issues for the cycle prefix network. This paper has two objectives. The first objective is to confirm an observation first made on the basis of computer experiment regarding the diameter change when certain links are removed. In a recent paper [**3**], Comellas and Fiol describe a routing scheme which leads to an upper bound on the diameter of the modified cycle prefix digraph. However, they left out the question of whether the bound is exact. We settle this question by exhibiting vertices that achieve the diameter bound. We also present a reachability theorem for the modified cycle prefix digraph. For some parameters, one may construct large symmetric networks with small degree and diameter based on these digraphs by using the method of Conway and Guy [**4**].

The second objective of this paper is concerned with the recently introduced notion of the wide diameter of a network. Mathematically, the notion of the wide diameter of a graph naturally stems from the classical theorem of Menger relating connectivity to disjoint paths. However, such a notion has not been studied in graph theory until very recently when it became relevant from an engineering point of view. Hsu has given a vivid account on the background of the wide diameter [**7**]:

"The concept of the wide diameter of a graph G arises naturally from the study of routing, reliability, randomized routing, fault tolerance, and other communication protocols (such as byzantive algorithm) in parallel architecture and distributed computer networks. By considering both the width and the length of a container, we are able to give a global and systematic treatment on the interconnection network for various distributed systems."

"Although the concept of a container and the notion of wide diameter have been discussed and used in practical applications, the graph theory questions suggested have not, at least until recently, been studied as extensively as the questions in the hardware and software design, development, and implementation of distributed computing systems."

In the case of the cycle prefix network, we completely determine its wide diameter through an explicit construction. It turns out that the wide diameter is very close to the ordinary diameter of the network. In other words, routing in parallel in the cycle prefix network costs little extra time compared to ordinary single path routing.

2. The Cycle Prefix Digraphs $\Gamma_\Delta(D)$ and $\Gamma_\Delta(D,-r)$

The cycle prefix digraph $\Gamma_\Delta(D)$ $(\Delta \geq D)$ is defined as a digraph whose vertex set consists of sequences $x_1x_2\cdots x_D$ over an alphabet $\{1,2,\ldots,\Delta+1\}$, where $x_1,x_2,\ldots,x_D$ are distinct. Such sequences are called partial permutations or D-permutations. The adjacency relations for a vertex $x_1x_2\cdots x_D$ are described as follows:

$$x_1x_2\cdots x_D \Rightarrow \begin{cases} x_kx_1\cdots x_{k-1}x_{k+1}\cdots x_D, & \text{for } 2\leq k\leq D, \\ yx_1x_2\cdots x_{D-1}, & \text{for } y\neq x_1,x_2,\ldots,x_D. \end{cases}$$

We say that the vertex $x_kx_1\cdots x_{k-1}x_{k+1}\cdots x_D$ is obtained from $x_1x_2\cdots x_D$ via a rotation on the prefix $x_1x_2\cdots x_k$, denoted,

$$x_kx_1\cdots x_{k-1}x_{k+1}\cdots x_D = R_k(x_1x_2\cdots x_D)\ .$$

In particular, the operation R_D is called a full rotation, and $x_Dx_1x_2\cdots x_{D-1}$ is called the full rotation of $x_1x_2\cdots x_D$. By comparison, the rotations R_k for $k<D$ are called partial rotations. We say that the sequence $yx_1x_2\cdots x_{D-1}$ is obtained from $x_1x_2\cdots x_D$ via a shift, denoted

$$yx_1x_2\cdots x_{D-1} = S_y(x_1x_2\cdots x_D).$$

Since the term "adjacent" is somewhat ambiguous for a directed graph, when (u,v) is an arc in a digraph we shall say that u is *adjacent to* v while v is *next to* u. The term "adjacent from" is used by some authors to distinguish it from "adjacent to". From the above sequence definition of $\Gamma_\Delta(D)$, it is easily seen to be vertex symmetric, a fact that immediately follows from the Cayley coset digraph definition of $\Gamma_\Delta(D)$ [**5**].

We notice that some authors prefer the shift operation in the direction from right to left. For consistency, we shall follow the notation in [**5**], that is, the shift from left to right. A right-to-left notation is however adopted in [**3**].

In the original study of cycle prefix digraphs $\Gamma_\Delta(D)$, it was observed through computational evidence that if one rules out the double arcs in $\Gamma_\Delta(D)$, then the diameter of the resulting digraph, denoted, $\Gamma_\Delta(D,-1)$, increases only by one. They then considered a new construction based on the cycle prefix digaph $\Gamma_\Delta(D)$. Suppose $\Gamma_\Delta(D,-r)$ is the digraph obtained from $\Gamma_\Delta(D)$ by deleting the arcs represented by the partial rotations $R_2,R_3,\ldots,R_{r+1}$. Formally speaking,

$\Gamma_\Delta(D,-r)$ has the same vertex set as $\Gamma_\Delta(D)$, and the adjacency relations for a vertex $x_1x_2\cdots x_D$ are given by

$$x_1x_2\cdots x_D \Rightarrow \begin{cases} x_kx_1\cdots x_{k-1}x_{k+1}\cdots x_D, & \text{for } r+2\leq k\leq D, \\ yx_1x_2\cdots x_{D-1}, & \text{for } y\neq x_1,x_2,\ldots,x_D. \end{cases}$$

Note that the degree of $\Gamma_\Delta(D,-r)$ decreases by r compared with $\Gamma_\Delta(D)$. It is intuitive to surmise that the diameter of $\Gamma_\Delta(D,-r)$ would increase by the same scale. Recently, Comellas and Fiol [**3**] have shown that in most cases the diameter increase of $\Gamma_\Delta(D,-r)$ is in fact bounded by r. However, they left out the problem of whether this bound is exact. We will fill this gap by proving the exactness of the diameter bound. Moreover, we shall study the reachability property of the digraph $\Gamma_\Delta(D,-r)$, while this property for $\Gamma_\Delta(D)$ has been studied in [**3**]. The details are presented in the next section.

3. Restricted Routing for $\Gamma_\Delta(D,-r)$

For the sake of clear presentation, we shall start with the reachability of the digraph $\Gamma_\Delta(D,-r)$. It is easy to see that $\Gamma_\Delta(D,-r)$ is vertex symmetric and that one may choose the *standard origin* $12\cdots D$ while considering routing for any two vertices.

DEFINITION 3.1 (k-REACHABLE DIGRAPHS). *A digraph is said to be k-reachable if for any two vertices u and v, which are not necessarily distinct, there exists a path (with repeated vertices and arcs allowed) from u to v of length k.*

Comellas and Fiol [**3**] have shown that the digraph $\Gamma_\Delta(D)$ is D-reachable for $\Delta\geq D\geq 3$. Here we will present a stronger result for $\Gamma_\Delta(D,-r)$.

THEOREM 3.1. *Suppose $r\geq 0$ and $\Delta\geq D\geq 2r+3$. Then the vertex symmetric digraph $\Gamma_\Delta(D,-r)$ is $(D+r)$-reachable.*

The main reason for the above theorem to hold lies in the following observation about *dead angles*. Given a vertex $X=x_1x_2\cdots x_D$, the prefix $x_1x_2\cdots x_{r+1}$ is called the dead angle of X in $\Gamma_\Delta(D,-r)$. We say that a letter z is in the dead angle of X if $z=x_i$ for some $1\leq i\leq r+1$. From now on, the vertex X is always referred to as $x_1x_2\cdots x_D$.

LEMMA 3.1 (DEAD ANGLE PRINCIPLE). *Let X be a vertex in $\Gamma_\Delta(D,-r)$. Then there exists a vertex Y next to X such that Y begins with a letter z, if and only if z is not in the dead angle of X.*

The proof of the above lemma is straightforward, which is based on a property of $\Gamma_\Delta(D)$ regarding how the rotation and shift operations complement each other: Suppose z is not in the dead angle of X. If z is indeed in X, then a rotation operation on X may put z back to the beginning of the sequence; otherwise, a shift operation can achieve the same goal with ease. For the above reason, one sees that the two operations are coherent with each other, although they look rather unrelated. Moreover, suppose Y is next to X in $\Gamma_\Delta(D,-r)$, then Y is determined by its first element. We now give a proof of Theorem 3.1.

Proof. Let $I = 12\cdots D$ be the standard origin and $X = x_1x_2\cdots x_D$ be the destination. It suffices to show that there is a directed path of length $D+r$ from I to X. We first consider the case when $x_D \neq 1$. Let A be the set $\{x_1, x_2, \ldots, x_{D-r-1}\}$. Since there are $r+1$ elements in the dead angle of X, and $D-(r+1) > r+1$, there exists an element y_1 in A that is not in the dead angle of I. By the dead angle principle, I is adjacent to a vertex P_1 with prefix $y_1 1\cdots(r+1)$. Let $A_1 = A\backslash\{y_1\}$. Considering the dead angle of P_1, the same condition on D and r ensures that there exists an element y_2 in A_1 that is not in the dead angle of P_1 (implying that y_1 and y_2 are distinct). Hence P_1 is adjacent to a vertex P_2 with prefix $y_2y_1 1\cdots r$. Repeating the above procedure, one may reach a vertex P_r having prefix $y_r\cdots y_2y_1\, 1$, where $y_1, y_2, \ldots, y_r$ come from A. Up to now, it has taken r steps to get to P_r from I. Since $x_D \neq 1$, we may construct a path of length D from P_r to X, and display it by showing the prefixes:

$$\begin{aligned} y_r\cdots y_2y_1 1 \quad &\Rightarrow \quad x_Dy_r\cdots y_2y_1 \quad \Rightarrow \quad x_{D-1}x_Dy_r\cdots y_2y_1 \quad \Rightarrow \cdots \\ &\Rightarrow \quad x_{D-r}\cdots x_{D-1}x_Dy_r\cdots y_2y_1 \quad \Rightarrow \quad x_{D-r-1}x_{D-r}\cdots x_{D-1}x_D \\ &\Rightarrow \quad \cdots \quad \Rightarrow \quad x_1x_2\cdots x_D\,. \end{aligned}$$

We next consider the case when $x_D = 1$. Let $A = \{x_1, x_2, \ldots, x_{D-r-2}\}$ and $B = \{2, 3, \ldots, r+1\}$. Since $D-r-2 > r$, there exists an element y_1 such that $y_1 \in A$ but $y_1 \notin B$. Note that $1 \notin A, B$. By the dead angle principle, I is adjacent to a vertex P_1 with prefix $y_1 1 2\cdots(r+1)$. Let $A_1 = A\backslash\{y_1\}$, $B_1 = B\backslash\{r+1\}$. The same condition on D and r ensures that there exists $y_2 \in A_1$, but $y_2 \notin B_1$. It follows that P_1 is adjacent to a vertex P_2 with prefix $y_2y_1\, 1 2\cdots r$. Repeating this procedure, one ends up with a vertex P_r having prefix $y_ry_{r-1}\cdots y_1 1$, where $y_i \in A$. For $D-r-2 > r$, there exists $x_j \in A$ and $x_j \neq y_i$ for $1 \le i \le r$. We continue with the following path of length $D-r$ starting from P_r (with only prefixes shown):

$$\begin{aligned} y_ry_{r-1}\cdots y_1 1 \quad &\Rightarrow \quad x_jy_ry_{r-1}\cdots y_1 1 \quad \Rightarrow \quad x_{D-1}x_jy_ry_{r-1}\cdots y_1 1 \\ &\Rightarrow \quad x_{D-2}x_{D-1}x_jy_ry_{r-1}\cdots y_1 1 \quad \Rightarrow \cdots \\ &\Rightarrow \quad x_{D-r}\cdots x_{D-1}x_jy_ry_{r-1}\cdots y_1 1 \\ &\Rightarrow \quad x_{D-r-1}x_{D-r}\cdots x_{D-1}x_jy_ry_{r-1}\cdots y_1 1\,. \end{aligned}$$

The last vertex is labeled P_{2r+2} according to the length. Note that $y_i \in A$, we claim that there is a path from P_{2r+2} to X of the following form:

$$\begin{aligned} P_{2r+2} &\Rightarrow x_{D-r-2}x_{D-r-1}\cdots x_{D-1}x_j\cdots 1 \Rightarrow \cdots \\ &\Rightarrow x_{j+1}\cdots x_{D-r}\cdots x_{D-1}x_j\cdots 1 \\ &\Rightarrow x_jx_{j+1}\cdots x_{D-r-1}\cdots x_{D-1}\cdots 1 \Rightarrow \cdots \\ &\Rightarrow x_1x_2\cdots x_{D-1}1 = X\,. \end{aligned}$$

Because $y_i \in A$, at each step it is impossible to bump 1 out of the sequence so that the last vertex has to be X. Summing up all the segments, we get a path of length $D+r$. □

Specializing the above theorem to $r=0$, it follows the reachability of $\Gamma_\Delta(D)$ first observed in [**3**]. The rest of this section is concerned with the diameter of $\Gamma_\Delta(D,-r)$. It is shown in [**4**] that the diameter of $\Gamma_\Delta(D,-r)$ does not exceed $D+r$ for $\Delta \geq D \geq 2r+2$. This upper bound is achieved by the following construction that is a much simpler version than the construction for the reachability of $\Gamma_\Delta(D,-r)$. Note that the reachability result requires a slightly stronger condition on the parameters of $\Gamma_\Delta(D,-r)$. Let us give an outline. Let $I = 12\cdots D$ be the standard origin, and $X = x_1x_2\cdots x_D$ be any vertex in $\Gamma_\Delta(D,-r)$. For the case $x_D \neq 1$, one may first try to reach from I a vertex with prefix $y_ry_{r-1}\cdots y_1 1$, where $y_i \neq x_{D-1}, x_{D-2}, \ldots, x_{D-r}$. Then one may continue with vertices having prefixes x_D, $x_{D-1}x_D$, etc. For the case $x_D = 1$, one may get to a vertex with prefix $y_ry_{r-1}\cdots y_1 1$ such that $y_i \in \{x_1, x_2, \ldots, x_{D-r-2}\}$. Then one may get to X via vertices with prefixes x_{D-1}, $x_{D-2}x_{D-1}$, etc. The last element $x_D = 1$ will eventually take care of itself.

To show that the above diameter bound is exact, we find a class of vertices that achieve the bound. A vertex X in $\Gamma_\Delta(D,-r)$ is called a *remote vertex* if $x_{D-1} = 1$, $x_D = D$, and there exists $x_i > D$ for some $1 \leq i \leq D-2$. We shall use the common notation $d(X,Y)$ to denote the distance from X to Y in a digraph. Then we have the following theorem:

THEOREM 3.2. *Let $\Delta \geq D \geq 2r+2$, and X be a remote vertex in $\Gamma_\Delta(D,-r)$, then the distance from the standard origin $I = 12\cdots D$ to X equals $D+r$.*

Proof. The diameter upper bound is already established, so it suffices to show that $d(I,X) \geq D+r$. Suppose $I \Rightarrow P_1 \Rightarrow P_2 \Rightarrow \cdots \Rightarrow P_m = X$ is a shortest path from I to X. Since $X = x_1x_2\cdots x_{D-2}1D$ and there exists $x_i > D$ for some i, to reach X from I requires at least one shift operation. Thus, element D in I cannot remain in the last position during the process to reach X from I. Since D is in the destination vertex X, it is either moved back to the beginning position at some point, or is removed out of the sequence by a shift operation and then is put back to the beginning by another shift operation; moreover this operation has to happen within the first $r+1$ steps in the shortest path routing because it takes at least $D-1$ more steps to move D to the last position. Since any operation (shift or rotation) on a vertex, say $Y = y_1y_2\cdots y_D$, moves the elements in the dead angle to the positions on the right hand side, the vertex P_{r+1} must have the prefix $z_{r+1}\cdots z_2 z_1 1$. Let $P_{r+1} = z_{r+1}\cdots z_2 z_1\, 1\, v_1 v_2 \cdots v_{D-r-2}$. Then one may notice that D is does not appear in the suffix $v_1 v_2 \cdots v_{D-r-2}$. If D does not appear in $z_{r+1}\cdots z_2 z_1$, then it will take at least D steps to reach X from P_{r+1} as far as the last element of X is concerned. This contradicts the upper bound $D + r$ on the diameter of $\Gamma_\Delta(D, -r)$. Furthermore, D has to remain in the vertices on the path from P_{r+1} to X. Observing that 1 precedes D in X while D precedes 1 in P_{r+1}, there has to be a step in the routing from P_{r+1} to P_m such that in the resulting vertex, say P_l, 1 precedes D instead. The only way for this to happen is that P_l has the form $P_l = 1 \cdots D \cdots$. Since it takes at least $D - 2$ steps to bring 1 to the second last position of X, it follows that

$$m \geq (D-2) + l \geq D + r\,,$$

because $l \geq r + 2$. $\square$

We remark that when $r \geq 1$ the digraph $\Gamma_\Delta(D, -r)$ does not have the unique shortest path property like $\Gamma_\Delta(D)$. For example, in $\Gamma_4(4, -1)$ there are two shortest paths from 1234 to 5214, illustrated below:

$$1234 \;\Rightarrow\; 4123 \;\Rightarrow\; 5412 \;\Rightarrow\; 1542 \;\Rightarrow\; 2154 \;\Rightarrow\; 5214,$$

$$1234 \;\Rightarrow\; 5123 \;\Rightarrow\; 4512 \;\Rightarrow\; 1452 \;\Rightarrow\; 2145 \;\Rightarrow\; 5214.$$

4. The Wide Diameter of $\Gamma_\Delta(D)$

Connectivity considerations of a network were primarily motivated by its fault tolerance capabilities, while the diameter is a measurement of routing efficiency along a single path. Interestingly, the recent notion of wide diameter is a kind of unification of both the diameter and the connectivity due to the classical theorem of Menger. Let G be a digraph of connectivity k and diameter D. By Menger's theorem, between any two distinct vertices x and y in G there are k vertex-disjoint paths. Such a set of disjoint paths, denoted by $C(x, y)$, is called a container, and its length is defined as the maximum length among the paths in the container. The wide distance from x to y is then defined to be the minimum length of the containers from x to y, and the wide diameter is the maximum wide distance among all the pairs of distinct vertices. The consideration of the wide diameter of a network has a solid practical background. Clearly, the wide diameter of a digraph is at least as large as the ordinary diameter. However, it is rather remarkable that for most of the popular interconnection networks the wide diameters are not significantly bigger than (actually, a small constant bigger than) the ordinary diameters, like the hypercube, de Bruijn, Kautz, and the star networks. The main result of this section is to show that such a remarkable phenomenon also occurs to the cycle prefix network. For a vertex X in $\Gamma_\Delta(D)$, we shall use $N(X)$ to denoted the set of vertices next to X, and $M(X)$ the set of vertices adjacent to X. We shall use the notation $i \circ X$ to denote the vertex next X that is obtained by rotating the element i to the beginning position if

i is X, or by shifting i into X and bumping the last element out of X, namely $S_i(X)$ by the previous notation. Note that $N(X)$ consists of vertices $X_i = i \circ X$ for $i \neq x_1$. For $Y = 12 \cdots D$, $M(Y)$ consists of vertices

$$(4.1) \qquad Y_i = \begin{cases} 2\,3 \cdots i\,1\,(i+1) \cdots D\,, & \text{if } \; 2 \le i \le D\,. \\ 2\,3 \cdots D\,i\,, & \text{if } \; D < i \le \Delta + 1\,. \\ Y\,, & \text{if } \; i = 1\,. \end{cases}$$

THEOREM 4.1. *The wide diameter of* $\Gamma_\Delta(D)$ *is at most* $D+2$. *It is exactly* $D+2$ *for* $D \ge 4$. *Specially, if* X *and* Y *are distinct vertices in* $\Gamma_\Delta(D)$, *then there is a bijection* θ *of* $N(x)$ *to* $M(Y)$ *such that the shortest paths from* Z *to* $\theta(Z)$ *are vertex disjoint and do not contain either* X *or* Y.

Since the proof of the above theorem heavily depends on the unique shortest path property, we here give a brief review of the shortest path routing in $\Gamma_\Delta(D)$. Given two vertices X and Y in $\Gamma_\Delta(D)$, a tail of Y with respect to X (as the origin) is a suffix $y_{k+1} \cdots y_D$ such that it forms a subsequence of X, say $x_{i_1} x_{i_2} \cdots x_{i_{D-k}}$, and all the elements $x_1, x_2, \ldots, x_{i_{D-k}}$ occur in Y. Note that tail can be the empty sequence. A header of Y with respect to X is a prefix $y_1 \cdots y_k$ such that the complementary suffix $y_{k+1} \cdots y_D$ is a tail. It is proved in [**5**] that the distance from X to Y equals the length of the shortest header of Y with respect to X. Suppose $y_1 y_2 \cdots y_k$ is the shortest header of Y with respect to X, then the shortest path from X to Y is determined by the following prefixes:

$$X \quad \Rightarrow \quad y_k \cdots \quad \Rightarrow \quad y_{k-1} y_k \cdots \quad \Rightarrow \quad \cdots \quad \Rightarrow \quad y_1 \cdots y_k \cdots = Y\,,$$

Without loss of generality Y can be assumed to be the standard origin. For clarity we list the following conditions which are equivalent to $d(X, Y) = k$ for $k < D$:

(a). y_D appears in X, say $x_j = y_D$.
(b). $x_1, x_2, \ldots, x_j$ are all in Y (but they do not necessarily form a subsequence).
(c). $y_{k+1} \cdots y_D$ is a subsequence of X, but $y_k y_{k+1}$ is not.

When $d(X, Y) = D$, the distance condition is equivalent to the following statement:

(a). Either y_D is not in X,
(b). or y_D is in X, say $x_j = y_D$, but there exists x_r with $r < j$ such that x_r is not in Y.

For example, let $X = 47285136$ and $Y = 82164753$, the shortest header of Y with respect to X is illustrated by $8216 \,|\, 4753$ The following property is helpful in understanding the routing scheme in $\Gamma_\Delta(D)$ and it implies the uniqueness of a shortest path.

PROPOSITION 4.1. Given $X = x_1 x_2 \cdots x_D$ and $Y = y_1 y_2 \cdots y_D$ in $\Gamma_\Delta(D)$, suppose $k = d(X, Y)$. Then $d(i \circ X, Y) \geq d(X, Y)$ unless $i = y_k$, in which case $d(i \circ X, Y) = d(X, Y) - 1$.

In order to reach the conclusion in the above theorem concerning the wide diameter of $\Gamma_\Delta(D)$, we shall start with the easiest case $x_1 = 1$, and we shall give a complete treatment of this case. For the other cases, we shall only give guidelines of the proof because the details are more or less of the same style in arguments and they are too tedious to be presented in full length. In this regard, we hope that a simpler construction will be found with a better understanding of the wide diameter of $\Gamma_\Delta(D)$. There is no doubt that the construction given in this paper is *ad hoc*, although it somehow does give the best bound.

From now on we assume that $Y = 12 \cdots D$, and Y_i's are given by (4.1). For the case $x_1 = 1$, the mapping θ is defined by

$$\theta(X_i) = Y_i\,, \quad (2 \leq i \leq \Delta + 1).$$

The following is an example for $\Delta = 5, D = 4$ and $X = 1325$:

$$\begin{array}{llllllllll}
X_2 = & 2135 & \to & \underline{4}213 & \to & \underline{34}21 & \to & \underline{134}2 & \to & \underline{2134} & = Y_2\,, \\
X_3 = & 3125 & \to & \underline{4}312 & \to & \underline{14}32 & \to & \underline{314}2 & \to & \underline{2314} & = Y_3\,, \\
X_4 = & 4132 & \to & \underline{3}412 & \to & & & & & \underline{23}41 & = Y_4\,, \\
X_5 = & 5132 & \to & \underline{4}513 & \to & \underline{34}51 & \to & & & \underline{234}5 & = Y_5\,, \\
X_6 = & 6132 & \to & \underline{4}613 & \to & \underline{34}61 & \to & & & \underline{234}6 & = Y_6\,.
\end{array}$$

The following lemma gives the distance from X_i to Y_i, from which the shortest path routing is determined in terms of the shortest header.

LEMMA 4.1. *Let $X = x_1 x_2 \cdots x_D$ be a vertex in $\Gamma_\Delta(D)$ such that $x_1 = 1$. Suppose that the distance from X to $Y = 12 \cdots D$ is k, then the distance from X_i to Y_i is given by*

$$d(X_i, Y_i) = \begin{cases} k, & \text{if } \ 1 < i < k, \\ k - 2, & \text{if } \ i = k, \\ i - 2, & \text{if } \ k < i \leq D, \\ D - 1, & \text{if } \ i > D. \end{cases}$$

Proof. We first consider the case when $k < D$. Note that $d(X, Y) = k$ is equivalent to the above conditions (a), (b) and (c) altogether. We need to check the same conditions for the corresponding distances in various cases.

For $1 < i < k$, the verification for $d(i \circ X, Y_i) = k$ is divided into the following three steps:

(a). $y_D = D$ appears in X: Suppose $i \circ X$ does not contain D. Since D is in X, it must be the last element in X and $i \circ X$ is obtained from X via a shift operation. By the condition (b) for X and Y, $x_1, x_2, \ldots, x_D$ are all in Y, implying that X is a permutation on $1, 2, \ldots, D$. Thus, $i \circ X$ is obtained from X via a rotation, which is a contradiction.

(b). Suppose $x_j = D$. If $i \circ X$ is next to X via a rotation, then every element prior to D in $i \circ X$ is in Y_i since Y_i is a permutation of Y. If $i \circ X$ is next to X via a shift operation, the above argument for (a) shows that D cannot be the last element of X. It also follows that every element prior to D in $i \circ X$ is still in Y_i.

(c). Since $(k+1, k+2, \ldots, D)$ is a subsequence of X, it follows that it is also a subsequence of $i \circ X$ because $i < k$ and D stays in $i \circ X$. Clearly, $(k, k+1)$ cannot be a subsequence of $i \circ X$ because it is not a subsequence of X.

Combining (a), (b) and (c) one sees that $d(X_i, Y_i) = k$, and the shortest header of Y_i with respect to X_i is illustrated as follows:

$$2\,3 \cdots i\,1\,(i+1) \cdots k \,|\, (k+1) \cdots D\,.$$

For the case $i = k$, $d(X_k, Y_k) = k - 2$: The verification of conditions (a) and (b) are the same as for the previous case. The only catch for condition (c) is that $(k, 1, k+1, \ldots D)$ is a subsequence of $k \circ X$. Since k is the first element of $k \circ X$, it follows that $d(X_k, Y_k) = k - 2$ and the shortest header of Y_k is illustrated below:

$$2\,3 \cdots (k-1) \,|\, k\,1\,(k+1) \cdots D\,.$$

For the case $k < i \leq D$, $d(X_i, Y_i) = i - 2$: The argument for conditions (a) and (b) remain the same. Noticing that the first two elements of $i \circ X$ is $i\,1$ and that $(i, 1, i+1, \cdots, D)$ is a subsequence of $i \circ X$, it follows that the shortest header of Y_i is given below:

$$2\,3 \cdots k \cdots (i-1) \,|\, i\,1\,(i+1) \cdots D\,.$$

Now comes the last subcase: $D < i \leq \Delta + 1$. The tail containing the single element i of Y_i is clearly the longest tail with respect to X_i, and it is illustrated below:

$$2\,3 \cdots D \,|\, i\,.$$

We finally finish up the main case $k = D$. In this case, we have $d(X, Y) = D$, or in the terminology of [**5**] we say that Y is not a closed vertex with respect to X. For $1 < i < D$, if D is not in X, then D is not in $i \circ X$ either. Suppose $x_j = D$ and there exists x_r with $r < j$ that is not in Y. If $i \circ X$ does not contain D, then we are done. If $i \circ X$ contains D, then x_r stays in $i \circ X$, but it is not in Y_i. Hence we still have $d(X_i, Y_i) = D$.

For $i = D$, we have $D \circ X = D\,1 \cdots$ and $Y_D = 2\,3 \cdots D\,1$. Clearly $d(X_D, Y_D) = D - 2$. For $i > D$, the argument is simply the same as for the case $k < D$. This completes all the cases. □

The shortest path routing from X_i to Y_i easily follows from the above lemma. Our next goal is to show that all the shortest paths from X_i to Y_i are vertex-disjoint. To this end, we need to define two statistics on a vertex so that they can be used to distinguish the vertices along the shortest paths from X_i to Y_i. Given $X = x_1x_2\cdots x_D$, suppose $d(X,Y) = k$, where $Y = 12\cdots D$. We define $\alpha(X)$ to be the first element x_i such that $x_i \notin \{k+1, k+2, \dots, D\}$, and $\beta(X,i)$ to be the minimum of $D+1$ and the smallest element x_j which is on the right of i such that $x_j > k$. For example, suppose $X = 531624$ and $Y = 123456$. Then $d(X,Y) = 4$, $\alpha(X) = 3$ and $\beta(X,3) = 6$. Let $\beta(X) = \beta(X, \alpha(X))$, we call $(\alpha(X), \beta(X))$ the *characteristic pair* of X.

It turns out that the characteristic pair of a vertex on the shortest path from X_i to Y_i can be easily determined in the routing process. The following table illustrates the shortest path P_i from X_i to Y_i in various cases, together with the characteristic pairs from which one sees that all the vertices are indeed distinct. The notation $j \rightarrow$ means the operation of getting $j \circ Z$ from $Z = z_1z_2\cdots z_D$ for $j \neq z_1$.

The cases other than $x_1 = 1$ are more tedious. The critical part is to construct the mapping θ from $N(X)$ to $M(Y)$. Recall that $d(X,Y) = k$.

For $x_1 = k+1$, we have

$$\theta(i \circ X) = \begin{cases} Y_k, & \text{if } \ i = 1\,, \\ Y_i, & \text{if } \ 1 < i < k\,, \\ Y_{\beta(X,1)-1}, & \text{if } \ i = k\,, \\ Y_{i-1}, & \text{if } \ k+1 \le i < \beta(X,1)\,, \\ Y_i, & \text{if } \ \beta(X,1) \le i \le \Delta + 1\,. \end{cases}$$

It is straightforward to see that θ is a bijection. The distance from $i \circ X$ to $\theta(i \circ X)$ is given below:

$$d(i \circ X, \theta(i \circ X)) = \begin{cases} k-1, & \text{if } \ i = 1\,, \\ k, & \text{if } \ 1 < i < k\,, \\ k-2, & \text{if } \ i = k\,, \\ i-1, & \text{if } \ k+1 \le i < \beta(X,1)\,, \\ i-2, & \text{if } \ \beta(X,1) \le i \le D\,, \\ D-1, & \text{if } \ D < i \le \Delta + 1\,. \end{cases}$$

We omit the detailed verification of the above distances. Based on these distances, we give Table 2 to illustrate the shortest path routing from $i \circ X$ to $\theta(i \circ X)$. In addition to the characteristic pairs, we need one more statistic to distinguish the vertices. Without loss of generality we may assume that

TABLE 1. Case 1: $x_1 = 1$.

i $\quad P_i$	$\alpha(V)$	$\beta(V)$	Notes
(a)$1 < i < k$:			
$i \to$ $k \to$ $\vdots$ $i+1 \to$	i	$k+1$	$i \neq 1$
$1 \to$ $i \to$ $\vdots$ $2 \to$	1	$i+1$	$i+1 \neq k+1$
(b) $i = k$:			
$k \to$ $\vdots$ $2 \to$	1	$k+1$	
(c)$k+1 \leq i \leq D$:			
$i \to$ $\vdots$ $2 \to$	1	$i+1$	
(d)$D < i \leq \Delta + 1$:			
$i \to$ $D \to$ $\vdots$ $2 \to$	i	$D+1$	

the elements in X that are greater than D occur in increasing order starting $D+1, D+2, \ldots$, because a permutation on the set $\{D+1, D+2, \ldots, \Delta+1\}$ can map the vertex X into this form without affecting the destination vertex or the other elements in X that do not exceed D.

For $x_1 \neq 1$ or $k+1$, we have

$$\theta(i \circ X) = \begin{cases} Y_{\beta(X,1)-1}, & \text{if } i = 1\,, \\ Y_i, & \text{if } 1 < i < k\,, \\ Y_{\alpha(X)}, & \text{if } i = k\,, \\ Y_{i-1}, & \text{if } k+1 \le i < \beta(X,1)\,, \\ Y_i, & \text{if } \beta(X,1) \le i \le \Delta+1\,. \end{cases}$$

In this case, the distances are given below:

$$d(i \circ X, \theta(i \circ X)) = \begin{cases} \beta(X,1)-2, & \text{if } i = 1\,, \\ k, & \text{if } 1 < i < k\,, i \neq x_1, \\ k-1, & \text{if } i = k\,, \\ i-1, & \text{if } k+1 \le i < \beta(X,1)\,, \\ i-2, & \text{if } \beta(X,1) \le i \le D\,, \\ D-1, & \text{if } D < i \le \Delta+1, i \neq k+1\,. \end{cases}$$

We note that in this case $x_1 < k$. When $k < D$, this claim follows from the condition $d(X,Y) = k$ (the detailed argument is omitted). For $k = D$, since $x_1 \neq 1$ or $D+1$, it is easy to see that $x_1 \neq D$. Hence, the assumption on the elements in X that are greater than D implies that $x_1 < D$ because $x_1 \neq D+1$. The detailed information on the shortest path routing from $i \circ X$ to $\theta(i \circ X)$ is given in Table 3, which also includes the statistic $\beta(V,1)$ to distinguish the vertices.

Having completed the above case by case studies, we arrive at the conclusion that the wide diameter of $\Gamma_\Delta(D)$ does not exceed $D+2$. To determine when the wide diameter is exactly $D+2$, let $X = (\Delta+1, \Delta, \ldots, \Delta-D+2)$ and $Y = 12\cdots D$. The distance from $i \circ X$ to Y is D except for $i = D$. If $1 < i < D$, the unique shortest path from $i \circ X$ to Y goes through $23\cdots D\,(\Delta+1)$. Provided that $D-2 \ge 2$, this implies that, of any Δ disjoint paths from X to Y, at least one must have length at least $D+2$. □

We have left open the problem of determining the wide diameter of $\Gamma_\Delta(D, -r)$.

TABLE 2. Case 2: $x_1 = k+1$.

i P_i	$\alpha(V)$	$\beta(V)$	$\beta(V,1)$	Notes
(a)$i=1$:				
$1 \to$ $k \to$ $\vdots$ $2 \to$	1	$k+1$	$k+1$	Ignore if $k+1 = 2$.
(b)$1 < i < k$:				
$i \to$ $k \to$ $\vdots$ $i+1 \to$	i	$k+1$	$\beta(X,1)$	
$1 \to$ $i \to$ $\vdots$ $2 \to$	1	$i+1$	$i+1$	$i+1 \neq k+1$
(c)$i=k$:				
$k \to$ $\vdots$ $2 \to$	$\neq k+1$	$> k+1$	$\beta(X,1)$	Ignore if $k+1 = 2$.
(d)$k+1 < i < \beta(X,1)$:				
$i \to$	$k+1$	$i+1$	$\beta(X,1)$	
$1 \to$ $i-1 \to$ $\vdots$ $2 \to$	1	i	i	
(e)$\beta(X,1) \leq i \leq D$:				
$i \to$ $\vdots$ $2 \to$		$i+1$	$i+1$	$i+1 \neq \beta(X,1)$
(f)$D < i \leq \Delta+1, i \neq k+1$:				
$i \to$ $D \to$ $\vdots$ $2 \to$	i	$D+1$	$D+1$	

TABLE 3. Case 3.

P_i	$\alpha(V)$	$\beta(V)$	$\beta(V,1)$	Notes
(a)$i=1$:				
$1 \to$	1	$k+1$	$k+1$	
$\beta(X,1)-1 \to$ $\vdots$ $2 \to$	1	$\beta(X,1)$	$\beta(X,1)$	
(b) $1 < i < k, i \neq x_1$:				
$i \to$ $k \to$ $\vdots$ $i+1 \to$	i	$k+1$	$\beta(X,1)$	$i \neq x_1$
$1 \to$ $i \to$ $\vdots$ $2 \to$	1	$i+1$	$i+1$	$i+1 \neq k+1$, $i+1 \neq x_1+1$
(c) $i=k$:				
$k \to$ $\vdots$ $x_1+1 \to$	x_1	$k+1$	$\beta(X,1)$	$d(V,Y) < k$
$1 \to$ $x_1 \to$ $\vdots$ $2 \to$	1	x_1+1	x_1+1	$x_1 < k$
(d) $k+1 \leq i < \beta(X,1)$:				
$i \to$	x_1	$i+1$	$\beta(X,1)$	$i+1 \neq k+1$
$1 \to$ $i-1 \to$ $\vdots$ $2 \to$	1	i	i	
(e) $\beta(X,1) \leq i \leq D$:				
$i \to$ $\vdots$ $2 \to$	x_1	$i+1$	$i+1$	
(f) $D < i \leq \Delta+1, i \neq k+1$:				
$i \to$ $D \to$ $\vdots$ $2 \to$	i	$D+1$	$D+1$	

References

1. S. B. Akers and B. Krishnamurthy, *A group-theoretic model for symmetric interconnection networks*, IEEE Trans. on Computers, 38 (1989), 555-566.
2. J.-C. Bermond, ed., Interconnection Networks, Special Issue, Discrete Applied Mathematics, Vol. 37/38 (1992).
3. F. Comellas and M. A. Fiol, *Vertex symmetric digraphs with small diameter*, preprint, 1992.
4. J. H. Conway and M. J. T. Guy, *Message graphs*, Ann. Discrete Math. 13 (1982), 61-64.
5. V. Faber, J. Moore and W. Y. C. Chen, *Cycle prefix digraphs for symmetric interconnection networks*, Networks, 23 (1993), 641-649.
6. D. F. Hsu, ed., Interconnection Networks and Algorithms, Special Issue, Networks, 23 (1993).
7. ———, *On container width and length in graphs, groups, and networks*, IEICE Trans. on Fundamentals of Electronics, Communications and Computer Sciences, V. E77-A, No.4 (1994), 668-680.
8. M. Jiang and F. Ruskey, *Determining the Hamilton-Connectedness of certain vertex-transitive graphs*, Discrete Math. 133 (1994), 159-169.
9. G. Sabidussi, *Vertex transitive graphs*, Monatsh. Math. 68 (1969), 426-438.

CIC-3, Mail Stop B265, Los Alamos National Laboratory, Los Alamos, NM 87545

E-mail address: chen@c3.lanl.gov, vxf@c3.lanl.gov, knill@c3.lanl.gov

DIMACS Series in Discrete Mathematics
and Theoretical Computer Science
Volume **21**, 1995

PERMUTATION ROUTING VIA CAYLEY GRAPHS WITH AN EXAMPLE FOR BUS INTERCONNECTION NETWORKS

GENE COOPERMAN AND LARRY FINKELSTEIN

ABSTRACT. Cayley graphs have been used extensively to design interconnection networks and provide a natural setting for studying point-to-point routing [1, 2, 3, 5, 6, 7, 12]. The extension of these techniques to the more important problem of permutation routing on interconnection networks presents fundamental problems. This is due to the potentially explosive growth in both the size of the graph and the number of generating permutations, referred to as *one-step permutation routes*, used to define the underlying graph. This paper describes a technique for moderating that growth so that the techniques in [8] can be applied for finding optimal permutation routes. In a particularly striking example, a bus interconnection architecture involving 1.0×10^{17} permutations (nodes of the Cayley graph) is reduced to a computation on a graph with only 3,950 nodes. Further, it is shown how many of the 58,624 generators (directed edges labelled by one-step permutation routes) at each node of the graph may be eliminated as locally redundant.

1. INTRODUCTION

There has been an extensive literature on the use of Cayley graphs to design interconnection networks [1, 2, 5, 6, 7, 12]. Cayley graphs have been used extensively to study point-to-point routing, and they have been particularly attractive for the degree-diameter problem [3, 5, 7]. Nevertheless, other patterns of routing, such as broadcast, permutation routing, and general many-to-one routing are at least as important for parallel computing. This paper describes the use of Cayley graphs to study permutation routing. There are also other approaches to permutation routing [1, 11] that concentrate more on heuristics and an overall framework.

A *Cayley graph* $\mathcal{G}$ is a directed labelled graph associated with a group, G, and generating set S. The nodes of $\mathcal{G}$ are elements of G and the edges are labelled by generators in S. Two nodes, g and h are connected by a directed edge, (g, h), and the edge is labelled by $s \in S$, if $h = gs$. In the case where $S = S^{-1}$, whenever g is connected to h by a directed edge labelled by s, h is simultaneously connected to g by a directed edge labelled by s^{-1}, and so $\mathcal{G}$ may be viewed as undirected. This is typically the case and will be assumed throughout this paper. Cayley graphs

---

1991 *Mathematics Subject Classification.* Primary 05C25, 68M10; Secondary 20-04, 68R10, 90B12.

Key words and phrases. permutation routing, Cayley graph, bus interconnection network.

The authors were partially supported by NSF Grant CCR-9204469.

have the property that they are *vertex-symmetric.* This means that for any two nodes, there is a graph isomorphism of the graph into itself that maps the first node into the second. In the case of Cayley graphs, this graph isomorphism also preserves the labels of the edges. In fact, this property can be easily established by simply embedding G into $\mathrm{Aut}(\mathcal{G})$ through the left regular action so that the resulting image is transitive on the nodes of the graph and preserves edge labels.

A spanning tree $\mathcal{T}$ for $\mathcal{G}$ rooted at the identity e allows one to find shortest paths between arbitrary nodes. Given elements g and h of G, one uses $\mathcal{T}$ to find a shortest path from e to $g^{-1}h$. If w is the word in the edge labels along this path, then $h = gw$. Since $\mathcal{G}$ is vertex-symmetric, it follows that w also defines a shortest path in $\mathcal{G}$ from g to h as well. An important interpretation is that w is a minimal length word in the generating set S that represents the group element $g^{-1}h$. In particular, this allows for the solution of the *minimal word* problem for G relative to the generating set S.

In [8], an efficient method of encoding Cayley graphs was developed, using standard techniques from computational group theory [13]. This method required $\log_2 3$ bits of storage per node of the graph to store a data structure for a spanning tree for the Cayley graph. The spanning tree was derived essentially through a method of breadth-first search. During computation of the data structure, $\log_2 5$ bits per node were temporarily required. Since this is approximately 2 bits, it was called the 2-bit method. Furthermore, it was shown how to use this data structure to find shortest paths in $\mathcal{G}$ between any two nodes. Applications of this method are presented in [9].

We will now show how to apply these ideas to permutation routing in an arbitrary interconnection graph $\mathcal{I}$. We first define a *one-step permutation route* π to be a permutation of the nodes Ω of $\mathcal{I}$ with the property that either $\pi(x)$ is connected to x for each node x or $\pi(x) = x$. The idea is that any one-step permutation route can be executed in parallel without any conflicts. A k-step permutation route is one that is obtained from a sequence of at most k one-step permutation routes. If $\mathcal{I}$ is connected, then a transposition of the nodes of any edge is a one-step permutation route. This leads to the following observation.

Theorem 1. *If $\mathcal{I}$ is connected, then the set S of one-step permutation routes generates* $\mathrm{Sym}(\Omega)$ *and any permutation of the nodes of Ω is a $\binom{|\Omega|}{2}$ permutation route.*

The basic problem we want to consider is given an arbitrary permutation π of $\mathcal{I}$, express π as a sequence of k one-step permutation routes where k is minimal. This can then be viewed as finding a word of minimal length in the set S of one-step permutations that represents π. It is precisely this problem which must be solved.

2. Bus Interconnection Networks

We now apply the previous ideas to bus interconnection networks. From this, it will be clear how to generalize to other examples. Specifically, let $\mathcal{C}$ be a set of chips and let $\mathcal{B}$ be a set of bus lines. Let $\Omega = \mathcal{C} \cup \mathcal{B}$ and let $\mathcal{I}$ be a graph on Ω such that (a, b) is an edge of $\mathcal{I}$ if and only if $a \in \mathcal{C}$, $b \in \mathcal{B}$, and chip a is connected to bus line b.

The following model corresponds to the case in which each chip has only one register. In each clock cycle, for each chip a, either the data in a is interchanged with the data on some unique bus line b or else the data on chip a remains fixed. Section 5 discusses how to generalizes this to multiple registers per chip. Essentially, one identifies $\mathcal{C}$ with the set of chip registers, rather than chips. The simpler model is analyzed in detail in section 3 to illustrate the concepts.

Let S be the set of all involutions $g \in \mathrm{Sym}(\Omega)$ with the property that g is a product of disjoint transpositions of the form $(a\ b)$ where $a \in \mathcal{C}$ and $b \in \mathcal{B}$. The elements of S are precisely the one-step permutation routes of $\mathcal{I}$. Let $M = \max(|\mathcal{C}|, |B|)$ and let $m = \min(|\mathcal{C}|, |\mathcal{B}|)$. Then, the size of $|S|$ is bounded above by $(M+1)^m$.

This is a restricted version of an interconnection network in which distinct data is both read and written to the bus in one clock cycle, thus insuring that the communication is indeed a permutation. This may be the physical situation if the setup time to connect chips and bus lines dominates the actual transfer time. The bus line and/or the chip can hold dummy data, in which case the important transfer of data is in one direction only.

3. Experimental Model

For simplicity in computational experiments, we also assume that each chip is connected to each bus line. This allows us to find experimental data dependent only on parameters $|\mathcal{C}|$ and $|\mathcal{B}|$. It is easy to see that if S is the set of one-step permutation routes for $\mathcal{I}$ defined in the previous section, then S generates $G = \mathrm{Sym}(\mathcal{C} \cup \mathcal{B})$. The technique can be easily applied to other interconnection patterns, and details of that generalization are contained in section 5.

The goal is to express each permutation route as a shortest possible sequence of one-step permutation routes. In the language of group theory, for each permutation g of Ω that setwise stabilizes $\mathcal{B}$, we must find a shortest word in S that represents g. Given the Cayley graph $\mathcal{G}$ for G with generating set S, this is equivalent to finding a shortest path in $\mathcal{G}$ from the identity node 1 to the node g.

Rather than work with the Cayley graph $\mathcal{G}$, we will work with a reduced labelled multigraph, defined below. This smaller graph greatly reduces the necessary computation. The reduced graph is constructed through two reductions. First, we construct a Cayley coset graph $\mathcal{G}_H$ from $\mathcal{G}$, and second, we construct a reduced labelled multigraph $\tilde{\mathcal{G}}$ from $\mathcal{G}_H$.

The first reduction is obtained by noting that we are concerned only with permutations of data on the chips, and not with permutations of whatever data may initially be on the bus lines. This can be best expressed by right cosets $Hg \in G/H$, where $g \in G$ and $H \cong \mathrm{Sym}(\mathcal{B})$ is the subgroup of G that fixes $\mathcal{C}$ pointwise. Formally, the *Cayley coset graph* of G/H with generating set S is a directed labelled multigraph $\mathcal{G}_H$, with nodes labelled by G/H. Two nodes Hg_1 and Hg_2 are connected by an edge with label s if $Hg_1 s = Hg_2$. Thus, the image of $g' \in Hg$ on $\mathcal{C}$ depends only on g and not on the choice of g'. So, it suffices to find a shortest word w in S representing an element of Hg. Many routing problems can be formulated in terms of Cayley coset graphs (also referred to as group action graph in [2]). In general, Cayley coset graphs are not vertex-symmetric.

The second reduction is obtained from the symmetries of the chosen bus interconnection architecture. Recall that a *group automorphism* is a permutation σ of G such that $\sigma(gh) = \sigma(g)\sigma(h)$ and $\sigma(g^{-1}) = (\sigma(g))^{-1}$ for all $g, h \in G$. For G

corresponding to the bus interconnection architecture $\mathcal{I}$, we define a *symmetry* to be an automorphism, σ, of G such that $\sigma(H) = H$ and $\sigma(S) = S$.

This definition insures that the symmetries of G act as graph automorphisms of the Cayley graph $\mathcal{G}$ (with generating set S), which map paths of minimal length in $\mathcal{G}$ to other paths of minimal length. If (g_1, g_2) is a directed edge of $\mathcal{G}$ with label $s \in S$, then σ maps (g_1, g_2) to the directed edge $(\sigma(g_1), \sigma(g_2))$ with label $\sigma(s) \in S$. This and the invertibility of σ imply that σ preserves the paths of minimal length in $\mathcal{G}$. Furthermore, as indicated above, we are actually interested in shortest words which represent a fixed coset Hg. But the symmetries of G also act as graph automorphisms of $\mathcal{G}_H$ in the same sense as for $\mathcal{G}$. To see this, note that if (Hg_1, Hg_2) is an edge of $\mathcal{G}_H$ with edge label $s \in S$ then $\sigma(H) = H$ and $\sigma(s) \in S$ implies that $(\sigma(Hg_1), \sigma(Hg_2)) = (H\sigma(g_1), H\sigma(g_2))$ is an edge of $\mathcal{G}_H$ with label $\sigma(s)$. Thus σ preserves paths of minimal length in $\mathcal{G}_H$ as well.

Conjugation provides a natural means of constructing such symmetries. Define the *conjugate* of g by u to be $g^u = u^{-1}gu$, and extend the definition to sets $H^U = \{h^u\colon\ h \in H, u \in U\}$. The mapping $\sigma_u\colon\ g \mapsto g^u$ is an automorphism of G, and is known in group theory as an *inner automorphism.* In the case of our bus interconnection network, we let $U \cong \mathrm{Sym}(\mathcal{C}) \times \mathrm{Sym}(\mathcal{B})$, and choose as the set of symmetries $\{\sigma_u\colon\ u \in U\}$. We often refer to a symmetry σ_u simply as u. Note that $H^U = H$ and $S^U = S$ for this choice of U, and so the definition of symmetry is satisfied. Hence, conjugation by U preserves shortest words. From group theory, it is known that if $|\mathcal{C} \cup \mathcal{B}| \neq 6$, then the set U contains all possible symmetries.

With this motivation, we now introduce the *reduced labelled multigraph* $\tilde{\mathcal{G}}$, derived from $\mathcal{G}$, in which nodes of $\tilde{\mathcal{G}}$ are identified with subsets $[Hx] = (Hx)^U = Hx^U \subseteq G$ for $x \in G$. Thus, $\{[Hx]\}_{x \in G}$ partitions G into disjoint subsets. In the language of group theory, $[Hx]$ can also be viewed as the union of the cosets in the orbit of Hx under the conjugation action of U on G/H.

It remains to define the edges of $\tilde{\mathcal{G}}$. For purposes of computation, the edge set of $\tilde{\mathcal{G}}$ is defined to be only a subset of the naturally induced edge set from $\mathcal{G}_H$. One would normally expect that for $[Hx] \neq [Hy]$, there is an edge between $[Hx]$ and $[Hy]$ if there is an edge in $\mathcal{G}$ between x' and y' for $x' \in [Hx]$ and $y' \in [Hy]$. For purposes of computation, it is more convenient to first choose an arbitrary transversal $T \subseteq G$ consisting of exactly one point from each distinct set $[Hx]$ for $x \in G$. Denote by $\overline{x}$ the unique element in $[Hx] \cap T$. We define $\tilde{\mathcal{G}}$ to have a directed edge between $[Hx]$ and $[Hy]$ labelled by $s \in S$ if and only if $[H\overline{x}s] = [Hy]$.

We next show that the edges in the two views of $\tilde{\mathcal{G}}$ really do correspond. There is an edge in the unlabelled graph if and only if there is at least one directed, labelled edge in the directed, labelled multigraph. The backward implication is easy. For the forward implication, suppose there is an edge in the unlabelled graph. So, there are $x' \in [H\overline{x}]$, $y' \in [Hy]$ and $s' \in S$ such that $x's' = y'$. Then there is a $u \in U$, such that $x'^u \in H\overline{x}$ and so $x'^u s'^u = y'^u \in [Hy]$. Note also that when S is closed under inverses, if there is a directed edge from $[Hx]$ to $[Hy]$, then there is at least one directed edge from $[Hy]$ to $[Hx]$.

Suppose now that g is an arbitrary element of G, and we want to construct a shortest word on S whose action on $\mathcal{C}$ is the same as the action of g on $\mathcal{C}$. This is the same problem as finding a shortest sequence, $g_1, \ldots, g_k$ of elements of S such that $g(g_1 \cdots g_k)^{-1} \in H$, or equivalently, that $g \in Hg_1 \cdots g_k$. The next result shows how to construct such a sequence from any path of length k from $[H]$ to $[Hg]$.

Theorem 2. *Let g be an arbitrary element of G. There exists a path in $\tilde{\mathcal{G}}$ from $[H]$ to $[Hg]$ of length k if and only if there exist $g'_1, \ldots, g'_k \in S$ such that $g \in Hg'_1 \cdots g'_k$.*

Proof. Suppose first that there exists a path of length k from $[H]$ to $[Hg]$ in $\tilde{\mathcal{G}}$. Then there exists $g_1, \ldots, g_k \in S$ and $w_0, \ldots, w_k \in T$ (hence $w_i = \overline{w_i}$) such that $w_0 = 1$, $w_k = \overline{g}$ and $[Hw_{i-1}g_i] = [Hw_i]$. Note that $[Hw_{i-1}g_i] = [Hw_i]$ implies that there are $h_i \in H$ and $u_i \in U$ such that $w_i = h_i w_{i-1}^{u_i} g_i^{u_i}$. Iterating this, shows that $w_k = hg_1^{u'_1} \cdots g_k^{u'_k}$ for $h = h_k(h_{k-1}(h_{k-2}\cdots)^{u_{k-2}})^{u_{k-1}} \in H$ and $u'_i = u_i \cdots u_k$. Letting $g = h_{k+1}w_k^{u_{k+1}}$ and setting $g'_i = g_i^{u'_i u_{k+1}} \in S$ proves the assertion in the forward direction.

Conversely, suppose $g \in Hg'_1 \cdots g'_k$. Define u'_i and h'_i so that $w_i = h'_i(g'_1 \cdots g'_i)^{u'_i}$ satisfies $\overline{w_i} = w_i$. Let $w_0 = 1$. Note that $w_k = \overline{g}$ and $[Hw_i(g'_{i+1})^{u'_{i+1}}] = [Hw_{i+1}]$. This then gives the required path of length k from $[H]$ to $[Hg]$ in $\mathcal{G}$. □

3.1. Encoding of nodes of reduced graph. Next, we consider how to produce a 1-1 mapping of the nodes of $\mathcal{G}$ into a "small" set of integers. The encoding serves as an almost perfect hash function of $\mathcal{G}$, in that the number of integers is not much larger than the number the number of nodes. The results in [8] show how to use such a 1-1 mapping to store a spanning tree using storage of only $\log_2(3)$ bits times the size of the "small" integer set. This is especially important for the larger graphs. For example, given 16 chips and 4 bus lines ($|\mathcal{C}| = 16$, $|\mathcal{B}| = 4$), the number of cosets, Hg, to explore is $|G/H| = (16+4)!/4! \approx 1.0 \times 10^{17}$. Yet the reduced graph has only 3,950 nodes. The following technique allows a unique encoding of those 3,950 nodes within the range [0, 5,822].

In order to construct a unique encoding of $[Hg]$, we require two auxiliary functions, $r(x,y)$ and $s(x,y)$. We denote by $r(x,y)$ the number of partitions of x indistinguishable elements, such that the largest set within the partition is of size no larger than y. So, for example, $r(5,2) = 3$ since the set of 5 elements can be partitioned into subsets of size $(2,2,1)$, $(2,1,1,1)$ and $(1,1,1,1,1)$ for which no subset has more than 2 elements. In general,

$$r(x,y) = \sum_{i=1}^{y} r(x-i, \min(i, y, x-i)).$$

Also, $s(x,y)$ is the number of ways in which $x+y$ indistinguishable elements can be partitioned, where the $x+y$ elements are first split into two distinguished subsets, the first of size at most x and the second of size at least y. So, for example, $s(1,2) = 5$ and the partitions are of the form $((1),(2))$, $((1),(1,1))$, $((),(3))$, $((),(2,1))$ and $((),(1,1,1))$. In general,

$$s(x,y) = r(x,x)r(y,y) + s(x-1, y+1)$$

for $x > 0$ and $s(0,y) = r(y,y)$.

We now give a unique characterization of a node of the reduced graph in terms of a sequence of invariants. The proof that this is a unique characterization is easy, but is not included here. In the restriction of g to the union of the cycles that fix $\mathcal{B}$, let c_i be the length of the i-th such cycle. Let b_i be the number of points $x \in \mathcal{C}$ such that $x^{g^j} \in \mathcal{C}$ for all $j < i$ and $x^{g^i} \notin \mathcal{C}$. Then $[Hg]$ is uniquely characterized by (a) the sequence $c_1, \ldots, c_{|\mathcal{C}|}$ and (b) the sequence $b_1, \ldots, b_{|\mathcal{B}|}$. Alternatively, one can say that the c_i and b_i are invariants of any permutation $h \in [Hg]$ and the collection of invariants identifies a unique $[Hg]$.

We can now describe the encoding of the equivalence class $[Hg]$ in terms of its invariants, c_i and b_i for fixed $|\mathcal{C}|$. Let $b = \sum_i b_i$ and let $c = \sum_i c_i$. Note that $b + c = |\mathcal{C}|$. Then the encoding is given by

$$\text{Encoding of } [Hg]\text{: } s(b-1, c+1) + t((b_1, \ldots, b_{|\mathcal{B}|}))r(c,c) + t((c_1, \ldots, c_{|\mathcal{C}|})).$$

(We set $s(b-1, c+1) = 0$ when $b = 0$.) Here, for fixed b, if $\sum_{j=1}^{i} x_i = b$, then $t((x_1, \ldots, x_i))$ is a unique encoding of the partition $(x_1, \ldots, x_i)$ among all partitions of b elements. The value of the function "t" is defined to be independent of the order of its arguments. We re-order the arguments so that $x_1 \geq \cdots \geq x_i$ and define

$$t((x_1, \ldots, x_i)) = t((x_2, \ldots, x_i)) + r(\sum_{j=1}^{i} x_i, x_1 - 1)$$

for $i > 0$ and $t(()) = 0$ for $i = 0$. Note that $t((x_1, \ldots, x_i)) \leq r(\sum_{j=1}^{i} x_i, \max_{1 \leq j \leq i} x_i) \leq r(\sum_{j=1}^{i} x_i, \sum_{j=1}^{i} x_i)$.

3.2. Optimization: Use of Reduced Generating Sets. Since the reduced labelled multigraph (defined in section 3) is a multigraph, an edge between nodes in the reduced graph may have multiple labels. One can recognize, a priori, that certain edge labels in the reduced graph are redundant. This allows us to moderate the exponential explosion in the number of generators as $|\mathcal{C}|$ and $|\mathcal{B}|$ increase. This is critical, since $|S| = \sum_{i=1}^{|\mathcal{B}|} \binom{|\mathcal{C}|}{i}\binom{|\mathcal{B}|}{i} i!$. To take an extreme example, for 16 chips and 4 bus lines, $|S| = 58{,}624$. Yet for $x = 1 \in G$, there are only 4 non-redundant generators, g, (each uniquely characterized by the number of transpositions in g) yielding distinct equivalence classes $[Hxg]$. All other elements of xS are conjugate to one of the 12 under U.

In order to discover these redundant edge labels, consider the the node $[Hx]$, $\overline{x} = x$, with outgoing edge labelled by g_0. One must eliminate those $g \in S$ such that $g \neq g_0$ and $[H\overline{x}g] = [H\overline{x}g_0]$. Equivalently, for a given $x \in G$ and $g_0 \in S$, one must find those $g \in S$ with $g \neq g_0$ such that $h(\overline{x}g)^u = \overline{x}g_0$ for some $h \in H$ and $u \in U$.

The following three heuristics are used to remove certain $g \in S$ from consideration. The statement assumes some arbitrary ordering of the points $\mathcal{C} \cup \mathcal{B}$, so that "smallest" and "<" are well-defined. The correctness of these heuristics follows from the results of section 4, which define a total ordering on S. So, if $g, g_0 \in S$, $g_0 < g$, and the heuristic can establish that $[H\overline{x}g] = [H\overline{x}g_0]$, then g is eliminated. Let $\mathcal{C}' = \{a\colon a^{x^i} \in \mathcal{C}\, \forall i\} \subseteq \mathcal{C}$.

(1) First, g can be removed if there is a point $a \in \mathcal{C}'$ moved by g that is not the smallest in the cycle of $\overline{x}$ containing a and for which g does not move the point that is first in that cycle. (There is a conjugation in U that "rotates" the points of the cycle until a is first in the cycle, while leaving $\overline{x}$ unchanged.)
(2) Second, g can be removed when there is a point of $\mathcal{C}'$ moved by g such that it is smallest in a cycle of $\overline{x}$ if there is an earlier cycle of points of the same length as the current cycle, and g does does not move the point that

# chips	# bus lines 1	2	3	4	5	6	7	8
4	6/12 0.05	3/17 0.2	3/19 0.7	3/20 0.7				
5	7/19 0.1	4/28 0.3	3/33 1.0	3/35 3.1	3/36 5.6			
6	9/30 0.2	5/47 0.6	4/57 3.5	3/62 11.8	3/64 30	3/65 56		
7	10/45 0.9	6/73 1.2	4/92 6.7	3/102 37	3/107 120	3/109 293	3/110 699	
8	12/67 0.4	6/114 2.2	5/147 17	4/167 104	3/177 460	3/182 1432	3/184 4259	3/185 13351

is smallest in that cycle. (In that case, there is a conjugation in U that interchanges the two cycles, and so leaves $\overline{x}$ unchanged.)

(3) Third, g can be removed if there are points $a, a' \in \mathcal{B}$ such that $a' < a$, $(a')^{\overline{x}} = a'$, $(a')^g = a'$, $a^{\overline{x}} = a$ and $a^g \neq a$. (The transposition $(a\ a') \in U$ leaves $\overline{x}$ unchanged.)

3.3. Experimental Results. The following table describes the maximum length permutation route, the CPU time, and the number of nodes in the reduced graph for various values of the number of chips and the number of buses. The computation was carried out under AKCL 1.615 (Common LISP). A SPARCstation-2 was used. For each tested combination of number of chips and bus lines, three numbers are reported. The first is the number of steps for an optimal permutation route for the worst case permutation (the diameter of the reduced labelled multigraph). The second is the number of nodes required to store the internal data structure. For large numbers of nodes, this will usually dominate the space requirements. Only two bits are required to store each node [8]. So, storage requirements are minimal for the examples shown. The third number is the number of seconds of CPU time to generate the data structure. From this data structure, an optimal permutation route can be computed in less than a millisecond for any desired permutation.

For 12 chips with 4 bus lines and 16 chips with 4 bus lines, two special tests were made. The results were a diameter of 5 with 919 nodes (5/919) and a diameter of 6 with 3,950 nodes (6/3950), respectively. The computation was carried out on a SPARCserver 670 in 43 minutes and 10 hours of CPU time, respectively.

The range of encodings is larger than, but usually close to the actual number of states. The range of encodings is independent of the number of bus lines. For example, at the beginning of section 3.1, we saw an example of 16 chips and 4 bus lines, in which the range [0, 5,822] was needed for 167 nodes. If the 2-bit encoding scheme is used, space must be allotted for the total range of encodings. Where the actual number of states used is much smaller, one can use hashing techniques to require space proportional to the actual number of states. Naturally, the constant of proportionality would be much higher than two bits.

Note that the length of the optimal permutation route for the worst case is always at least three. The reason for this is easy to see, since a one-step permutation route is required to move the data from the chips to the bus lines. A second one-step permutation route moves the data from the bus lines to other chips. A third one-step route then places some of the data back on the bus lines. In practice, one is only interested in permutation routes that leave all data on the chip, and none on the bus lines. Hence, for permutation routes of interest, with sufficiently many bus lines, there is always a permutation route of length at most two.

4. Pruning Locally Redundant Generators

Although $\tilde{\mathcal{G}}$ was previously defined as the reduced labelled multigraph, we also identify it with the labels the nodes of the graph, which are the equivalence classes $[Hx]$ for $x \in G$. For $[Hx] \in \tilde{\mathcal{G}}$ and $y \in S$, we define $[Hx]y = [H(\overline{x}y)]$ as in section 3.

The next theorem characterizes a set of representatives, $\overline{x}S_x$, for the distinct equivalence classes $[Hx]S \subseteq \tilde{\mathcal{G}}$. Equivalently, this characterizes representatives for the nearest neighbors in the reduced labelled multigraph.

Theorem 3. *Let ρ be a total ordering on S. Let $H \subseteq U$, $H^U = H$ and $S^U = S$. For $x \in G$, let $S_x = \{g \in S\colon\ g \leq_\rho h\ \forall h \in (\overline{x}^{-1}(\overline{x}g)^U H) \cap S\}$. Then $[Hx]S = [Hx]S_x$ and $|[Hx]S| = |S_x|$ — i.e. the number of nearest neighbors of $[Hx]$ in $\tilde{\mathcal{G}}$ is $|S_x|$.*

Proof. $[Hx]S = [Hx]S_x$ follows from the definitions of $[Hx]$ and S_x, along with the properties of H. Let $g, h \in S$. Then

$$\begin{aligned}[Hx]h = [Hx]g \quad &\Leftrightarrow \quad H(\overline{x}h)^U = H(\overline{x}g)^U \Leftrightarrow (\overline{x}h) \in H(\overline{x}g)^U \\ &\Leftrightarrow \quad h \in \overline{x}^{-1}H(\overline{x}g)^U = \overline{x}^{-1}(\overline{x}g)^U H.\end{aligned}$$

Thus if $g, h \in S_x$ and $[Hx]h = [Hx]g$ then g and h must both be lexically least, and so $g = h$. □

Define the *centralizer* of $x \in G$ in U to be $\mathrm{Cent}_U(x) = \{g \in U\colon\ x^g = x\}$.

Corollary 4. *Let ρ be a total ordering on S. Let $H \subseteq U$, $H^U = H$ and $S^U = S$. For $x \in G$, let $S'_x = \{g \in S\colon\ g \leq_\rho h\ \forall h \in (g^{\mathrm{Cent}_U(\overline{x})}H) \cap S\}$. Then $[Hx]S = [Hx]S'_x$ and $|[Hx]S| \leq |S'_x|$ for $[Hx]S \subseteq \tilde{\mathcal{G}}$.*

Proof. Note that since $\mathrm{Cent}_U(\overline{x}) \subseteq U$, $g^{\mathrm{Cent}_U(\overline{x})}H = \overline{x}^{-1}(\overline{x}g)^{\mathrm{Cent}_U(\overline{x})}H \subseteq \overline{x}^{-1}(\overline{x}g)^U H$. So, $S_x \subseteq S'_x \subseteq S$. □

The advantage of S'_x over S_x is that, given $\mathrm{Cent}_U(\overline{x})$, there are well-known techniques for enumerating the elements of S'_x [4]. For U equal to $\mathrm{Sym}(\mathcal{C}) \times \mathrm{Sym}(\mathcal{B})$, $\mathrm{Sym}(\Omega)$ and many other special cases of interest, computing $\mathrm{Cent}_U(\overline{x})$ is also efficient [10].

The heuristics of section 3.2 follow from the special case for S'_x. The order, ρ, is a variation of lexical ordering.

$$g >_\rho h \text{ iff } i^g = i^h\ \forall i < j, j^g \neq j^h, \text{ and either } j^h = j \text{ or } j^g < j^h \neq j$$

The use of a lexical ordering means that one can generate all elements of S'_x in lexical order and if a permutation $g \in X'_x$ satisfies $i^g \leq i^h$ or $i^h = i\ \forall i \leq j$, $\forall h \in g^{\mathrm{Cent}_U(\overline{x})}H$, then the subtree $\{h \in g^{\mathrm{Cent}_U(\overline{x})}H\colon\ j^g \neq j^h, i^g = i^h\ \forall i < j\} \cap S'_x = \emptyset$ and so the entire subtree can be eliminated.

Where it is feasible to compute S_x or where S'_x is not too much larger than S_x, the entire computation of a data structure for the spanning tree of the reduced labelled multigraph has a nice complexity characterization. Since $|S_x| \leq |\tilde{\mathcal{G}}$, the algorithm examines at most $|\tilde{\mathcal{G}}|^2$ directed edges in its breadth-first search. Note that enumeration of S'_x in lexical order costs $O(|S'_x|n)$ for $n = |\Omega|$, since on the current branch, either the entire subtree will be eliminated, or a new element of S'_x will be found before the necessity of backtracking. Computing the product xg and its encoding cost $O(n)$. So, the total cost of the algorithm is $O(|\tilde{\mathcal{G}}|^2(n+E)+|\tilde{\mathcal{G}}|(A+Rn))$, where A is the cost of computing the automorphism group and R is the cost of enumerating the elements of S_x of S'_x. Further, for the bus interconnection networks of section 3, $O(|\tilde{\mathcal{G}}|^2 n)$ is the dominant term.

5. Generalization to Other Models

The previous section was described under the assumption that each chip had only one register (and hence hold only one piece of data at a time). Naturally, most implementations will allow multiple registers or memory cells on a single chip. Hence, shortest paths derived from previous calculations serve only as an upper bound on the shortest length path when multiple registers are allowed.

The model can be extended to k registers per chip, by including k copies of each chip. Every generator must also be replicated to include variation in which each bus is connected to an arbitrary register. Finally, the subgroup H must be extended to include generating elements that transpose arbitrary registers from the same chip.

The previous section was described in a situation with full connections between chips and registers. If there are fewer connections, in general, one will need to choose a subgroup of $U = \mathrm{Sym}(\mathcal{C}) \times \mathrm{Sym}(\mathcal{B})$ that preserves the architecture under conjugation. Thus, the subgroup U can be viewed as the automorphism group of a labelled graph, where each node of the graph is labelled either by $\mathcal{C}$ or by $\mathcal{B}$. Usually ease of manufacturing dictates a uniform architecture, and so most real-world example are still likely to have a fairly large automorphism group.

Finally, a few remarks about generalizations to other models besides bus interconnection networks is worthwhile. There is a common difficulty in translating routes in physical networks to permutation routes. Suppose one has nodes A, B and C with edges between A and B, and between B and C. One would like to add a transposition between A and B as a one-step permutation route. Similarly, one would like to add a transposition between B and C. Finally, one would like to to move data from A to B while other data is moved from B to C, in parallel, in a single one-step permutation route. However, a naive product of transpositions would send data from A to C, which cannot be done in one step. The solution is to split each node into two nodes corresponding to two registers/ports: a sending port and a receiving port. Thus, one can move data from A to B in parallel with distinct data moving from B to C, and this can be expressed as a permutation.

6. Conclusion

It has been demonstrated how to find permutation routes for larger data structures than would be possible by a direct approach. The solution involves pre-computing a special data structure in seconds. One can then derive shortest permutation routes achieving an arbitrary permutation from the pre-computed data structure. Finding that solution is a computation that can be carried out in less than a millisecond (and much shorter times are possible with hardware support).

In implementations, one would pre-compute the data structure for a given hardware architecture off-line. The required space is very small, although the CPU time for the pre-computation can be appreciable. Since the pre-computation is done once only, long CPU times may be acceptable. Specific permutation routes would either be computed at compile-time for an individual application, or else it would be be computed as part of a set-up routine for a route determined at run-time. In the latter case, one would expect the computed permutation route to be used many times to justify the overhead of computing such a route at run-time.

7. Acknowledgements

The authors wish to thank Richard Draper, Chuck Fiduccia and Bryant York for stimulating discussions.

References

1. F. Annexstein and M. Baumslag, "A Unified Framework for Off-line Permutation Routing in Parallel Networks", Math. Syst. Theory **23** (1991), pp. 233–252.
2. F. Annexstein, M. Baumslag, and A.L. Rosenberg, "Group Action Graphs and Parallel Architectures", SIAM J. Computing **19** (1990), pp. 544–569.
3. J-C. Bermond, C. Delome, and J-J. Quisquater, "Strategies for interconnection networks: Some methods from graph theory", *Journal of Parallel and Distributed Computing* **3** (1986), pp. 433-449.
4. G. Butler, *Fundamental Algorithms for Permutation Groups*, Lecture notes in computer science **559**, Springer-Verlag, New York (1991).
5. L. Campbell, G.E. Carlsson, and M.J. Dinneen, "Small Diameter Symmetric Networks from Linear Groups", *IEEE Transactions on Computers* **41**, no. 2, (1992), pp. 218–220.
6. G.E. Carlson, J.E. Cruthirds, H.B. Sexton, and C.G. Wright, "Interconnection networks based on a generalization of cube-connected cycles", *I.E.E.E. Trans. Comp.* **C-34** (1985), pp. 769–777.
7. D.V Chudnovsky, G.V. Chudnovsky and M.M. Denneau, "Regular Graphs with Small Diameter as Models for Interconnection Networks", *I.E.E.E. Trans. Comp.* (1988), pp. 232–239.
8. G. Cooperman and L. Finkelstein, "New Methods for Using Cayley Graphs in Interconnection Networks", *Discrete Applied Mathematics* **37/38** (special issue on Interconnection Networks) (1992), pp. 95–118.
9. G. Cooperman, L. Finkelstein and N. Sarawagi, "Applications of Cayley Graphs", *Applied Algebra, Algebraic Algorithms and Error-Correcting Codes (AAECC-8, Tokyo, 1990)* Springer-Verlag Lecture Notes in Computer Science **508**, S. Sakata (ed.) (1991), pp. 367–378.
10. E. Luks, Permutation Groups and Polynomial-Time Computation, *Proceedings of DIMACS Workshop on Groups and Computation*, DIMACS-AMS **11**, L. Finkelstein and W.M. Kantor (eds.), AMS, Providence, RI, (1993), pp. 139–175.
11. M. Ramras, "Routing Permutations on a Graph", *Networks* **23**, no. 4, (1993), pp. 391–398.
12. S.T. Schibell and R.M. Stafford, "Processor Interconnection Networks from Cayley Graphs", Disc. Appl. Math. **40** (1992), pp. 333–362.
13. C.C. Sims, "Computation with Permutation Groups", in *Proc. Second Symposium on Symbolic and Algebraic Manipulation*, S.R. Petrick (ed.), ACM, New York (1971).

College of Computer Science, Northeastern University, Boston, MA 02115
E-mail address: gene@ccs.neu.edu

College of Computer Science, Northeastern University, Boston, MA 02115
E-mail address: laf@ccs.neu.edu

DIMACS Series in Discrete Mathematics
and Theoretical Computer Science
Volume **21**, 1995

Using Helpful Sets to Improve Graph Bisections

RALF DIEKMANN, BURKHARD MONIEN AND ROBERT PREIS

November 1, 1994

ABSTRACT. We describe a new, linear time heuristic for the improvement of graph bisections. The method is a variant of local search with sophisticated neighborhood relations. It is based on graph-theoretic observations that were used to find upper bounds for the bisection width of regular graphs.

Efficiently implemented, the new method can serve as an alternative to the commonly used local heuristics, not only in terms of the quality of attained solutions, but also in terms of space and time requirements. We compare our heuristic with a number of well known bisection algorithms. Extensive measurements show that the new method is a real improvement for graphs of certain types.

1. Introduction

The problem of assigning large process networks or graph structured data items to massively parallel computing systems is one of the most important tasks in the field of parallel programming. The related mapping problem is *NP*-complete and there exist almost no efficient sequential or parallel heuristics that solve this problem sufficiently.

With the growing performance of interconnection networks and especially with the establishment of independent routing networks, it is appropriate to reduce the mapping problem to the task of partitioning the graph into as many equal sized (or weighted) clusters as there are numbers of processors and to minimize the number of edges crossing the partition boundaries. Because such partitioning problems are still hard to solve and there exist, again, only a very small number of efficient heuristics (first parallel approaches can be found in [**9, 24**]), they are often handled by recursive bisection.

1991 *Mathematics Subject Classification.* Primary 68R10; Secondary 05C10, 94C15, 90C35.

This work was partly supported by the German Research Foundation (DFG Forschergruppe "Effiziente Nutzung massiv paralleler Systeme") and by the ESPRIT Basic Research Action No. 7141 (ALCOM II).

A number of well-known bisection algorithms do exist. The most important are the Kernighan-Lin heuristic (KL) [**15**] with its improvements by Fiduccia and Mattheyses [**8**], Spectral Bisection [**10, 11, 22**], Inertial [**27**], Simulated Annealing [**14, 17**] and the geometric approach by Miller et al. [**2, 21**], which generalizes a number of earlier results on the size of graph separators starting with Lipton and Tarjan [**20**].

As the mapping process is only a preprocessing step to the parallel application that has to be executed, heuristics have to be accurate, fast and as little space-consuming as possible.

In this paper, we introduce a new method for improving graph partitions that is more powerful than most of the commonly used bisection heuristics, not only if solution quality is considered, but also with regard to space and time requirements. Just as the KL algorithm, our heuristic is based on the local search method: it repeatedly tries to improve a given bisection with the help of local rearrangements. As follows, we classify the nodes of a graph according to their number of external and internal edges (i.e. edges crossing the cut). Based on this classification we search for k-helpful sets, i.e. sets of nodes that improve the cut size by k and for balancing sets that increase the cut size by not more than $k-1$ edges. Whenever such sets are found, the corresponding nodes are exchanged. The idea of this heuristic is based on a proof technique used in [**13**] to show upper bounds on the bisection width of 4-regular graphs.

We compare our algorithm to results of the Spectral, Multilevel and Inertial method with and without additional KL that were obtained using the Chaco[1] partitioning library [**12**] and a self-implemented version of KL with its improvements by Fiduccia and Mattheyses. As test suite we used several self-generated regular networks, especially De Bruijn networks, regular random graphs with an existing small edge separator [**3**], a number of graphs from the Harwell-Boeing collection [**5**] that are also used in [**22**] and some finite element discretizations from [**7**], [**9**] and [**23**].

Extensive measurements show that the new heuristic is an improvement for a variety of different graphs. The implementation of the new method is very efficient, not only for time, but also for space requirements. Currently, we are able to bisect graphs with 1M nodes and 2M edges on a Sparc Station SS10/30 with 96MB memory in about 6 minutes, without affecting the solution quality.

The following section gives an overview of existing graph partitioning heuristics. In section 3, we give a detailed description of our algorithm. Section 4 compares the results of our heuristic to those obtained with the Chaco library and with a self-implemented version of KL.

[1]Chaco is copyrighted by Sandia National Laboratories. Its use is granted under License Agreement No. 93-N00053-021.

2. Graph Partitioning

The graph partitioning problem can be defined in the following way:

DEFINITION 2.1. *Let $G = (V, E)$ be a graph and $\pi : V \to \{1, 2\}$ a mapping that cuts the nodes into two sets V_1 and V_2. For a node $v \in V$ let* deg(v) *be its degree,* ext$(v) = |\{w \in V; \{v, w\} \in E, \pi(v) \neq \pi(w)\}|$ *its number of* ***external edges*** *and* int(v) = deg(v) − ext(v) *its number of* ***internal edges****. Define* ext$(\pi) = \sum_{v \in V_1}$ ext(v) *to be the* ***cut size*** *of π and* bal$(\pi) =| (| V_1 | - | V_2 |) |$ *its* ***balance****.*

The task of the ***partitioning problem*** *is to find a mapping π that minimizes* ext(π) *while keeping* bal(π) *below a given constant c. If $c \leq 1$ is required, the problem is called the* ***bisection problem****.*

Throughout this paper the bisection problem is considered. Remember that some of the methods from the literature calculate partitions with weaker constraints on the balance.

The bisection problem is *NP*-complete (see e.g. [**19**]) but there exist a number of efficient heuristic solution methods [**12, 14**]. Usually, they are divided into *global* and *local* methods [**9, 18**]. Global methods are sometimes called *construction heuristics* because they take the graph description as input and generate a valid bisection. Local methods are called *improvement heuristics.* They take the graph and a valid bisection as input and try to improve the partition.

Commonly used global heuristics are simple *Front Techniques* [**6**], *Coordinate-* and *Graph Bisection* [**26**], *Inertial Bisection* [**27**], *Geometric Partitioning* [**2, 21**], *Spectral Bisection* [**22**] with different eigensolvers [**12**] and combined with edge-contraction schemes [**1**] and *Multilevel Partitioning* [**11**]. In addition to these more or less sophisticated heuristics, there exist many simple node numbering schemes [**12**] which construct valid solutions to the bisection problem.

These valid solutions are used as input to local methods like *Simulated Annealing* [**4, 14, 17**] or *Kernighan-Lin* [**15**] which are mainly based on local search.

Our method belongs to the local heuristics. Thus it compares directly to the KL-algorithm.

The *Kernighan-Lin* heuristic [**15**] is the most frequently used local partitioning method. Starting with a given initial bisection, it tries to find a sequence of node-pair exchanges that leads to an improvement of the cut size.

For a description of the KL algorithm, we need the following definition:

DEFINITION 2.2. (*diff*-value)
Let $G = (V, E)$ be a graph. For a node $v \in V$ define diff(v) = ext(v) − int(v) *to be the* ***diff-value*** *of this node.*

Note that the cut size decreases by *diff*(v) if v changes the side of the cut. For a pair $(u, v), u \in V_1, v \in V_2$ of nodes let $b(u, v) = diff(u) + diff(v) - e(u, v)$ with $e(u, v) = 2$ if $\{u, v\} \in E$, 0 otherwise. $b(u, v)$ describes the change in cut size if u and v are exchanged. The KL algorithm repeats choosing pairs (u, v) of unmarked nodes with $b(u, v)$ maximal, exchanges them logically and marks them. Note that $b(u, v) \geq 0$ must not necessarily hold throughout the choice of pairs. If all nodes are marked, the algorithm chooses the sequence of pair exchanges that leads to a maximal decrease in cut size and exchanges the corresponding nodes physically. If there is no such improving sequence, i.e. the algorithm is not able to improve the bisection, it terminates.

The KL heuristic as originally described in [**15**] is not very efficient because of the $O(n^2)$ steps for choosing node pairs. There are several possible improvements, among which the linear-time version of Fiduccia and Mattheyses [**8**] is the most important. Instead of performing pair exchanges it only moves nodes with maximal *diff*-value while keeping the balance below a certain constant. Generalizations of the KL algorithm to k-partitioning are also known [**10, 16**], just as a randomized version as well as several other improvements [**12**].

In general, the results of KL are convincing, provided it is started with a fairly satisfactory initial bisection.

3. The New Heuristic

Just as the KL algorithm, our heuristic is based on local search. It starts with a given initial bisection π and tries to improve the cut size with the help of local rearrangements. The choice of these rearrangements is the main difference to KL. The algorithm starts on nodes of a certain type to search for *k-helpful sets*, i.e. subsets of nodes from either V_1 or V_2 that decrease the cut size by k if moved to the other side. If such a set is found in one part of the graph, say V_1, it is moved to V_2. The algorithm then starts to search for a set of nodes in the overweighted part V_2 that is large enough to rebalance the bisection but increases the cut size by not more than $(k - 1)$ edges. If such a *balancing set* is found, it is moved to V_1 and the whole process is started again.

Before we describe our algorithm in more detail, we have to define helpful and balancing sets:

DEFINITION 3.1. (Helpful Set)
Let $S \subset V_i$, $i \in \{1, 2\}$ be a subset of nodes from one partition. For each $v \in S$ define $int_s(v) = |\{w \in S; \{v, w\} \in E\}|$ as the number of internal edges w.r.t. S.

$$H(S) = \sum_{v \in S} (ext(v) - int(v) + int_s(v))$$

*is the **helpfulness** of S. S is called $H(S)$-**helpful**.*

A move of a set S to the other side decreases the cut size by $H(S)$. Thus the helpfulness plays the same role for a set as the *diff*-value does for a single node. Throughout the rest of the paper we will use 'helpfulness' and '*diff*-value' synonymously for sets and nodes. Examples of 2-helpful sets are shown in Fig. 1 (nodes are marked with their *diff*-value).

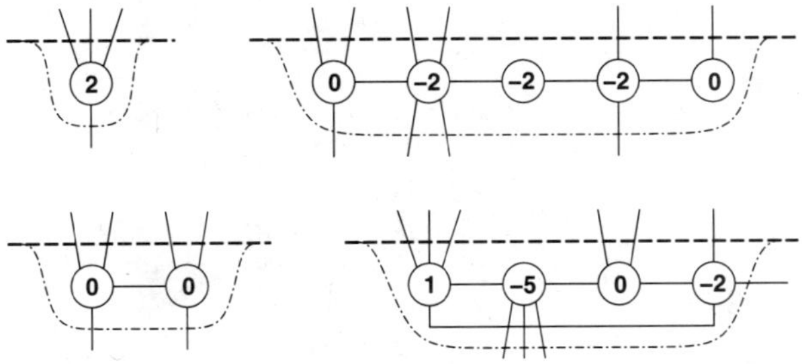

FIGURE 1. 2-helpful sets.

DEFINITION 3.2. (Balancing Set)
Let $S \subset V_i$ be a k-helpful set. A set $\bar{S} \subset V_j \cup S$, $j \neq i$ is called a ***balancing set*** *of S, if $|\bar{S}| = |S|$ and $\bar{S}$ is at least $(-k+1)$-helpful.*

Note that the cut size increases by not more than $(k-1)$ if $\bar{S}$ is moved from one side of the cut to the other.

3.1. The algorithm. Figure 2 shows the main algorithm. The size of the helpful sets to be searched for is controlled by *limit*. It is initialized with $cutsize/2$ and changed in each round, depending on a successful or unsuccessful search.

In each round, the algorithm starts to search for k-helpful sets in both parts of the graph. If no k-helpful set S with $k \geq limit$ is found, the set with the highest helpfulness is taken and *limit* is set to this value. If no k-helpful set with $k > 0$ is found, then *limit* is set to 0, which leads to a termination of the algorithm.

If any k-helpful set S with $k > 0$ is found, *limit* stays positive and the algorithm moves S from one side of the cut to the other, thereby decreasing the cut

```
PROCEDURE Improve Bisection
  limit = cutsize/2;
  REPEAT
    search for a k-helpful set S with k ≥ limit;
    IF (no k-helpful set S with k ≥ limit found)
      IF (any k-helpful set with k > 0 found)
        S = set with highest helpfulness found; limit = H(S);
      ELSE
        S = ∅; limit = 0;
    IF (S ≠ ∅)
      move S to the other side;
      search for a balancing set S̄ of S;
      IF (successful)
        move S̄ to the other side; limit = limit ·2;
      ELSE
        move S back to its original position; limit = ⌊limit/2⌋;
  UNTIL (limit = 0)
```

FIGURE 2. The main algorithm.

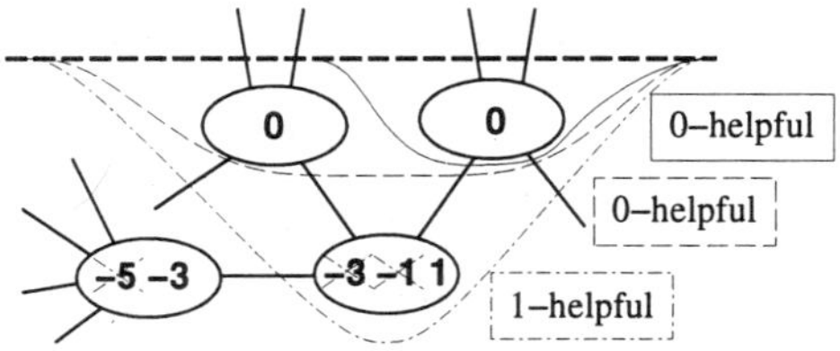

FIGURE 3. Changing *diff*-values of neighbors.

size by k. It then starts to search for a balancing set $\bar{S}$. If such a set is found, the bisection is rebalanced and *limit* is increased, assuming that it may now be possible to find a more helpful set in the next round. If the algorithm fails to rebalance, S is moved back to its origin and the value of *limit* is decreased, assuming that a less helpful set will probably have a smaller size and be easier to balance.

Termination is reached if *limit* is equal to 0. This happens if either no k-helpful set with $k > 0$ was found or the balancing of a 1-helpful set was not successful (*limit* decreased to 0). The net improvement of the cut size in each successful round of the algorithm (move of S, rebalance with $\bar{S}$) is at least one edge.

The crucial steps of the algorithm are searching for helpful and balancing sets. Both use the building of helpful sets, which will be described in the next section. A closer look at the two steps will be taken in sections 3.3 and 3.4.

3.2. Building helpful sets. The main search function of this heuristic proceeds in a BFS manner comparable to Kruskal's algorithm for calculating minimum spanning trees (see e.g. [**25**]).

It starts with an empty set (helpfulness 0) and builds it up with nodes from one side of the cut. In each step it chooses a node from a certain range of *diff*-values, moves it into the set (the helpfulness of which increases by the *diff*-value), and increases the *diff*-values of its neighbors that lie in the same part of the bisection (these nodes have one internal edge changed to an external, therefore, their *diff*-value increases by 2, cf. Fig. 3).

This increase of *diff*-values and the movement of nodes is performed logically. Considering only neighbors on the same side enables the search function to be applied independently on both sides of the cut. If a set of nodes is physically moved to the other side of the cut, the *diff*-values of neighbors on the target side have to be changed, too.

During the search the size of the set increases in each step. There are certain cases that stop the search process:

- The helpfulness of the set reaches a specified *limit.*
- Only nodes with a certain *diff*-value are considered (e.g. ≥ 0) and none of those are left.
- The size of the set reaches a specified value.

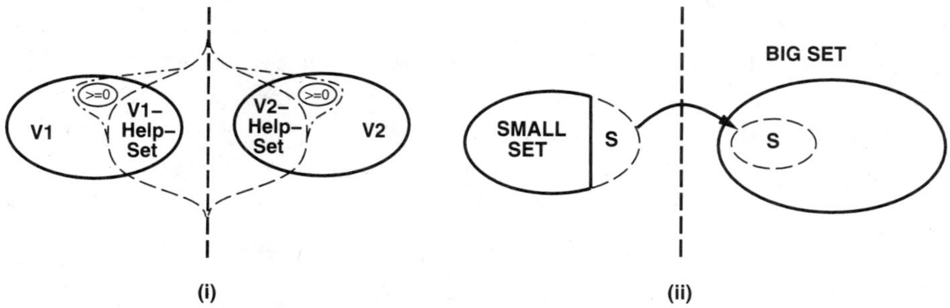

FIGURE 4. Searching for a k-helpful set by moving nodes with *diff*-value ≥ 0 logically to the Help-Sets (i); Moving S physically to the other side (ii).

Note that if only nodes with a positive *diff*-value are considered, the helpfulness of the set always stays the same or increases, but never decreases.

The BFS manner of this search can be observed if we consider the way the nodes are taken into the set. A node can be chosen if its *diff*-value is higher than a specified value. This happens if either the value is initially in that range or if it has increased beyond that value during the search. In the latter case the node has at least one neighbor which has already been taken into the set.

3.3. Search for a k-helpful set with $k \geq$ *limit*. On each side of the cut a helpful set is built up according to the described building process (cf. Fig. 4(i)). We will call these sets *Help-Sets*. Their cardinality increases dynamically during the search. By only updating the *diff*-values of the neighbors of moved nodes in the source-set, the two search processes run independently.

Throughout the search only nodes with *diff*-values greater than or equal to 0 are considered. This can lead to an unsuccessful termination if none of them is left, but it also guarantees that the helpfulness of the Help-Sets never decreases. The search terminates if the helpfulness of either of the Help-Set reaches the intended value of *limit* or if on both sides there exist no further nodes with a *diff*-value of at least 0.

After a successful termination, S is set to the Help-Set with the highest helpfulness and is moved to the other side physically. The other Help-Set is no longer considered. In Figure 4(ii), BIG SET represents either V_1 or V_2 unified with S. The SMALL SET is the remaining part which was reduced by S.

3.4. Balancing. Finding a balancing set $\bar{S}$ is more difficult because of the additional constraint concerning its size. Remember that a set $\bar{S}$ has to be of size $|S|$ and should be as helpful as possible, but not less than $(-k+1)$-helpful (if S was k-helpful).

The basic idea is to start with an empty set $\bar{S}$ and to choose subsets of nodes from BIG SET that increase its size as much as possible while decreasing its helpfulness as little as possible. The algorithm proceeds in three phases. The first phase adds those nodes to $\bar{S}$ that have a positive *diff*-value. The second phase tries to find subsets of nodes that, added to $\bar{S}$, keep its helpfulness unchanged, i.e. it searches for 0-helpful sets. Each of these two phases terminates if $\bar{S}$ becomes large enough. If they are not sufficient, i.e. they do not manage to fill $\bar{S}$ to the size of S, a third phase uses a greedy strategy trying to finish the balancing.

Note that, as S is part of BIG SET, the three phases return some nodes from S. But note also that the procedure is not able to take the whole set S because in this case $H(\bar{S})$ would be equal to $-k$.

In the following, we describe the three phases in more detail.

3.4.1. *Phase 1.* The first phase is straightforward. As long as there are nodes in the BIG SET with *diff*-values greater than or equal to 0, the node with the highest *diff*-value is taken into $\bar{S}$. As already described in Section 3.2, the nodes are considered to have moved to the other side and the *diff*-values of their neighbors in BIG SET are increased.

During this phase, the size of $\bar{S}$ increases step by step and $H(\bar{S})$ stays the same or increases, too. Phase 1 terminates if it succeeds in balancing or if there are no more nodes in BIG SET left whose *diff*-values are at least 0.

3.4.2. *Phase 2.* This phase tries to find 0-helpful sets, i.e. subsets of nodes from BIG SET that, when moved to $\bar{S}$, do not decrease its helpfulness. The main idea is to perform several searches, starting on single -1- or -2-helpful nodes. Each search tries to complete its subset of nodes to a 0-helpful set. Notice that the BIG SET does not contain nodes with a positive *diff*-value at the beginning of this phase.

Figure 5 shows some possible 0-helpful sets. They can, for example, consist of rings of -2-helpful nodes or lines of -2-helpful nodes with -1-helpful nodes at each end. But there are also more complex arrangements possible, which may include $-d$-helpful nodes with arbitrary d. Phase 2 tries to find as many of these 0-helpful sets as possible.

Figure 6 shows the second phase. As in the first phase, the main search function is used (cf. sec. 3.2). But instead of moving the nodes directly to $\bar{S}$, they are (logically) kept in a set called -2-SET. -2-SET contains the present subset of nodes that is tried to be completed to a 0-helpful set. Let $k' = k + H(\bar{S})$ be the improvement in cut size that is caused by S and the present $\bar{S}$ as being moved to the other side. For the start of each search, only nodes with a *diff*-

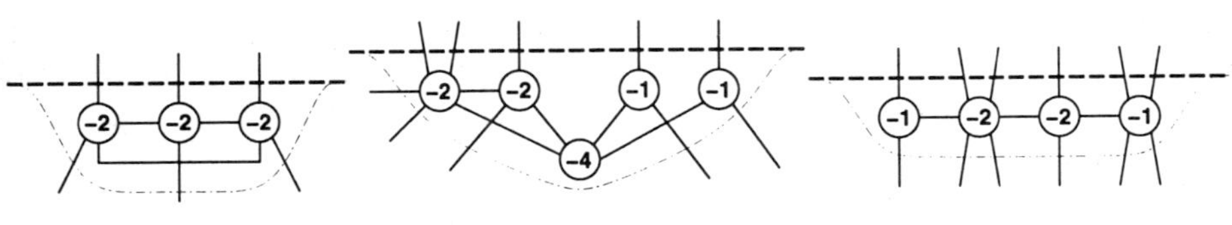

FIGURE 5. Some 0-helpful sets.

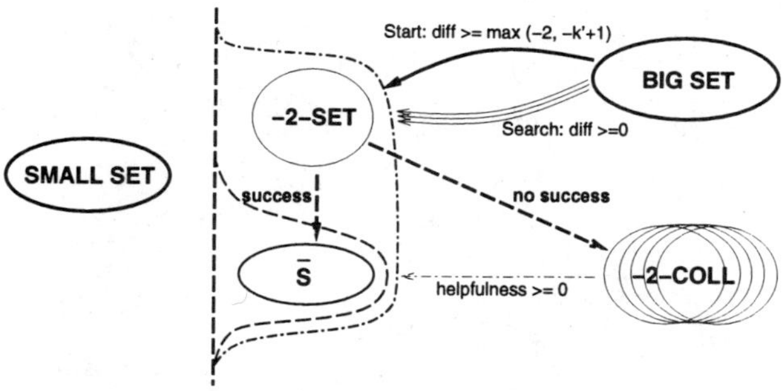

FIGURE 6. The second Phase.

value of at least $max(-2, -k'+1)$ are considered. Throughout the search only nodes that are at least 0-helpful can be moved to −2-SET. This guarantees that a −2-SET can move to $\bar{S}$ at any stage throughout the search without decreasing its helpfulness too much. This is especially important if a −2-SET becomes large enough to finish the balancing. We will come to this later. The restriction has several implications:

- only −1- and −2-helpful nodes from the BIG SET are considered during the search process;
- the helpfulness of −2-SET is at least −2 and never decreases;
- if k' is 2, only −1-helpful nodes can be taken and
- if k' is 1, phase two does not start at all.

If the search for the present −2-SET terminates because there are no more nodes in BIG SET that could be moved and if its helpfulness is still below 0, it is stored for later use. The collection of such sets will be called −2-*COLL* (cf. Fig. 6). It includes subsets of nodes that are at least −2-helpful. Many of them may become 0-helpful if some other nodes from BIG SET are moved via −2-SET to $\bar{S}$.

The update of *diff*-values of neighbors during the search happens only with those neighbors that are part of the BIG SET. The sets in −2-COLL are not considered to have changed sides. Therefore, if a −2-SET can not be completed to be 0-helpful and is moved to −2-COLL, the *diff*-values of its neighbors in BIG SET are restored. The *diff*-values of neighbors in −2-COLL are not changed in this stage of the search because the nodes in −2-COLL are not considered for the building of −2-SETs. Only after a −2-SET becomes 0-helpful and is moved to $\bar{S}$ the *diff*-values of its neighbors in −2-COLL are changed.

Figure 7 shows the search procedure of phase 2. The search for the present −2-SET terminates if one of the following 3 cases occurs:

1. Its helpfulness becomes 0 or larger. In this case the whole set is moved to $\bar{S}$. The *diff*-values of neighbors in −2-COLL are updated.
   Each such update in −2-COLL increases the helpfulness of one of its sets by 2. Therefore, some of them may become 0- or even more helpful and can be moved to $\bar{S}$. The algorithm takes the set with the highest

```
PROCEDURE Phase 2
  WHILE ((|S̄| < |S|) AND BIG SET contains nodes
          with diff-value ≥ max(−2, −k + 1 − H(S̄)))
    start building a −2-SET with the max. helpful node;

    WHILE (H(−2-Set) < 0 AND (| − 2-SET| + |S̄| < |S|) AND
              BIG SET has node with diff-value ≥ 0)
       move node with the highest diff-value to −2-SET;

    IF (H(−2-SET) ≥ 0)
       move −2-SET to S̄; /* SUCCESS */
       while ((|S̄| < |S|) AND −2-COLL contains sets S′ with H(S′) ≥ 0)
         take set S′ ∈ −2-COLL with max. helpfulness;
         IF (|S′| + |S̄| < |S|)
            move S′ to S̄;
         ELSE
            reduce S′ to |S| − |S̄| and move it to S̄;
    ELSE IF ((| − 2-SET| + |S̄| = |S|))
       move −2-SET to S̄; /* SUCCESS */
    ELSE
       move −2-SET to −2-COLL; /* NO SUCCESS */
```

FIGURE 7. Phase 2.

helpfulness and moves it to $\bar{S}$, as long as such sets are left in -2-COLL. If one of these sets is larger than $|S| - |\bar{S}|$, it is reduced to the needed size by removing the last added nodes. The way the sets in -2-COLL are constructed guarantees that this reduction is valid, i.e. that its helpfulness stays above $-k'$. As a result of the reduction, $|\bar{S}|$ becomes equal to $|S|$ and the whole balancing procedure terminates with success.

2. The partition becomes balanced, i.e. $|-2\text{-SET}| + |\bar{S}| = |S|$. In this case, the helpfulness of -2-SET is still at least $max(-2, -k'+1)$ (guaranteed by its construction) and it can be moved to $\bar{S}$. The partition is balanced and the search terminates with success.
3. There are no nodes left in BIG SET with a *diff*-value of at least 0. In this case the present -2-SET can not be enlarged any further and is stored in -2-COLL for later use.

3.4.3. *Phase 3.* This phase is straightforward again. It chooses nodes from BIG SET and sets from -2-COLL to be moved to $\bar{S}$. It selects the node/set with the highest helpfulness as long as the move does not diminish the helpfulness of $\bar{S}$ to a value below $-k+1$. If a chosen set from -2-COLL has a useful helpfulness but the number of nodes is larger than needed, i.e. larger than $|S| - |\bar{S}|$, its size is reduced as in Phase 2. Phase 3 terminates if every move of a node or a set would result in a helpfulness of $\bar{S}$ below $-k+1$ or if it manages to fill up $\bar{S}$ to the size of S.

3.5. Efficiency. We estimate the time the algorithm needs to carry out one round. The number of rounds the algorithm performs depends on the characteristic of the individual problem and especially on the initial solution the method

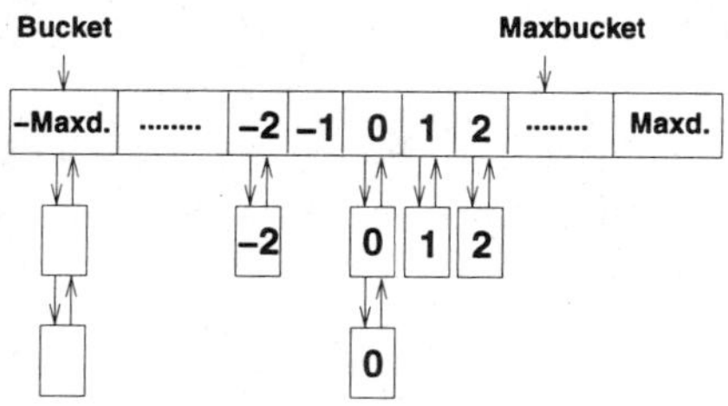

FIGURE 8. The bucket structure.

is started on. This effect can be observed with all local partitioning heuristics.

The main functions that are used throughout the search are choosing a node with the highest *diff*-value and changing the *diff*-value of its neighbors. Both operations, choosing a node and updating a single neighbor, can be done in constant time by using buckets to store the nodes of V_1 and V_2. Within each bucket, nodes with the same *diff*-value are stored as a doubly linked list (cf. Fig. 8). This enables the insertion and deletion of nodes in constant time; operations which have to be performed if *diff*-values are changed. $2 * Maxdegree + 1$ buckets are needed to store the nodes on one side of the cut. A pointer *Maxbucket* guarantees the extraction of a node with the highest *diff*-value in constant time.

LEMMA 3.1. *Finding a k-helpful set takes* $O(|V| + |E|)$ *steps.*

PROOF. The number of nodes moved to the Help-Sets is bounded by $|V|$ and the number of *diff*-values updated at adjacent nodes is no more than $|E|$. The nodes are stored as doubly linked lists in buckets, sorted according to their *diff*-values. The insertion and deletion of a node from a single bucket takes $O(1)$ time. The pointer *Maxbucket* enables the fast identification of the first nonempty bucket (starting from *Maxdegree*). This pointer is moved if either higher-valued buckets get filled throughout the search or if the bucket it points to becomes empty. In the first case, the pointer moves no more than two steps to the right, as *diff*-values increase by no more than 2. If its bucket empties, the pointer has to move a number of steps (at most $Maxdegree = |V|$ steps) to the left to find the next non-empty bucket (cf. Fig. 8). But as each edge of the graph increases the pointer by no more than 2 (because each edge is considered at most once), it can move at most $2 \cdot |E| + 2 \cdot |V|$ steps in total to the left (including the maximal distance between its start and end position). Thus the amortized time for finding helpful sets is $O(|V| + |E|)$.

□

LEMMA 3.2. *Finding a balancing set takes* $O(|V| + |E|)$ *steps.*

PROOF. The nodes of the BIG SET are stored in a bucket structure, as already described (cf. Fig. 8). Nodes move only a small constant number of times from one side of the cut to the other (more precisely they can change sides no more than 4 times). Thus the number of times an edge changes its status from internal to external is also bounded by a constant and the conjectures of Lemma 3.1 hold for all bucket operations.

The sets in the −2-COLL are also stored in buckets sorted according to their helpfulness. Thus the same observations for nodes apply for the sets, too.

□

COROLLARY 3.1. *One round of the new improvement heuristic terminates within* $O(|V| + |E|)$ *steps.*

4. Comparison to Other Methods

In this section, we present our results and compare them to those obtained with different bisection heuristics. Thanks to Bruce Hendrickson and Robert Leland from Sandia, who provided the code of their Chaco library of partitioning methods [**12**], we were able not only to compare results of cut sizes to similar results in the literature, but also to compare running times and space requirements. The Chaco library includes implementations of Inertial, Spectral, Multilevel and some simple partitioning algorithms as global methods and the KL with its improvements by Fiduccia and Mattheyses [**8, 19**] as a local method.

4.1. Test-graphs. As graphs in our benchmark set, we chose De Bruijn graphs $DB_a(d)$ of varying basis a and dimension d, some random graphs with an upper bound on the bisection width generated with an algorithm from [**3**] and graphs that result from numerical modeling of physical phenomena. Some of them were extracted from the Harwell-Boeing Collection [**5**], the others are mainly finite element discretizations obtained from [**7, 9, 18, 23**].

Table 1 lists the test-graphs with their main characteristics. Main characteristics are not only the number of nodes and the number of edges, but also the minimum, average and maximum degree of the nodes in the graph. The random and the De Bruijn graphs are regular, the others have large deviations in

graph	$\|V\|$	$\|E\|$	degree			geom. dim.	best bis. found
$DB_2(15)$	32768	65533	2	4.00	4	–	3606
$DB_2(16)$	65536	131069	2	4.00	4	–	6646
$DB_2(17)$	131072	262141	2	4.00	4	–	12470
$DB_4(8)$	65536	262134	6	8.00	8	–	25616
BCSSTK13	2003	40940	4	40.88	94	–	2355
NASA4704	4704	50026	5	21.27	41	–	1292
airfoil1	4253	12289	3	5.78	9	2	74
whitaker3	9800	28989	3	5.92	8	2	127
crack	10240	30380	3	5.93	9	2	184
brack2	62631	366559	3	11.71	32	3	731
rnd-100000-3	100000	150000	3	3.00	3	–	112
rnd-100000-4	100000	200000	4	4.00	4	–	112

TABLE 1. Test-graphs: *degree* means min., av. and max. node degree (self-loops are extracted).

graph	SP	+KL(c)	+HS	ML	KL-5	HS-5
$DB_2(15)$	4095 (17.69)	4055 (23.63)	3682 (28.30)	4018 (12.48)	3820 (31.27)	3784 (64.67)
$DB_2(16)$	8192 (37.25)	8116 (53.01)	7416 (67.59)	7588 (31.85)	7862 (72.64)	7284 (163.02)
$DB_2(17)$	– –	– –	– –	13512 (71.08)	14772 (197.48)	13244 (267.91)
$DB_4(8)$	29136 (16.76)	28236 (49.12)	27998 (25.48)	28010 (77.86)	27484 (80.73)	26560 (48.09)
BCSSTK13	3591 (2.42)	2880 (2.99)	2668 (2.62)	2920 (1.54)	2386 (2.15)	2386 (0.87)
NASA4704	1319 (5.83)	1313 (6.28)	1313 (5.93)	1292 (1.96)	1313 (4.12)	1652 (0.86)
airfoil1	142 (23.03)	85 (23.39)	88 (23.08)	79 (1.46)	85 (1.64)	83 (0.56)
whitaker3	137 (32.89)	128 (33.30)	133 (33.00)	127 (1.46)	130 (5.68)	129 (1.16)
crack	233 (65.22)	185 (66.09)	221 (65.35)	200 (2.08)	212 (5.37)	219 (1.52)
brack2	– –	– –	– –	872 (12.01)	762 (62.23)	2674 (13.77)
rnd-100000-3	112 (41.45)	112 (43.51)	112 (42.95)	114 (18.60)	3692 (82.77)	19234 (15.39)
rnd-100000-4	112 (34.99)	112 (37.87)	112 (36.81)	112 (21.68)	112 (53.93)	112 (20.12)

TABLE 2. Results of Spectral Partitioning (SP) (with additional KL(c) or the new heuristic (HS) as clean-up), Multilevel (ML), KL and HS. KL and HS are done with 5 runs on different simple global partitions. Numbers in brackets indicate running times in seconds on SS10/30.

their node degrees (especially BCSSTK13, NASA4704 and brack2). Coordinate information is provided with some of the test-graphs, i.e. they are embedded in either $\mathbb{R}^2$ or $\mathbb{R}^3$. 'geom. dim.' refers to the space dimension they are embedded in. In addition, Table 1 provides information about the best bisection we found for each of the graphs. These are results obtained from numerous test-runs with different global and local partitioning methods. The random graphs have an existing small edge separator of 112 that is guaranteed by their construction [**3**].

4.2. Bisection Results. Table 2 shows bisection results of several methods. The following methods are compared:

- The Spectral method (SP) from the Chaco library [**12**] together with Kernighan–Lin (+KL(c)) and our new heuristic (+HS) as a local clean-up.
- The Multilevel method (ML) from the Chaco library [**11**]. It includes the Spectral method as coarse graph partitioner and the Kernighan–Lin method as local refiner on different levels.

- The Kernighan–Lin method (KL-5) with our own implementation of the improvements by Fiduccia and Mattheyses [**8**]. The results include 5 runs on different simple global partitionings.
- Our algorithm (HS-5), which is also used with 5 runs on different simple global partitionings.

KL(c) refers to the implementation of the Kernighan-Lin heuristic in the Chaco library [**12**] and KL to our own implementation. The Spectral method offers the use of different eigensolvers. These are discussed in more detail in [**12**]. For our measurements we used selective orthogonalization. It turned out to be faster than full orthogonalization and still is very robust. Other solvers recommended for very large graphs are more likely to misconverge, resulting in an extremely differing performance when used for the different test-graphs of our benchmark suit. The Multilevel method refers to the Multilevel partitioner from [**11**] that is included in the Chaco library [**12**]. The graph was coarsed down to 200 nodes.

The local heuristics are not ergodic, i.e. their results depend very much on the global solution they are started on. Choosing only one run on a single global partition would possibly lead to a bad final solution. Choosing a large number of runs would be too time-consuming and not practical for a potential user. We choose 5 runs on different initial bisections for the KL and our method. They are constructed with the following simple global partitioning methods (cf. Sec. 2): *Linear*, *Scattered*, *Random*, *Binary* (based on the relation of 0's to 1's in the binary representation of nodes) and *Greedy*.

The results in Table 2 show the obtained cut sizes and the time needed. This time includes the calculation time of the global and the local partitioning method. Not included is the time for data input, evaluation and data output. If local methods are used as clean-up in addition to SP, the calculation time of the global method is included. The columns of KL-5 and HS-5 show the best partitioning result of the 5 runs and the total time needed for all 5 runs. All measurements are performed on a stand-alone Sparc Station SS10/30 with 96MB main memory.

Several observations from Table 2 can be made:

- Spectral Partitioning runs out of memory on very large graphs and is very time-consuming for most of the graphs. The use of the ML solver is recommended for very large graphs and performs satisfying on many graphs. But in the case of De Bruijn graphs, the ML solver has a very high running time and, in addition, computes disappointing bisection results. Thus the ML solver is not very robust and the use of the selective orthogonalization instead is more convincing.
- KL and our method, used as clean-up, significantly improve the results of Spectral Partitioning with an appropriate increase in time.
- The Multilevel method and the KL produce good results for the graphs from physical applications needing only little running time.
- Our method is faster on those graphs but also computes worse results.

$DB_2(X)$	13	14	15	16	17	18	19	20
best bis. found	1032	1928	3606	6646	12470	23296	43756	81804

TABLE 3. Bisection results for De Bruijn graphs of varying size obtained with the new method.

- For De Bruijn graphs our method produces the best solutions of all methods. This results in a slight increase in running time.
- Local heuristics have problems to produce good results for 3-regular graphs. This was already observed in [**3**].

Note that if the algorithms are randomized and allowed to perform several runs, or if the local methods are allowed to run on a large number of different random global bisections, all of them are able to find better solutions (e.g. the best solution for the $DB_2(17)$ that we found with our new algorithm is 12470). Although for practical applications such a policy is often not possible, it is still of theoretical interest. Table 3 shows results for the De Bruijn graph that can be obtained if the global methods perform many randomized runs or if the local methods are started on many different random bisections. They were all found with random global bisection and the new algorithm as local refinement. As they are the best bisection results of large De Bruijns ever published, to our knowledge, they are of interest in their own right. Note that the new heuristic is able to bisect $DB_2(20)$ in about 6 minutes on a Sparc Station SS10.

Some of the graphs in our benchmark set are provided with coordinate information. The Inertial method is based on this information. Bisection results of Inertial partitioning are shown in Table 4. Again, the use of KL or our method as a local clean-up improves the results significantly. The Inertial method itself is very fast (even with additional clean-up it is faster than all other methods we tested). As an additional clean-up, our method keeps the used time lower than KL and produces similar results. As the Inertial method does not make use of adjacency information it is possible to construct examples where it produces fairly bad results. But for the geometric graphs we used it computes results comparable to those of the Spectral method in a fraction of time.

graph	IN	+KL(c)	+HS
airfoil1	94	83	83
	(0.04)	(0.17)	(0.08)
whitaker3	137	133	133
	(0.06)	(0.39)	(0.16)
crack	377	194	214
	(0.06)	(1.04)	(0.17)
brack2	817	783	782
	(0.42)	(3.52)	(1.29)

TABLE 4. Results of Inertial (IN) partitioning with additional KL(c) or HS as clean-up. Numbers in brackets indicate running times in seconds on SS10/30.

5. Conclusions

With our heuristic we have presented a new linear-time algorithm for graph partitioning that is very powerful with a variety of different graphs. The current implementation of our method not only runs faster than versions of the KL with several improvements included, but often also finds better solutions.

We presented bisection results and running times for a number of very large graphs. Especially for large De Bruijn graphs we were able to obtain highly convincing results.

Acknowledgements

Many people have helped to make this work possible. We would especially like to thank Bruce Hendrickson and Robert Leland from Sandia, who provided the code of the Chaco library and made many helpful comments. Thanks also to Anja Feldmann (CMU) for many stimulating discussions and to Alex Pothen (Old Dominion Univ.) and Steve Hammond (RIACS), who provided some of their sample input data.

Without the help of the Sun administration group at the University of Paderborn and especially the assistance by Dietmar Neuwöhner, all the measurements that were necessary to perform this work would not have been possible. We also want to thank many of our colleagues, especially Ralf Klasing, Reinhard Lüling, Jürgen Schulze, Derk Meyer and Carsten Spräner for many inspiring discussions.

References

1. S.T. Barnard, H.D. Simon: *A fast multilevel implementation of recursive spectral bisection for partitioning unstructured problems.* Proc. 6th SIAM Conf. Par. Processing for Scientific Computing, SIAM, 1993, pp. 711-718
2. G.E. Blelloch, A. Feldmann, O. Ghattas, J.R. Gilbert, G.L. Miller, D.R. O'Hallaron, E.J. Schwabe, J.R. Shewchuk, S.-H. Teng: *Automated Parallel Solution of Unstructured PDE problems.* CACM, to appear
3. T.N. Bui, S. Chaudhuri, F.T. Leighton, M. Sisper: *Graph Bisection Algorithms with good average case behaviour.* Combinatorica 7 (2), 1987, pp. 171-191
4. R. Diekmann, R. Lüling, J. Simon: *Problem Independent Distributed Simulated Annealing and its Applications.* R.V.V. Vidal (ed.): *Applied Simulated Annealing*, Lecture Notes in Economics and Mathematical Systems, Springer LNEMS 396 (1993), pp. 17-44
5. I.S. Duff, R.G. Grimmes, J.G. Lewis: *Sparse Matrix Test Problems.* ACM Trans. on Math. Software 15(1), 1989, pp. 1-14
6. C. Farhat: *A Simple and Efficient Automatic FEM Domain Decomposer.* Computers & Structures, Vol. 28(5), 1988, pp. 579-602
7. A. Feldmann, personal communication
8. C.M. Fiduccia, R.M. Mattheyses: *A linear-time heuristic for improving network partitions.* Proc. of the 19th IEEE Design Automation Conference 1982, pp. 175-181
9. S.W. Hammond: *Mapping Unstructured Grid Computations to Massively Parallel Computers.* Technical Report No. 92.14, RIACS, NASA Ames Research Center, June 1992
10. B. Hendrickson, R. Leland: *Multidimensional Spectral Load Balancing.* Technical Report SAND93-0074, Sandia National Laboratories, Jan. 1993
11. ______: *A Multilevel Algorithm for Partitioning Graphs.* Technical Report SAND93-1301, Sandia National Laboratories, Oct. 1993

12. ______: *The Chaco User's Guide.*
Technical Report SAND93-2339, Sandia National Laboratories, Nov. 1993
13. J. Hromkovič, B. Monien: *The Bisection Problem for Graphs of Degree 4 (Configuring Transputer Systems).* In: Buchmann-Ganzinger-Paul: (Festschrift zum 60. Geburtstag von Günter Hotz), B.G. Teubner, Stuttgart-Leipzig 1992, pp. 215-234
14. D.S. Johnson, C.R. Aragon, L.A. McGeoch, C. Schevon: *Optimization by Simulated Annealing: An Experimental Evaluation; Part 1, Graph Partitioning.*
Operations Research, Vol. 37, No. 6, November-December 1989, pp. 865-893
15. B.W. Kernighan, S. Lin: *An Effective Heuristic Procedure for Partitioning Graphs.*
The Bell Systems Technical Journal, Feb. 1970, pp. 291-308
16. J. De Keyser, D. Roose: *Distributed mapping of SPMD Programs with a Generalized Kernighan-Lin Heuristic.* High-Performance Computing and Networking, Proceedings 1994, Springer LNCS 797, pp. 227-232
17. S. Kirkpatrick, C.D. Gelatt, M.P. Vecchi: *Optimization by Simulated Annealing.*
Science, Vol. 220, No. 4598, May 1983, pp. 671-680
18. R. Leland, personal communication
19. T. Lengauer: *Combinatorial Algorithms for Integrated Circuit Layout.* B.G. Teubner, 1990
20. R.J. Lipton, R.E. Tarjan: *A Separator Theorem for Planar Graphs.*
SIAM J. of Appl. Mathematics 36, 1979, pp. 177-189
21. G.L. Miller, S.H. Teng, W. Thurston, S.A. Vavasis: *Automatic mesh partitioning.* Graph Theary and Sparse Matrix Computation. A. George, J.R. Gilbert, J.W.H. Liu, eds. Springer-Verlag, 1993
22. A. Pothen, H.D. Simon, K.P. Liu: *Partitioning Sparse Matrices with Eigenvectors of Graphs.* SIAM J. on Matrix Analysis and Applications 11/3, 1990, pp. 430-452
23. A. Pothen, personal communication
24. J.E. Savage, M.G. Wloka: *Parallelism in Graph-Partitioning.*
Journal of Parallel and Distributed Computing 13, 1991, pp. 257-272
25. R. Sedgewick: *Algorithms.* Addison-Wesley, 1984
26. H.D. Simon: *Partitioning of unstructured problems for parallel processing.* Par. Methods on Large Scale Structural Analysis and Physics Applications, Pergamon Press, 1991
27. R.D. Williams: *Performance of Dynamic Load Balancing Algorithms for Unstructered Mesh Calculations.* Concurrency 3, 1991, pp. 457-481

DEPARTMENT OF MATHEMATICS AND COMPUTER SCIENCE, UNIVERSITY OF PADERBORN, WARBURGER STR. 100, D-33095 PADERBORN, GERMANY
E-mail address: diek@uni-paderborn.de

DEPARTMENT OF MATHEMATICS AND COMPUTER SCIENCE, UNIVERSITY OF PADERBORN, WARBURGER STR. 100, D-33095 PADERBORN, GERMANY
E-mail address: bm@uni-paderborn.de

DEPARTMENT OF MATHEMATICS AND COMPUTER SCIENCE, UNIVERSITY OF PADERBORN, WARBURGER STR. 100, D-33095 PADERBORN, GERMANY
E-mail address: robsy@uni-paderborn.de

DIMACS Series in Discrete Mathematics
and Theoretical Computer Science
Volume **21**, 1995

Modification of Consecutive-d Digraphs

DING-ZHU DU, D. FRANK HSU, AND DANIEL J. KLEITMAN

ABSTRACT. The concept of a consecutive-d digraph was proposed by Du, Hsu and Hwang as a generalization of many digraphs, such as de Bruijn digraphs, Kautz digraphs, and Imase-Itoh digraphs, which contain many hamiltonian digraphs with near-minimum diameter and near-maximum connectivity. In this paper we show sufficient and necessary conditions for various ways to modify consecutive-d digraphs to reach maximum line-connectivity.

1. Introduction

Many interconnection networks require to have hamiltonian circuit, small diameter, and large connectivity [**1, 2**]. de Bruijn graphs [**7**], Kautz graphs [**23**] and their generalizations given by Imase and Itoh [**20**] and Reddy, Pradhan and Kuhl [**19, 25**] are good cabdidates for the networks. To study those digraphs uniformly, Du, Hsu, and Hwang [**11**] introduced the concept of consecutive-d digraphs. A consecutive-d digraph $G(d,n,q,r)$ has n nodes, labeled by integers mod n, with edges from each node i to d consecutive nodes, namely those with label $qi + r + k \pmod{n}$ for $0 \le k < d \le n$, where r and q are integers and $-n/2 < q \le n/2, q \ne 0$. The generalized de Bruijn digraphs [**19, 25**] and the generalized Kautz digraphs [**20**] are its two subclasses consisting of $G_B(d,n) = G(d,n,d,0)$ and $G_I(d,n) = G(d,n,n-d,n-d)$, respectively.

The consecutive digraph $G(d,n,q,r)$ is hamiltonian when $d \ge 5$ [**11**] or $d \ge \max(3,q)$ [**10**]. Its line-connectivity was determined in [**13**] that $G(d,n,q,r)$ is d-line-connected iff it has no loop and every node has in-degree d (i.e. $g \mid d$, where $g = gcd(n,q)$) and moreover, if every node has in-degree d then $G(d,n,q,r)$ is

1991 *Mathematics Subject Classification.* 05C40,05C20.

The first author was supported in part by the NSF under grant CCR-9208913.
The second author was supported in part by Komatsu Endowed Chair.
The third author was supported in part by contracts AF OSR-86-0078 and NSF DMS-86-06225.

at least $(d-1)$-line-connected.. This result confirms that consecutive-d digraphs have very good line-connectivity, especially if no loop exists.
How do we modify consecutive-d digraphs to reach the maximum line-connectivity if loops exist? Essentially, the modification is to replace all loops by some cycles. A digraph is called a *modified* $G(d,n,q,r)$ if it is constructed from $G(d,n,q,r)$ by connecting all loop-nodes into disjoint cycles of size at least two and deleting all loops. The modification is said to be *cyclic* if all loop-nodes are connected into a single cycle. The modified $G(d,n,q,r)$ is said to be *simple* if there is no multiple edge in the graph. Not every consecutive-d digraph has a simple and cyclic modification. For example, all cyclic modifications of $G(3,4,3,0)$ must contains a multiple edges. Since multiple edges contribute nothing to diameter, avoiding multiplicity is of some interests.

Our main results in this paper are as follows.

THEOREM 1.1. *Let $d \geq 2$. There exists a modified $G(d,n,q,r)$ having line-connecitivity d if and only if g divides d. If g divides d, then the following holds:*
(a) Every cyclically modified $G(d,n,q,r)$ is d-line-connected.
(b) Every modified $G(d,n,q,r)$ is d-line-connected unless the following occurs
(b1) $g=1$, $d \leq 3$ and $q \equiv \pm 1 \pmod n$, or
(b2) $G(d,n,q,r)$ is isomorphic to $G(3,6,3,1)$, or
(b3) $g=d=2$ and $n=((q)_n+1)2^s$ for some natural number s.

THEOREM 1.2. *Let q^* be the magnitude of q. Suppose g divides d. Then for $d \geq 2$ and $n > (q^*+1)d$, there always exists a d-line-connected simple modified $G(d,n,q,r)$.*

2. Loops

Let $\lambda = gcd(q-1,n)$. (note: $\lambda = n$ if $q=1$.) Denote by $(x)_n$ the residue of x modulo n, represented by a number in $\{0,1,\ldots,n-1\}$. An edge is said to be with k-value i if it is contained in the subgraph $G(1,n,q,r+i)$. The following lemma is about the distribution of loops in the consecutive-d digraph $G(d,n,q,r)$.

LEMMA 2.1. *$G(d,n,q,r)$ has the following properties:*
(a) Each node has at most one loop.
(b) If $d \geq 2$ then $G(d,n,q,r)$ has either no loop or at least two loops.
(c) $G(d,n,q,r)$ has no loop iff $d < \lambda$ and $(r)_\lambda < \lambda - d$.
(d) If $d \leq \lambda$, then all loops of $G(d,n,q,r)$ are with the same k-value.
(e) If $\lambda = 1$, then for each k-value there exists exactly one loop with the k-value. If $\lambda > 1$, then for each k-value, either there is no loop or there are exactly λ loops with the k-value.
(f) If $|q-1| \leq d$ and x is a loop-node, then either $x + \lfloor n/(q-1) \rfloor$ or $x + \lceil n/(q-1) \rceil$ is a loop-node.

Proof. (a) and (b) were was proved in [13].

To prove (c), note that there is a loop at node i iff for some $k \in \{0, \ldots, d-1\}$,

$$(q-1)i + r + k \equiv 0 (mod\ n). \tag{1}$$

Thus, $G(d,n,q,r)$ has no loop iff $r+k$ is not divisible by λ for all k. This happens iff the two conditions hold.

For (d), note that when $d \leq \lambda$, there exists at most one k-value such that $r+k$ is divisible by λ. Therefore, loops could exist only for such a k-value.

To see (e), note that if $\lambda = 1$, the equation (2.1) has exactly one solution i for each k; if $\lambda > 1$, then i, $i + n/\lambda$, $\cdots$, $i + (\lambda - 1)n/\lambda$ satisfy (2.1) simultaneously.

Finally, we prove (f). If $q = 1$, then $\lambda = n$. So, either every node has a loop or every node has no loop. Thus, (f) holds trivially in this case. Next, consider $q \neq 1$. Suppose that the loop at x is $x \rightarrow qx + r + k$. Denote $k(x) = qx + r + k$. Then $k(x+1) - (x+1) = k(x) - x + (q-1)$. Denote $y = \lfloor n/(q-1) \rfloor$ and $z = \lceil n/(q-1) \rceil$. If $y = z$, then $k(x+y) - (x+y) = k(x) - x + n$, so that $x+y$ is a loop-node. If $y \neq z$, then $z = y+1$ and $(q-1)y < n < (q-1)z$ (if $q > 0$) or $(q-1)y > n(q-1)z$ (if $q < 0$). Since $|q-1| \leq d$, we have either $k + n - (q-1)y$ or $k + n - (q-1)z$ is in between 0 and $d-1$. It follows that either $x+y$ or $x+z$ has a loop. □

Let q^* denote an integer between 0 to $n/2$ such that $q \equiv q^*$ or $-q^* (mod\ n)$. q^* is called the *magnitude* of q. If $q^* < d$, then certainly $\lambda \leq d$, so that a loop must exist by Lemma 2.1(c).

3. Proof of Main Results

The proofs of Theorems 1.1 and 1.2 are based on investigating line-cuts of size $d-1$. If $G(d,n,q,r)$ is not d-line-connected, then we must have a line-cut of size $d-1$. That is, the nodes can be partitioned into two disjoint nonempty sets A and B such that at most $d-1$ edges go from A to B. When g divides d, the in-degree and the out-degree are equal at each node. Thus, the number of edges from B to A equals the number of edges from A to B. This fact always holds in the following lemmas.

A subset of Z_n is called a *consecutive run* if its elements can be consecutively numbered mod n.

Lemma 3.1. *Suppose that the nodes of $G(d,n,q,r)$ can be partitioned into two disjoint nonempty sets A and B such that at most $d-1$ edges go from A to B. If $g = 1$, then A and B both are consecutive runs unless $G(d,n,q,r)$ is isomorphic to $G(3,4,1,0)$ or $G(3,4,-1,0)$.*

Proof. Without loss of generality, assume $|A| \leq |B|$. If $|A| = 1$ then the conclusion holds trivially. Next, we consider $|A| \geq 2$. Suppose that either A or B does not consist of consecutive nodes. Then, we can find at least four different consecutive pairs of nodes, $\{a, a+1\}$, $\{b, b+1\}$, $\{c, c+1\}$ and $\{e, e+1\}$ such that in each pair, one node is in A and another is in B. Since $|A| \geq 2$, we have $n \geq 4$. Thus, at least two consecutive pairs are disjoint. We now consider a pair $\{a, a+1\}$. Note that each node has outedges to d consecutive nodes. Since $g = 1$, there is exactly one node linking to any d consecutive nodes. Thus, there are exactly $d-1$ nodes going to both a and $a+1$. Among these $2(d-1)$ edges, exactly $d-1$ edges going between A and B. Since at most $2(d-1)$ edges go between A and B, there are at most two disjoint pairs among the four. Without loss of generality, assume that the first two pairs are joint with $b \equiv a+1 \pmod{n}$. Since there are exactly $d-2$ nodes having edges going to all of a, $a+1$ and $a+2$, there are exactly d nodes linking to either both a and $a+1$ or both $a+1$ and $a+2$. This contributes at least d edges going between A and B. Since there are at most $2(d-1)$ edges between A and B, each of other two pairs cannot be disjoint from $\{a, a+1, a+2\}$. This happens only if $\{c, e\} = \{(a-1)_n, (a+2)_n\}$. In this case, we must also have $a+3 \equiv a-1 \pmod{n}$ since, otherwise, $\{a-1, a\}$ and $\{a+1, a+2, a+3\}$ are disjoint. $a+3 \equiv a-1 \pmod{n}$ implies $n = 4$. For $n = 4$, there are few cases so that we can find the exceptions easily. □

LEMMA 3.2. *Suppose that $g = 1$ and that the nodes of $G(d, n, q, r)$ can be partitioned into disjoint nonempty sets A and B such that at most $d-1$ edges go from A to B. Then either A or B contains only one node unless $d \leq 3$ and $q \equiv \pm 1 \pmod{n}$.*

PROOF. Let us first consider the case $q \not\equiv \pm 1 \pmod{n}$. By Lemma 3.1, both A and B are consecutive runs. Let q^* be the magnitude of q. Look at edges from A. Since at most $d-1$ of them go to B, there must exist a k-value such that all edges with it coming from A will stay within A. Let k' be such a k-value. That is, for any $i \in A$, $qi+r+k' \in A$. Consider any consecutive pair $i, i+1 \in A$. Note that $|B| \geq n/2 > q^*-1$. The q^*-1 nodes between $qi+r+k'$ and $q(i+1)+r+k'$, $\{qi+r+k'+j \cdot sign(q) \mid j = 1, \ldots, q^*-1\}$, must belong to A. If A contains x nodes, we find in this way that A has at least $x+(x-1)(q^*-1)$ nodes. Thus, $x+(x-1)(q^*-1) \leq x$. Since $q^* > 1$, we must have $x = 1$.

Now, assume that $q \equiv \pm 1 \pmod{n}$ and $d \geq 4$. Then $n \neq 4$. By Lemma 3.1, A and B are consecutive runs. Suppose to the contrary that $|B| \geq |A| \geq 2$. We first consider the cases $d \geq 4$. If A contains only two nodes, then at most 4 of the $2d$ edges emanating from A to A, so that there are at least $2d-4 \geq d$ edges from A to B, contradicting the assumption on A and B. Thus, A has at least three elements. Write $A = \{a, a+1, \ldots, a+k\}$ $(k \geq 2)$. From the proof of Lemma 3.1, we know that there are exactly $2(d-1)$ edges between A and B among $4d$ edges emerging into nodes $a-1$, a, $a+k$ or $a+k+1$. However, since $d \geq 4$, there exists at least one node whose edges go to both $a-2$ and $a+1$, which contributes an additional edge between A and B, contradicting the assumption on A and B. □

LEMMA 3.3. *Suppose that $g > 1$ and that g divides d. If the nodes of $G(d, n, q, r)$*

can be partitioned into two disjoint nonempty set A and B such that at most $d-1$ edges go from A to B, then either A or B has only one element unless
(i) $G(d,n,q,r)$ is isomorphic to $G(3,6,3,1)$, or
(ii) $g=d=2$ and $n=((q)_n+1)2^s$ for some natural number s.

PROOF. Since g divides n, the nodes of $G(d,n,q,r)$ can be partitioned into $n'=n/g$ groups $\bar{i}=\{i,i+n',\ldots,(g-1)n'\}$. In each group, all nodes have same successors. Let $\bar{G}$ be the digraph with the nodes $\bar{1},\ldots,\bar{n'}$ such that an edge from $\bar{i}$ to $\bar{j}$ exists iff an edge from i to x for some x in $\bar{j}$ exists in $G(a,n,q,r)$. It is easy to see that $\bar{G}$ is isomorphic to $G(d,n',q,r)$. However, we have to pay attention to the fact that n' may not be bigger than d. Note that $gcd(n',q)\mid g$ and $g\mid d$ imply $gcd(n',d)\mid d$. When $n'>d$, we know that $\bar{G}$ is $(d-1)$-line-connected (Corollary of Theorem 3.3 in [13]). When $n'\le d$, we may look at $\bar{G}$ as a multigraph. This multigraph is the union of $G(n',n',q,r),\cdots,G(n',n',q,r+(\ell-1)n'),G(d-\ell n',n',q,r+\ell n')$ where $\ell=\lfloor d/n'\rfloor$ and $G(n',n',q,r)$, $\cdots$, $G(n',n',q,r+(\ell-1)n')$ are the complete graph of n' nodes and hence $(n'-1)$-line-connected. Moreover, $G(d-\ell n',n',q,r+\ell n')$ is $(d-\ell n'-1)$-line-connected since $gcd(n',q)|g$ implies $gcd(n',q)|(d-\ell n')$. So, $\bar{G}$ is at least $d-\lceil d/n'\rceil$ $(\ge n'-1)$ line-connected. Since $n>d\ge g$, we have $n'\ge 2$ and $\lceil d/n'\rceil\le\min\{g,d-1\}$. So, each node of $\bar{G}$ has at most $\min\{g,d-1\}$ loops.

Define $\bar{A}=\{\bar{i}\mid\bar{i}\subseteq A\}$ and $\bar{B}=\{\bar{i}\mid\bar{i}\subseteq B\}$. Note that each edge of $\bar{G}$ contains g edges of $G(d,n,q,r)$. If $\bar{A}\cup\bar{B}$ contains all nodes of $\bar{G}$, then there are at most $\lfloor(d-1)/g\rfloor=d/g-1$ edges from $\bar{A}$ to $\bar{B}$ and $\bar{A}$ and $\bar{B}$ are nonempty. Since $g>1$, $d/g-1\le d-2$. Note that if $n'>d$, then $\bar{G}$ is $(d-1)$-line-connected. Thus, we have $n'\le d$. So, $\bar{G}$ is at least $(n'-1)$-connected. It follows that $d/g-1\ge n'-1$, that is, $d\ge n$, contradicting our assumption that $n>d$. Therefore, there exists a node $\bar{x}$ of $\bar{G}$ not in $\bar{A}\cup\bar{B}$, that is, $\bar{x}\cap A\ne\emptyset$ and $\bar{x}\cap B\ne\emptyset$. We claim that there exists only one such node. To see this, consider a node $a\in\bar{x}\cap A$ and a node $b\in\bar{x}\cap B$. Since a and b have the same d successors, there exist d out-edges of a or b going between A and B. If there are two nodes of $\bar{G}$ not in $\bar{A}\cup\bar{B}$, then we will obtain at least $2d$ edges between A and B. If there are two nodes of $\bar{G}$ not in $\bar{A}\cup\bar{B}$, then we will obtain at least $2d$ edges between A and B, contradicting the choice of A and B. Similarly, if $\bar{x}\cap A$ and $\bar{x}\cap B$ both contain at least two nodes, then we can consider another pair of nodes $a'\in(\bar{x}\cap A)\setminus\{a\}$ and $b'\in(\bar{x}\cap B)\setminus\{b\}$ to obtain other d edges between A and B, contradicting the choice of A and B. Therefore, either $\bar{x}\cap A$ or $\bar{x}\cap B$ contains only one element. Without loss of generality, assume that $\bar{x}\cap A$ has only one element. We next show that if $g>2$ then $\bar{A}=\emptyset$ unless $G(d,n,q,r)$ is isomorphic to $G(3,6,3,1)$.

In fact, if $\bar{A}$ is not empty, then at least $d-\lceil d/n'\rceil$ edges of $\bar{G}$ come to $\bar{A}$ from $\bar{B}\cup\{\bar{x}\}$ since $\bar{G}$ is $(d-\lceil d/n'\rceil)$-line-connected. These edges contains at least $(g-1)(d-\lceil d/n'\rceil)$ edges of $G(d,n,q,r)$, going from B to A. Since $n>d\ge g$, we have $n'\ge 2$. If $n'>2$ or d is even, then $d\ge 2\lceil d/n'\rceil\ge\frac{g-1}{g-2}\lceil d/n'\rceil$. Hence,

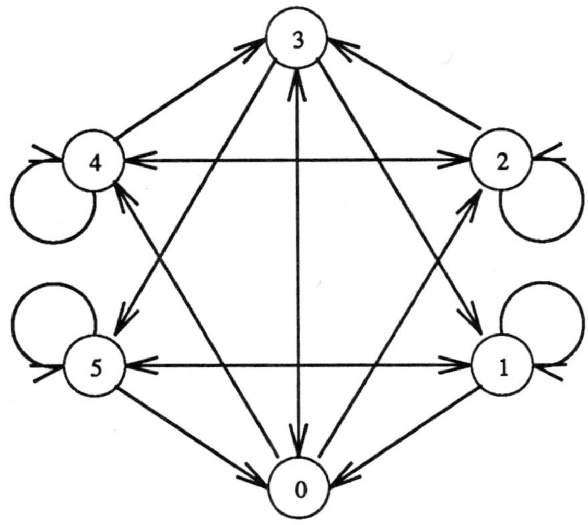

FIGURE 1. $G(3,6,3,1)$ $(A = \{0,1,3,5\}, B = \{2,4\})$

$(d - \lceil d/n' \rceil)(g-1) \geq d$. It contradicts the hypothesis on A and B. If $n' = 2$ and d is odd, then $|\bar{A}| = 1$, $\bar{B} = \emptyset$ and $n = 2g$. Since $n \geq d$ and $g|d$, we have $d = n$ or g. If $d = n$, then $G(d,n,q,r)$ is the complete digraph of order n. Since $g \geq 3$ and $|A \cap \bar{x}| = 1$, $|B| = g - 1 \geq 2$. Thus, there are at least $2(d-2)$ edges going from B to A. Because $d = 2g \geq 6$, we have $2(d-2) > d$, a contradiction. If $g = d$, then note that $\bar{x}$ consists of either all odd nodes or all even nodes. Thus, A receives at least $(d-1)/2$ edges from each node. If $g > 3$, then B has at least three nodes and hence A receives at least $3(d-1)/2$ edges from B, contradicting the assumption on A and B. Therefore, $g = d = 3$ and $n = 6$. Thus, $|A| = 4$ and $|B| = 2$. There is only one node in A receiving edges from B. (Otherwise, there are at least four edges from B to A.) Therefore, all in-edges of two nodes in B come from nodes in $\bar{x}$ and all in-edges of two nodes in A which do not receive edges from B come from $A \setminus \bar{x}$. It turns out that $G(d,n,q,r)$ is isomorphic to $G(3,6,3,1)$ as shown in Fig. 1.

When $g = 2$ and $d > 2$, both $\bar{x} \cap A$ and $\bar{x} \cap B$ have only one element. To show that either A or B has only one element, it suffices to prove that either $\bar{A}$ or $\bar{B}$ is empty. Suppose to the contrary that they both are nonempty. We first claim that no edge exists between $\bar{A}$ and $\bar{B}$. In fact, suppose that there is an edge from $\bar{A}$ to $\bar{B}$. This edge will contributes g edges from A to B. If at least $d-1$ edges exist from $\bar{A} \cup \{\bar{x}\}$ to $\bar{B}$, then at least $g + (d-2)(g-1) = d$ edges exist from A to B, which is impossible. Moreover, $\bar{G}$ is $(d - \lceil d/n' \rceil)$-line-connected and $\lceil d/n' \rceil \leq \lceil n/n' \rceil = g = 2$. So, $\bar{G}$ is $(d-2)$-line-connected. Thus, there are exactly $d-2$ edges from $\bar{A} \cup \{\bar{x}\}$ to $\bar{B}$. This implies that $n' < d$ and there are at most $(d-2)/2 \leq n' - 1$ edges from $\bar{A} \cup \{\bar{x}\}$ to $\bar{B}$ in multigraph $\bar{G}$ which contains the complete digraph $G(n', n', q, r)$. Hence, $\bar{B}$ must be a singleton. Moreover, since there are exactly $d-2$ edges from $\bar{A} \cup \{\bar{x}\}$ to $\bar{B}$, there are exactly $g + (g-1)(d-3) = d-1$ edges from A to $B \setminus \bar{x}$. Thus, all in-edges of $B \cap \bar{x}$ come from $B \setminus \bar{x}$ and hence $d = g = 2$, a contradiction. Similarly, it is also impossible that there exists an edge from $\bar{B}$ to $\bar{A}$. Thus, at least $d-1$ outedges from the node in $A \cap \bar{x}$ go to B and at least $d-1$ outedges from the node in $B \cap \bar{x}$ go to A. It follows that $\bar{x}$ has at least $d-1$ edges going to $\bar{A}$ and at least $d-1$ edges going to $\bar{B}$. Thus, at least $2(d-1)$ edges go out from $\bar{x}$. It implies $2(d-1) \leq d$. Hence $d = 2$.

Now, we consider the case $d = 2$ and hence $g = 2$ since $1 < g$ and g divides d. In the above proof, we suppose that $\bar{A}$ and $\bar{B}$ both are nonempty and have proved that no edge exists between $\bar{A}$ and $\bar{B}$ and that $\bar{x}$ has at least $d-1$ edges going to $\bar{A}$ and at least $d-1$ edges going to $\bar{B}$. The latter fact implies that $\bar{x}$ has exactly one edge going to $\bar{A}$ and exactly one edge going to $\bar{B}$. Therefore, only one edge goes from $\bar{A} \cup \{\bar{x}\}$ to $\bar{B}$ and only one edge goes from $\bar{B} \cup \{\bar{x}\}$ to $\bar{A}$. First, consider $gcd(q, n') = 1$. If $q \not\equiv \pm 1 (mod\ n')$, then by Lemma 3.2, either $\bar{A} \cup \{\bar{x}\}$ or $\bar{B}$ is a singleton and either $\bar{B} \cup \{\bar{x}\}$ and $\bar{A}$ is a singleton. This happens only if both $\bar{A}$ and $\bar{B}$ are singletons. Thus, $n' = 3$. This is impossible since $q \not\equiv \pm 1 (mod\ n')$ and $gcd(q, n') = 1$. If $q \equiv 1 (mod\ n')$, we notice that $\bar{x}$ has no loop, so that $\bar{G}$ has no loop and is d-line-connected, hence the above $(d-1)$ size cut cannot exists. If $q \equiv -1 (mod\ n')$, then $n = 2((q)_n + 1)$, which is the exceptional case. Now, we consider the case that $gcd(q, n') > 1$. Assume by induction that the lemma holds for $G(d, n', q, r)$. Note that $n' = ((q)_n + 1)2^{s'}$ implies $n = ((q)_n + 1)2^{s'+1}$. Thus, if $n \neq ((q)_n + 1)2^s$, then $n' \neq ((q)_n + 1)2^{s'}$. By the induction hypothesis, either $\bar{A} \cup \{\bar{x}\}$ or $\bar{B}$ is a singleton and either $\bar{B} \cup \{\bar{x}\}$ or $\bar{A}$ ia a singleton. It follows that both $\bar{A}$ and $\bar{B}$ are singletons. Hence $n' = 3$. Since $gcd(q, n') > 1$ implies $n' > 3$, we obtain a contradiction. Therefore, either $\bar{A}$ or $\bar{B}$ is empty, so that either A or B is singleton. □

LEMMA 3.4. *Suppose that g divides d and that the nodes of $G(d, n, q, r)$ can be partitioned into two disjoint nonempty sets A and B such that at most $d-1$ edges go from A to B. Then, both A and B contain loop-nodes.*

PROOF. For $g = 1$, if $d \geq 4$ or $q \not\equiv \pm 1$ (mod n) then either A or B contains only one node; such a node must have a loop. By Lemma 2.1(a)(b), both A and B have a loop. If $q = 1$, the existence of A and B implies that every node has a loop. For $q = -1$, we note that A and B are consecutive runs. Clearly, there is a k-value such that the edges with this k-value do not go from A to B. As shown in Fig. 2, all edges with such a k-value provide a symmetric reflection inside of A (Fig. 3.2). If A contains odd number of nodes, then the middle node will be fixed under the reflection and hence get a loop. If A contains even number of nodes, then two consecutive nodes a and $a+1$ situated at the middle positions of A will reflect to each other. Consider a k-value which is different in one from the former k-value. Such a k-value will give a loop at either node a or node $a+1$. Thus, A always has a node with a loop. Similarly, we can show that for B.

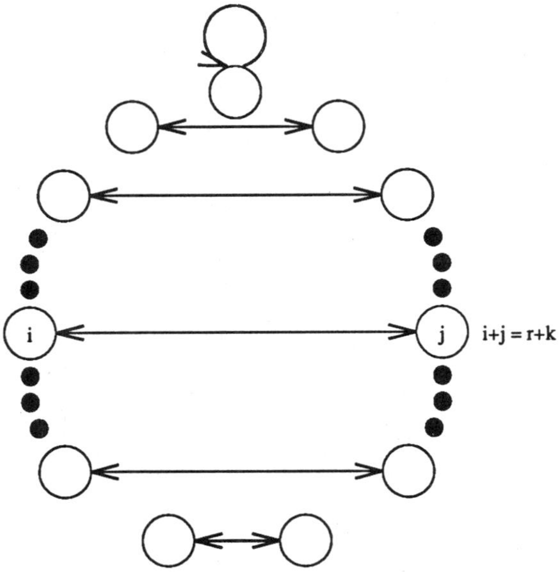

FIGURE 2. $G(1, n, -1, r+k)$

Now, we consider the case $g > 1$. If either A or B is a singleton then it is obvious. If nether A nor B is singleton, then by Lemma 3.2, we have either (1) $G(d, n, q, r)$ is isomorphic to $G(3, 6, 3, 1)$, or (2) $g = d = 2$ and $n = ((q)_n + 1)2^s$ for some natural number s. If (1) occurs, then it is easy to see that A and B both contain loops. If (2) occurs, then let $\bar{A}$, $\bar{B}$ and $\bar{x}$ be the sets and the point in the proof of Lemma 3.3. By induction, we may assume that the current lemma holds for $G(d, n', q, r)$. Thus, $\bar{A}$ and $\bar{B}$ both have loops. This imply that A and B both have loops since a node of $\bar{G}$ has a loop iff this node contains a node of $G(d, n, q, r)$ having a loop. □

PROOF. **(Theorem 1.1)** The first half of the theorem follows from Lemma 3.4 and the fact that the modification does not change the in-degree at each node. The second half follows from Lemmas 3.4, 3.2, and 3.3. □

PROOF. **(Theorem 1.2)** If $\lambda = gcd(q-1, n) = 1$ then by Lemma 2.1(e), for every k-value, there is exactly one node with a loop corresponding to the k-value. Such a node is denoted by i_k. Now, we give edges going from i_{k-1} to i_k for $k = 0, \ldots, d-2$ and from i_{d-1} to i_0 and claim that the resulting digraph has no multiple edge. For otherwise, suppose that the edge from i_{k-1} to i_k has been in $G(d, n, q, r)$. That is, $i_k \equiv qi_{k-1} + r + k' \pmod{n}$ for some $k' \in \{0, \ldots, d-1\}$. It follows that $i_k \equiv i_{k-1} + k' - k + 1 \pmod{n}$. Multiplying q in both sides and simplifying it, we obtain that $1 + i_k \equiv i_{k-1} + q(k' - k + 1) \pmod{n}$. Therefore, $1 \equiv (q-1)(k' - k + 1) \pmod{n}$. This is impossible since $n > (q' + 1)d$. Similarly, we can show that the edge from i_{d-1} to i_0 is not in $G(d, n, q, r)$. Thus, we have a cyclically-modified $G(d, n, q, r)$ without multiple edges, which, by Theorem 1.1, is d-line-connected.

If $\lambda = gcd(q-1, n) > 1$, then the digraph has either no loop or exactly λ loops for a given k-value. For each loop-node i, $i + n/\lambda$ must also be a loop-node. We edge i to $i + n/\lambda$. This edge is not in $G(d,n,q,r)$ if $q \not\equiv 1(mod\ n)$. In fact, if the edge is in $G(d,n,q,r)$, then $i + n/\lambda \equiv qi + r + k \pmod n$ for some k in $\{0, \ldots, d-1\}$. That is, $n/\lambda \equiv k - k' \pmod n$ for some $k, k' \in \{0, \ldots, d-1\}$. Thus, $\lambda(k - k') \equiv 0 \pmod n$. This is impossible when $q \not\equiv 1 \pmod n$. Therefore, when the two exceptional cases stated in Theorem 1.1 does not occur, a modified $G(d,n,q,r)$ exists without multiple edges. By Theorem 1.1, it is d-line-connected.

Now, we consider the exceptional cases. If $q \equiv 1 \pmod n$, then either 1 or $n-1$ is not in $\{(r)_n, \ldots, (r+d-1)_n\}$. Thus, we can edge i to either $i+1$ or $i-1$ for all i. This gives a cyclically-modification. If $q \equiv -1 \pmod n$ and $d \le 3$, then n is even and $\lambda = d = 2$ since $\lambda > 1$. In this case, we have exactly two loop nodes with the same k-value. As we proved in the last paragraph, the edges between these two loop nodes are not in $G(d,n,q,r)$. They give a cyclically-modified $G(d,n,q,r)$ which is d-line-connected by Theorem 1.1. Finally, if $g = d = 2$ and $n = ((q)_n + 1)2^s$ then we can prove it by induction on s. For $s = 0$, we have $q \equiv -1(mod\ n)$. For $s \ge 1$, $G(2, ((q)_n + 1)2^s, q, r)$ is the line digraph of $G(2, ((q)_n + 1)2^{s-1}, q, r)$. Each loop of the former graph is induced by a loop of the latter graph. There is a one-one correspondence between them. The modification for the former graph is inherited from the latter one through this correspondence. □

4. Discussion

de Bruijn graphs [**7**], Kautz graphs [**23**] and their generalizations have been extensively studied [1-6, 8-13, 15-21]. Imase, Soneoka and Okada [**21**] showed that for $n \ge d^3$, $G_B(d,n)$ and $G_I(d,n)$ are $(d-1)$-connected. Du and Hwang [**14**] showed that $G_I(d,n)$ is d-line-connected iff it has no loop, i.e. $(d+1) \mid n$. Homobono and Peyrat [**17**] showed that for $n \ge d^4$, $G_I(d,n)$ is d connected iff $(d+1) \mid n$ and $gcd(d,n) > 1$. Our results in this paper and [**13**] showed that the consectutive-d digraph $G(d,n,q,r)$ and its various modifications also have large connectivity. So, they can be candidates for certain interconnection networks.

References

1. J.C. Bermond, N. Homobono, and C. Peyrat, *Large fault-tolerant interconnection networks*, Graphs and Combinatorics, (1989).
2. J.C. Bermond, F. Comellas, and D.F. Hsu, *Distributed loop computer networks: a survey*, Journal of Parallel and Distributed Computing, to appear.
3. J.C. Bermond and C. Peyrat, *de Bruijn and Kautz networks: a competitor for the hypercube?* preprint.
4. J.C. Bermond and C. Peyrat, *Broadcasing in de Bruijn networks*, Proc. 19-th Southeastern Conference on Combinatorics, graph Theory and Computing, 1988.

5. J.C. Bermond, N. Homobono, and C. Peyrat, *Connectivity of Kautz networks*, French-Israel Conference on Combinatorics, Jerusalem, 1988.
6. J. Bond and A. Ivanyi, *Modeling of interconnection networks using de Bruijn graphs*, Proc. Thied Conference of Program Designers (A. Vitanyi ed.), Budapest, (1987) 75-88.
7. N.G. de Bruijn, *A combinatorial problem*, Koninklijke Nederlandse Academie van Wetenschappen Proc., A49 (1946) 758-764.
8. F.K. Chung, *Diameter of communication networks*, Proceedings of Symposia in Applied Mathematics, 34 (1986) 1-18.
9. D.Z. Du, D.F. Hsu, and F.K. Hwang, *Doubly-linked ring networks*, IEEE Trans. on Computers, C-34 (1985) 853-855.
10. D.Z. Du and D.F. Hsu, *On hamiltonian consecutive-d digraphs*, in *Combinatorics and Graph Theory*, (Banach Center Publication, vol. 25, Warsaw, 1989) 47-55.
11. D.Z. Du, D.F. Hsu, and F.K. Hwang, *Hamiltonian property of d-consecutive digraphs*, Math. Comput. Modeling Vol. 17, NO. 11 (1993) 61-63.
12. D.Z. Du, D.F. Hsu, F.K. Hwang and X.M. Zhang, *The hamiltonian property of generalized de Bruijn digraphs*, J. Comb. Theory, Series B, 52 (1991) 1-8.
13. D.Z. Du, D.F. Hsu and G.W. Peck, *Connectivity of consecutive-d digraphs*, Discrete Applied Mathematics, 37/38 (1992) 169-177.
14. D.Z. Du and F.K. Hwang, *Generalized de Bruijn digraphs*, Networks, 18(1988) 27-33.
15. M.A. Fiol, J.L.A. Yebra, and I. Alegre, *Line digraph iterations and the (d,k) digraph problem*, IEEE Trans. on Computers, C-33 (1984) 702-713.
16. N. Homobono, *Connectivity of generalized de Bruijn and Kautz graphs*, Proc. 11-th British Conference, Ars Combinatoria, (1988).
17. N. Homobono and C. Peyrat, *Connectivity of Imase and Itoh digraphs*, IEEE Trans. on Computers, (1988).
18. F.K. Hwang, *The hamiltonian property of linear functions*, Operations Research Letters, 6 (1987) 125-127.
19. M. Imase and M. Itoh, *Design to minimize a diameter on Building block network*, IEEE Trans. on Computers, C-30 (1981) 439-443.
20. M. Imase and M. Itoh, *A design for directed graph with minimum diameter*, IEEE Trans. on Computers, C-32 (1983) 782-784.
21. M. Imase, T. Soneoka, and K. Okada, *Connectivity of regular directed graphs with small diameter*, IEEE Trans. on Computers, C-34 (1985) 267-273.
22. M. Imase, I. Soneoka, and K. Okada, *A fault tolerant processor interconnection network*, Systems and Computers in Japan, 17:8 (1986) 21-30.
23. W.H. Kautz, *Bounds on directed (d,k) graphs*, Theory of cellular logic networks and machines, AFCRL-68-0668 Final Report (1968) 20-28.
24. V.P. Kumar and S.M. Reddy, *A class of graphs for fault-tolerant processor interconnections*, IEEE 1984 Int. Conf. Distributed Computing Systems, (1984) 448-460.
25. S.M. Reddy, J.G. Kuhl, S.H. Hosseini, and H. Lee, *On digraphs with minimum diameter and maximum connectivity*, Proc. of 20-th Annual Allerton Conference, (1982) 1018-1026.
26. S.M. Reddy, D.K. Pradhan, and J.G. Kuhl, *Directed graphs with minimal diameter and maximal connectivity*, School of Engineering Oakland Univ. Tech. Rep., 1980.
27. T. Soneoka, H. Nakada, and M. Imase, *Design of d-connected digraph with a minimum number of edges and a quasiminimal diameter*, Discrete Applied Mathematics, 27 (1990) 255-265.
28. T. Soneoka, M. Imase, and Y. Manabe, *Design of d-connected digraph with a minimum number of edges and a quasiminimal diameter II*, preprint.

COMPUTER SCIENCE DEPARTMENT, UNIVERSITY OF MINNESOTA, MINNEAPOLIS, MN 55455, U.S.A.

E-mail address: dzd@cs.umn.edu

DEPARTMENT OF COMPUTER AND INFORMATION SCIENCE, FORDHAM UNIVERSITY, BRONX, NEW YORK, U.S.A.

E-mail address: hsu@murray.fordham.edu

DEPARTMENT OF MATHEMATICS, MASSACHUSETTS INSTITUTE OF TECHNOLOGY, CAMBRIDGE, U.S.A.

DIMACS Series in Discrete Mathematics
and Theoretical Computer Science
Volume **21**, 1995

Highly Adaptive Wormhole Routing Algorithms for N-Dimensional Torus

JOSÉ DUATO AND PEDRO LÓPEZ

ABSTRACT. Deadlock avoidance is a key issue in wormhole networks. A first approach consists of removing the cyclic dependencies between channels. Many deterministic and adaptive routing algorithms have been proposed based on that approach. The absence of cyclic dependencies is a necessary and sufficient condition for deadlock-free deterministic routing. However, it can be relaxed for adaptive routing. A more powerful approach was proposed by us. It only requires the absence of cyclic dependencies on a connected channel subset. The remaining channels can be used in almost any way.

In this paper, we show that there exists a more relaxed condition for deadlock-free adaptive routing. This condition is the key for the design of more powerful adaptive routing algorithms. We apply this condition to the design of adaptive routing algorithms for n-dimensional torus. In particular, we propose a partially adaptive routing algorithm which doubles the throughput achieved by the deterministic algorithm without increasing the hardware complexity significantly.

1. Introduction

A lot of research effort has been dedicated during the last decade to improve the performance of multicomputers [**2**]. A key architectural issue is the interconnection network. Wormhole routing [**8**] has become the most efficient flow-control technique for multicomputers. In wormhole routing, each message is serialized into a sequence of parallel data units, referred to as flow control units, or flits [**7**]. The flit at the head of a message determines the route. As the header flit is routed, the remaining flits follow it in a pipeline fashion. When the header is routed, the router can offer one or more alternative paths, depending on whether the routing algorithm is deterministic or adaptive. If all the channels offered by the router are busy, the header is blocked until one of those channels

1991 *Mathematics Subject Classification.* Primary 54C40, 14E20; Secondary 46E25, 20C20.
This work was supported by the Spanish CICYT under Grant TIC91–1157–C03–03

is freed; the flow control within the network blocks the trailing flits. Wormhole routing avoids using storage bandwidth in the nodes through which messages are routed. Also, it makes the message latency largely insensitive to the distance in the network.

A more efficient flow control technique known as mad postman has been proposed in [**24**]. It forwards the message header even before the routing operation takes place. It is very well suited for meshes with narrow channels.

Deadlocks may appear if the routing algorithms are not carefully designed. A deadlock in the interconnection network of a multicomputer occurs when no message can advance toward its destination because the queues of the message system are full. As routing decisions must be taken very fast in wormhole networks, the only practical way to avoid deadlock is to design deadlock-free routing algorithms.

Many deadlock-free routing algorithms have been developed for store-and-forward computer networks [**20, 23, 33**]. These algorithms are based on a structured buffer pool. However, with wormhole routing or mad postman, buffer allocation cannot be restricted, because flits have no routing information. Once the header of a message has been accepted by a channel, the remaining flits must be accepted before the flits of any other message can be accepted. So, routing must be restricted to avoid deadlock. As mad postman has the same deadlock properties as wormhole routing, in what follows we will only refer to wormhole.

Dally [**8**] proposed the necessary and sufficient condition for a deterministic routing algorithm to be deadlock-free. Based on this condition, he developed a design methodology, defining a channel dependency graph and establishing a total order among channels. Routing is restricted to visit channels in decreasing order to eliminate cycles in the channel dependency graph. If the routing restrictions disconnect the network, physical channels are split into virtual channels to connect the network again. This methodology has been applied to the design of routing chips for multicomputers [**7**], multicomputer nodes with integrated communication support [**4**] and systolic communication [**26, 4**]. Also, the use of virtual channels can increase throughput considerably [**10**].

Although Dally's theorem was proposed for deterministic routing, it has been extensively applied to adaptive routing [**24, 29, 5, 21, 11**]. Also, it has been applied to multicast routing [**27**]. In these cases, Dally's theorem becomes a sufficient condition for deadlock-freedom. An important study about deadlock-free adaptive routing is the turn model [**21**]. In this model, a message produces a turn when it changes direction, usually changing from one dimension to another. Turns are combined into cycles. Prohibiting just enough turns to break all the cycles prevents deadlock. However, the turn model is also based on Dally's theorem, requiring the absence of cycles in the channel dependency graph. This approach is too restrictive, limiting the number of alternative paths that can be used.

A different approach consists of allowing the existence of cyclic dependencies between channels, providing fully adaptive routing with a modest number of virtual channels. The *-channels routing algorithm [**3**] prevents deadlock by dynamically disallowing paths based on occupancy of buffers in neighboring nodes.

However, the theory proposed by us [**12, 17**] is not restricted to any particular topology or routing algorithm. Cyclic dependencies between channels are allowed, provided that there exists a connected channel subset free of cyclic dependencies. To simplify the application of this theory, we also proposed two design methodologies. The routing algorithms designed with them achieve very good performance because the channels belonging to the cyclic subset can be used almost without restrictions. Several authors have proposed and evaluated routing algorithms based on our theory [**13, 17, 19, 30, 31, 32**]. More recently, Lin et al. [**28**] have proposed the message flow model which also allows the existence of cyclic dependencies between channels. The authors also propose adaptive routing algorithms for meshes and hypercubes, proving that they are deadlock-free. These algorithms can also be proved to be deadlock-free by applying the theory proposed in [**12**].

In this paper, we extend the theory proposed in [**12, 17**], giving a more relaxed condition for deadlock-free adaptive routing. Basically, we define routing subfunctions in a more flexible way. This definition produces two new kinds of channel dependency. Also, we apply the relaxed condition to the design of adaptive routing algorithms for k-ary n-cubes. Section 2 deals with the significance of this work. Then, we introduce the new theory and the new kinds of channel dependency in section 3. Section 4 proposes the new theory formally, showing its application to k-ary n-cubes in section 5. The resulting routing algorithm is evaluated by simulation in section 6. Finally, some conclusions are drawn.

2. Significance of this work

This paper improves our previous results by proposing a more relaxed condition to prove that an adaptive routing algorithm for a wormhole network is deadlock-free. Recently, it has been shown that virtual channels are expensive, increasing node delay considerably [**6**]. Thus, there is a growing interest on the design of adaptive routing algorithms with minimum routing restrictions and as few virtual channels per physical channel as possible [**22**]. The theorem proposed in this paper is the key for the design of those routing algorithms.

Recently, some manufacturers, including Cray, are moving towards low dimensional torus for their massively parallel computers. We will present an adaptive routing algorithm for the n-dimensional torus (k-ary n-cube) in section 5, which cannot be proved to be deadlock-free using our previous theory. Deadlock-freedom will be proved by using the new theory. The performance evaluation results will show the benefits of using routing algorithms designed with the new theory.

3. Informal description of the theory

With deterministic routing, messages have a single routing option at each node. Thus, it is necessary to remove all the cyclic dependencies between channels to prevent deadlocks [**8**]. When adaptive routing is considered, messages usually have several choices at each node. Thus, it is not necessary to eliminate all the cyclic dependencies, provided that every message can always find a path to its destination (connected subset) whose channels are not involved in cyclic dependencies.

For store-and-forward networks, it suffices with the existence of a connected channel subset C_1 whose channels are not involved in cyclic dependencies [**12**]. The channels belonging to C_1 can be considered as *escape paths* from the cycles. However, for wormhole networks, it is more complex. As each message can reserve several channels simultaneously, it is possible for a message to reserve a channel belonging to C_1 and then another channel not belonging to C_1, which may be involved in a cyclic dependency. In other words, a message involved in a cyclic dependency may occupy channels belonging to the escape paths, thus blocking other messages which try to use the channels of C_1 to reach their destination. We solved this problem by introducing indirect dependencies [**12**]. When a message uses a channel belonging to C_1, then another channel not belonging to C_1 and then it tries to use a channel belonging to C_1, there is an indirect dependency between both channels belonging to C_1. If cyclic dependencies are not allowed between channels belonging to C_1, even when indirect dependencies are considered, then we guarantee that the escape paths will not block forever.

Labeling each channel as belonging or not belonging to the escape paths is simple, but it can be improved. In this paper, we go one step further by conditionally assigning each channel to the escape paths, depending on the destination of the message. By doing so, we achieve the maximum flexibility in the definition of the escape paths, but we also introduce new problems. Effectively, a message M can reserve a channel c_i which does not belong to any escape path for the destination of M, and then another channel belonging to the escape path for the destination of M. If c_i belongs to the escape paths for other destinations, other messages may need it to reach their destination. However, that channel may be occupied forever by M, which in turn may be blocked by another message requesting c_i. It must be noticed that the problem arises when there are cyclic dependencies among channels belonging to the escape paths and channels which do not belong to the escape path for the message occupying them, but belong to the escape paths for other destinations. Again, the solution consists of defining a new kind of dependency (direct cross dependency). Moreover, if we combine this dependency with indirect dependencies, we obtain the indirect cross dependencies. By including all the new dependencies and prohibiting cycles in the escape paths, we can guarantee the absence of deadlock.

4. Definitions and theorem

This section proposes a more relaxed sufficient condition for deadlock-free adaptive routing in wormhole networks. This theory is also valid when messages are split into packets. However, as it has been shown that packetization does not improve performance significantly when virtual channels are used [**25**], we will consider that messages are not split. Otherwise, the following assumptions and definitions should refer to packets.

The basic assumptions are identical to the ones proposed in [**17**]. They are repeated here to make the paper self-contained:

(i) A node can generate messages destined for any other node at any rate.
(ii) A message arriving at its destination node is eventually consumed.
(iii) Wormhole routing is used. So, once a queue accepts the first flit of a message, it must accept the remainder of the message before accepting any flits from another message. Also, a message is routed as soon as the routing information reaches a node. Thus, a message may occupy several channels simultaneously.
(iv) A node can generate messages of arbitrary length. Messages will generally be longer than a single flit.
(v) An available queue may arbitrate between messages that request that queue, but may not choose among waiting messages.
(vi) A queue cannot contain flits belonging to different messages. After accepting a tail flit, a queue must be emptied before accepting another header flit. Thus, when a message is blocked, its header flit will always occupy the head of a queue.
(vii) The route taken by a message depends on its destination and the status of output channels (free or busy). At a given node, the routing function supplies a set of output channels based on the current and destination nodes. A selection from this set is made based on the status of output channels at the current node. So, *adaptive* routing will be considered.

Before proposing the theorem, some definitions are needed:

DEFINITION 4.1. *An* interconnection network I *is a strongly connected directed multigraph,* $I = G(N, C)$. *The vertices of the multigraph* N *represent the set of processing nodes. The arcs of the multigraph* C *represent the set of communication channels. More than a single channel is allowed to connect a given pair of nodes. Each channel* c_i *has an associated queue with capacity* $cap(c_i)$. *The source and destination nodes of channel* c_i *are denoted* s_i *and* d_i, *respectively. For the sake of simplicity, an enumeration of arbitrary channels is denoted* $c_1, c_2, \ldots, c_k$ *instead of* $c_{n_1}, c_{n_2}, \ldots, c_{n_k}$. *An enumeration does not imply any channel ordering.*

DEFINITION 4.2. *Let* F *be the set of* channel status, $F = \{free, busy\}$.

DEFINITION 4.3. *An* adaptive routing function $R : N \times N \rightarrow \mathcal{P}(C)$, *where* $\mathcal{P}(C)$ *is the power set of* C, *supplies a set of alternative output channels to send a message from the current node* n_c *to the destination node* n_d, $R(n_c, n_d) = \{c_1, c_2, \ldots, c_p\}$, *where* $s_i = n_c \ \forall i$. *In general,* p *will be less than the number of output channels per node to restrict routing and obtain deadlock-free algorithms. As a particular case,* $p = 1 \ \forall (n_c, n_d) \in N \times N, n_c \neq n_d$, *defines a deterministic routing function. Also,* $R(n,n) = \emptyset$, $\forall n \in N$. *The domain of* R *has been defined as* $N \times N$ *instead of* $C \times N$ *as in* [**8**]. *Also, defining the result of* R *as a set of channels (an element of* $\mathcal{P}(C)$*) allows us the use of the operations defined on sets. In particular, the* $\in$ *operator will be extensively used throughout this paper.*

DEFINITION 4.4. *A* selection function $S : \mathcal{P}(C) \times F^C \rightarrow C$ *selects a free output channel (if any) from the set supplied by the routing function. From the definition,* S *takes into account the status of all the channels belonging to the set supplied by the routing function. The selection can be random or based on static or dynamic priorities. Also, in the same way the result of a deterministic routing function may be a busy channel, if all the output channels are busy, any of them may be selected. The decomposition of the adaptive routing into two functions (routing and selection) will be critical while proving the theorem, because only the routing function determines whether a routing algorithm is deadlock-free or not. Thus, the selection function will only affect performance. Moreover, it is possible to extend the definition of the selection function by taking into account additional information, either local to the node or remote.*

DEFINITION 4.5. *A routing function* R *for a given interconnection network* I *is* connected *iff*

$$\forall x, y \in N, x \neq y, \exists c_1, c_2, \ldots, c_k \in C / \begin{cases} c_1 \in R(x, y) \\ c_{m+1} \in R(d_m, y), \quad m = 1, \ldots, k-1 \\ d_k = y \end{cases}$$

In other words, it is possible to establish a path $P(x, y) \subset \mathcal{P}(C)$ *between* x *and* y *using channels belonging to the sets supplied by* R. *Notice that the interconnection network is strongly connected, but it does not imply that the routing function must be connected.*

DEFINITION 4.6. *A* routing subfunction R_1 *for a given routing function* R *is a routing function that supplies a subset of the channels supplied by* R

$$R_1(x, y) \subseteq R(x, y) \quad \forall x, y \in N$$

Thus, R_1 *restricts the routing options supplied by* R. *It must be noticed that this definition allows the restriction of channel routing capability instead of simply removing channels. In other words, it is possible to restrict the use of a channel* c_i *when* R_1 *routes messages to some destinations while still allowing the use of* c_i *when routing messages to other destinations. This definition is more general than the one proposed in* [**12, 17**], *because now it is possible to remove a channel from* R_1 *only for some destinations. The set of all the channels supplied by* R_1 *is* $C_1 = \bigcup_{\forall x,y \in N} R_1(x,y)$. *Also, the complementary routing subfunction of* R_1 *will be denoted as*

$$R_1^R(x,y) = R(x,y) - R_1(x,y) \quad \forall x,y \in N$$

DEFINITION 4.7. *Given an interconnection network* I, *a routing function* R *and a pair of channels* $c_i, c_j \in C$, *there is a* direct dependency *from* c_i *to* c_j *iff*

$$\exists x \in N / c_i \in R(s_i, x) \text{ and } c_j \in R(d_i, x)$$

That is, c_j *can be used immediately after* c_i *by messages destined to some node* x.

DEFINITION 4.8. *Given an interconnection network* I, *a routing function* R, *a routing subfunction* R_1 *and a pair of channels* c_i, c_j *supplied by* R_1 *for some destinations, such that* $d_i \neq s_j$, *there is an* indirect dependency *from* c_i *to* c_j *iff*

$$\exists x \in N,\ \exists c_1, c_2, \ldots, c_k \in C / \begin{cases} c_i \in R_1(s_i, x) \\ c_1 \in R_1^R(d_i, x) \\ c_{m+1} \in R_1^R(d_m, x), \quad m = 1, \ldots, k-1 \\ c_j \in R_1(d_k, x) \end{cases}$$

That is, it is possible to establish a path from s_i *to* d_j *for messages destined to some node* x. c_i *and* c_j *are the first and last channels in that path and the only ones supplied by* R_1. *Thus,* c_j *can be used after* c_i *by some messages. As* c_i *and* c_j *are not adjacent, some other channels not supplied by* R_1 *are used between them.*

DEFINITION 4.9. *Given an interconnection network* I, *a routing function* R, *a routing subfunction* R_1 *and a pair of channels* $c_i, c_j \in C$, *there is a* direct cross dependency *from* c_i *to* c_j *iff*

$$\exists x, y \in N / \begin{cases} c_i \in R_1(s_i, x) \\ c_i \in R_1^R(s_i, y) \\ c_j \in R_1(d_i, y) \end{cases}$$

That is, c_j *can be used immediately after* c_i *by messages destined to some node* y. c_j *is supplied by* R_1 *and* c_i *cannot be supplied by* R_1 *for that destination. However,* c_i *is supplied by* R_1 *for some other destination(s).*

DEFINITION 4.10. *Given an interconnection network* I, *a routing function* R, *a routing subfunction* R_1 *and a pair of channels* $c_i, c_j \in C$, *such that* $d_i \neq s_j$, *there is an* indirect cross dependency *from* c_i *to* c_j *iff*

$$\exists x, y \in N,\ \exists c_1, c_2, \ldots, c_k \in C / \begin{cases} c_i \in R_1(s_i, x) \\ c_i \in R_1^R(s_i, y) \\ c_1 \in R_1^R(d_i, y) \\ c_{m+1} \in R_1^R(d_m, y), \quad m = 1, \ldots, k-1 \\ c_j \in R_1(d_k, y) \end{cases}$$

That is, it is possible to establish a path from s_i to d_j for messages destined to some node y. c_i and c_j are the first and last channels in that path. c_j is the only channel supplied by R_1. c_i cannot be supplied by R_1 for that destination. However, c_i is supplied by R_1 for some other destination(s). Thus, c_j can be used after c_i by some messages. As c_i and c_j are not adjacent, some other channels not supplied by R_1 are used between them.

DEFINITION 4.11. *A* channel dependency graph *D for a given interconnection network I and routing function R, is a directed graph, $D = G(C, E)$. The vertices of D are the channels of I. The arcs of D are the pairs of channels (c_i, c_j) such that there is a direct dependency from c_i to c_j. Notice that there are no 1-cycles in D, because channels are unidirectional.*

DEFINITION 4.12. *An* extended channel dependency graph *D_E for a given interconnection network I and routing subfunction R_1 of a routing function R, is a directed graph, $D_E = G(C_1, E_E)$. The vertices of D_E are the channels that can be supplied by the routing subfunction R_1 for some destinations. The arcs of D_E are the pairs of channels (c_i, c_j) such that there is either a direct, indirect, direct cross or indirect cross dependency from c_i to c_j.*

DEFINITION 4.13. *A* sink *channel for a given interconnection network I and routing function R is a channel c_i such that*

$$\forall x \in N, c_i \in R(s_i, x) \Rightarrow x = d_i$$

In other words, all the flits that enter a sink channel reach their destination in a single hop. As a result, there are no outgoing arcs from a sink channel in any channel dependency graph, as can be easily seen from the definitions.

DEFINITION 4.14. *A* configuration *is an assignment of a set of flits to each queue. All of the flits in any one queue belong to the same message (assumption* 6*). The number of flits in the queue for channel c_i is denoted $size(c_i)$. If the first flit in the queue for channel c_i is destined for node n_d, then $head(c_i) = n_d$. If the first flit is not a header and the next channel reserved by its header is c_j, then $next(c_i) = c_j$. Let $C_h \subseteq C$ be the set of channels containing a header flit at their queue head. Let $C_d \subseteq C$ be the set of channels containing a data or tail flit at their queue head. A configuration is* legal *iff*

$$\forall c_i \in C \begin{cases} size(c_i) \leq cap(c_i) \\ size(c_i) > 0 \Rightarrow c_i \in R(s_i, head(c_i)) \end{cases}$$

For each channel, the queue capacity is not exceeded and all the flits stored in the queue (if any), which have the same destination, can reach the channel from the previous node using the routing function.

DEFINITION 4.15. *A* deadlocked configuration *for a given interconnection network I and routing function R is a nonempty legal configuration verifying the following conditions:*

$$1)\ \forall c_i \in C_h \begin{cases} head(c_i) \neq d_i \\ size(c_j) > 0 \qquad \forall c_j \in R(d_i, head(c_i)) \end{cases}$$

$$2)\ \forall c_i \in C_d \begin{cases} head(c_i) \neq d_i \\ size(next(c_i)) = cap(next(c_i)) \end{cases}$$

In a deadlocked configuration there is not any flit one hop from its destination. Header flits cannot advance because the queues for all the alternative output channels supplied by the routing function are not empty (see assumption 6*). As a particular case (for disconnected routing functions), the routing function may not supply any output channel. Data and tail flits cannot advance because the next channel reserved by their message header has a full queue. No condition is imposed on empty channels. It must be noticed that a data flit can be blocked at a node even if there are free output channels to reach its destination. Also, in a deadlocked configuration, there is no message whose header flit has already arrived at its destination.*

DEFINITION 4.16. *A routing function R for an interconnection network I is* deadlock-free *iff there is not any deadlocked configuration for that routing function on that network.*

For comparison purposes, we include below the theorems proposed in [**12, 17**].

THEOREM 4.1. *A connected and adaptive routing function R for an interconnection network I is deadlock-free if there are no cycles in its channel dependency graph D.*

THEOREM 4.2. *A connected and adaptive routing function R for an interconnection network I is deadlock-free if there exists a subset of channels $C_1 \subseteq C$ that defines a routing subfunction R_1 which is connected and has no cycles in its extended channel dependency graph D_E.*

The following theorem proposes a more relaxed sufficient condition for an adaptive routing function to be deadlock-free. It is an extension of theorem 4.2. Although the conditions proposed by both theorems seem very similar, it must be noticed that the definition of extended channel dependency graph has been modified.

THEOREM 4.3. *A connected and adaptive routing function R for an interconnection network I is deadlock-free if there exists a routing subfunction R_1 which is connected and has no cycles in its extended channel dependency graph D_E.*

PROOF. Suppose that there exists a routing subfunction R_1 of the routing function R and that R_1 is connected and there are no cycles in D_E. If $R_1(x,y) = R(x,y) \;\; \forall x,y \in N$ then $D_E = D$. Thus, there is not any cycle in D and R is deadlock-free by theorem 4.1. Otherwise $R_1(x,y) \subset R(x,y) \;\; \forall x,y \in N$. Let $C_1 = \bigcup_{\forall x,y \in N} R_1(x,y)$. As there are no cycles in D_E, one can assign an order to the channels of C_1 so that if $(c_i, c_j) \in E_E$ then $c_i > c_j$. Consider the channel(s) $c_i \in C_1$ such that

$$\forall c_j \in C_1, (c_i, c_j) \notin E_E$$

Such a channel c_i is minimal in the order. Let us prove that it is a sink. If it were not a sink, as the routing subfunction R_1 is connected, for any legal configuration with a header flit stored in the queue head of c_i

$$d_i \neq head(c_i) \Rightarrow \exists c_k \in C_1 / c_k \in R_1(d_i, head(c_i))$$

As the configuration is legal then

$$c_i \in R(s_i, head(c_i))$$

Now, there are two possible cases:

a) $c_i \in R_1(s_i, head(c_i))$. In this case, there is a direct dependency from c_i to c_k.

b) $c_i \notin R_1(s_i, head(c_i))$. Taking into account that c_i can be supplied by R_1 for some destinations, there is a direct cross dependency from c_i to c_k.

In both cases $(c_i, c_k) \in E_E$, contrary to the assumption that c_i is minimal. So, $d_i = head(c_i)$ and c_i is a sink.

Suppose that there is a deadlocked configuration for R. There are two possible cases:

a) The queues for channels belonging to C_1 are empty. Thus, there will be channels belonging to $C - C_1$ with header flits at their queue heads. Let c_i be one of those channels. As R_1 is connected then

$$head(c_i) \neq d_i \Rightarrow \exists c_j \in C_1 / c_j \in R_1(d_i, head(c_i)) \Rightarrow \\ \exists c_j \in C / c_j \in R(d_i, head(c_i))$$

Also $size(c_j) = 0$ and R does not have a deadlock.

b) The queues for channels belonging to C_1 are not empty. Let c_i be a channel belonging to C_1 with a nonempty queue such that all the channels less than c_i have empty queues. Again, there are two possible cases:

b1) c_i is minimal. As shown above, it is also a sink and then, all the flits stored in its queue will be destined for d_i and the flit at the head of the queue for c_i is not blocked.

b2) c_i is not minimal. Thus

$$size(c_j) = 0 \ \ \forall c_j \in C_1 / c_i > c_j$$

existing three possible cases:

b2.1) c_i has a header at the queue head. Taking into account that R_1 is connected

$$\begin{aligned} head(c_i) \neq d_i \Rightarrow \exists c_j \in C_1 / c_j \in R_1(d_i, head(c_i)) \Rightarrow \\ \exists c_j \in C / c_j \in R(d_i, head(c_i)) \end{aligned}$$

Now, there are two possible cases:

b2.1.1) $c_i \in R_1(s_i, head(c_i))$. In this case, there is a direct dependency from c_i to c_j.

b2.1.2) $c_i \notin R_1(s_i, head(c_i))$. As the configuration is legal, then $c_i \in R(s_i, head(c_i))$. In this case, there is a direct cross dependency from c_i to c_j.

In both cases:

$$c_i > c_j \Rightarrow size(c_j) = 0$$

and R does not have a deadlock.

b2.2) c_i has a data flit at the queue head, not destined for d_i, and $next(c_i)$ has been supplied by R_1. Thus, there is either a direct or direct cross dependency from c_i to $next(c_i)$, depending on whether c_i was supplied by R_1 or not. Thus,

$$c_i > next(c_i) \Rightarrow size(next(c_i)) = 0$$

and R does not have a deadlock.

b2.3) c_i has a data flit at the queue head, not destined for d_i, and $next(c_i)$ has not been supplied by R_1. Let $c_1, c_2, \ldots, c_k \in C$ be the set of channels reserved by the message after reserving c_i, c_k containing the message header. Those channels cannot be supplied by R_1, because if a channel c_m, $m = 2, \ldots, k$ were supplied by R_1, then there would be an indirect or indirect cross dependency between c_i and c_m, implying that c_m has an empty queue and flits can advance. Thus,

$$\exists c_1, c_2, \ldots, c_k \in C / \begin{cases} c_1 \in R_1^R(d_i, head(c_k)) \\ c_{m+1} \in R_1^R(d_m, head(c_k)), \quad m = 1, \ldots, k-1 \end{cases}$$

As R_1 is connected

$$\begin{aligned} head(c_k) \neq d_k \Rightarrow \exists c_j \in C_1 / c_j \in R_1(d_k, head(c_k)) \Rightarrow \\ \exists c_j \in C / c_j \in R(d_k, head(c_k)) \end{aligned}$$

Again, we have two possible cases:

b2.3.1) $c_i \in R_1(s_i, head(c_k))$. In this case, there is an indirect dependency from c_i to c_j.

b2.3.2) $c_i \notin R_1(s_i, head(c_k))$. As the configuration is legal, then $c_i \in R(s_i, head(c_k))$. In this case, there is an indirect cross dependency from c_i to c_j.

In both cases:

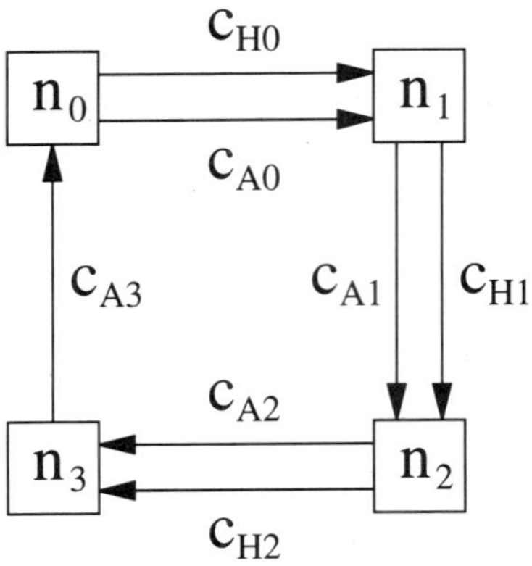

FIGURE 1. Channel labeling for a unidirectional ring

$$c_i > c_j \Rightarrow size(c_j) = 0$$

Thus, the header at the queue head for c_k is not blocked and R does not have a deadlock. □

The considerations given in [**17**] are also valid here.

5. Adaptive routing on k-ary n-cubes

In this section, we show the application of the theorem proposed in section 4 to the design of two adaptive routing algorithms for k-ary n-cubes, presenting two algorithms: partially adaptive and fully adaptive. First, we will propose a routing algorithm for unidirectional rings. Then, we will extend it for k-ary n-cubes. The routing algorithms proposed in this section are deadlock-free, but it is not possible to prove it by using theorem 4.2.

Without loss of generality, consider a unidirectional ring with four nodes denoted $n_i, i = \{0, 1, 2, 3\}$ and two channels connecting each pair of adjacent nodes, except nodes n_3 and n_0 which are linked by a single channel. Let $c_{Ai}, i = \{0, 1, 2, 3\}$ and $c_{Hi}, i = \{0, 1, 2\}$ be the outgoing channels from node n_i. The routing algorithm can be stated as follows: If the current node n_i is equal to the destination node n_j, store the message. Otherwise, use either c_{Ai}, $\forall j \neq i$ or c_{Hi}, $\forall j > i$. In other words, c_{Ai} channels can be used to forward messages to all the destinations. However, c_{Hi} channels can only be used if the destination is higher than the current node. Figure 1 shows the channel labeling.

This algorithm can be proved to be deadlock-free by applying theorem 4.3. Consider a routing subfunction R_1 which is equal to R, except that c_{A0} cannot be used and c_{A1}, c_{A2} can only be used to forward messages to destinations lower than the current node. In other words, we have removed from each channel c_{Ai} exactly the same routing capability supplied by the channel c_{Hi} in parallel with it. Obviously, R_1 is connected because messages at node n_i destined for a higher

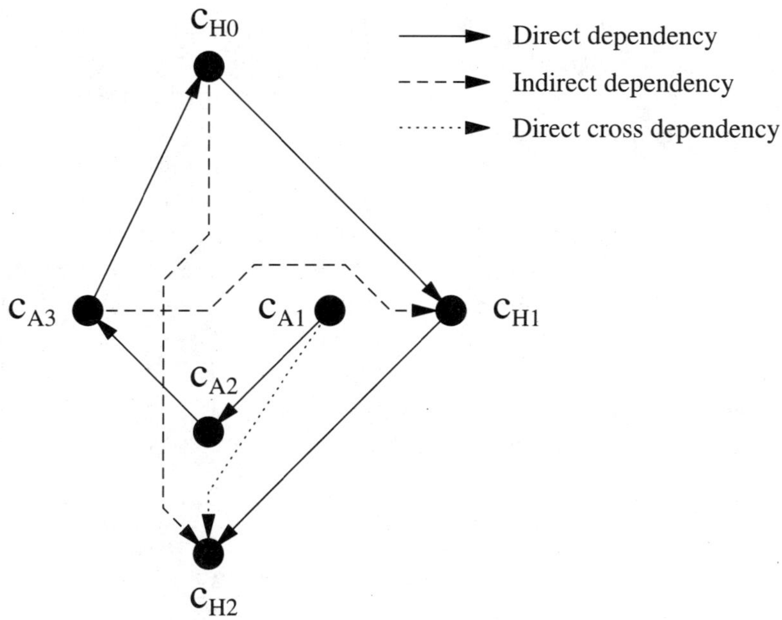

FIGURE 2. New extended channel dependency graph

node will be forwarded through c_{Hi} and messages destined for a lower node will be sent across c_{Ai}. Figure 2 shows the new extended channel dependency graph for R_1. As there are no cycles in the graph, we can conclude that R is deadlock-free. The same conclusion can be obtained if an additional channel c_{W3} is added in parallel with c_{A3} with the same routing capability as c_{A3}. This extension is useful to have the same number of channels connecting each pair of nodes. Also, if channels were connected from higher numbered nodes to lower numbered nodes, the routing algorithm would be identical, except that c_{Hi} channels would be labeled as c_{Li}, now being used to reach lower numbered nodes.

Now, we will extend this routing algorithm for bidirectional rings. As channels are bidirectional, they can be split into two unidirectional channels. Thus, both versions of the algorithm (with c_{Hi} and c_{Li} channels) are used, always forwarding messages through a minimal path. As messages follow a minimal path, once a message starts crossing channels in one direction, it cannot turn and continue in the opposite direction. Thus, the use of bidirectional channels does not introduce any additional channel dependency and the routing algorithm is deadlock-free.

The partially adaptive routing algorithm for k-ary n-cubes uses the algorithm for bidirectional rings (with c_{Wi} channels) inside each dimension. Channels can only be used crossing dimensions in ascending order. This routing algorithm only requires two virtual channels per physical channel. As dimensions are crossed in ascending order, adding dimensions does not introduce any additional cyclic dependency and the routing algorithm is still deadlock-free.

To design a fully adaptive routing algorithm based on the new theory we use the first design methodology proposed in [**12, 17**]. This methodology starts from a basic routing algorithm, adding virtual channels in a regular way. In this case, we start from the partially adaptive routing algorithm. Then, we add a third virtual channel to each physical channel. The third virtual channel is used for fully adaptive routing, crossing dimensions in any order following a minimal path. Obviously, the first and second virtual channels are used exactly in the same way as in the partially adaptive algorithm. The result is a fully adaptive minimal routing algorithm with three virtual channels per physical channel. Similarly to the example proposed in [**17**], it is easy to prove deadlock-freedom by including the additional virtual channels in $C - C_1$. In other words, the routing subfunction R_1 cannot use the fully adaptive virtual channels for any destination. As in the example proposed in [**17**], the fully adaptive virtual channels introduce new indirect dependencies but they cannot produce cyclic dependencies.

6. Performance evaluation of the algorithms

We have evaluated the behavior of the new adaptive algorithms by simulation. The evaluation methodology used is based on the one proposed in [**17**]. The most important performance measures are delay (additional latency required to deliver a message with respect to an idle network) and throughput (maximum traffic accepted by the network). Traffic is the flit reception rate. Delay is measured in clock cycles. Traffic is measured in flits per node per cycle.

Taking into account the sizes of current multicomputers and the studies about the optimal number of dimensions [**1**], we have evaluated the performance of the new algorithms on a 8-ary 3-cube network (512 nodes). In addition, the deterministic algorithm proposed in [**8**] for the k-ary n-cube, modified so that it uses bidirectional channels, has been evaluated for comparison purposes. It uses two virtual channels per physical channel as the partially adaptive algorithm. Our simulator models the network at the flit level. A single clock cycle is needed to route a header or to transmit one flit through the switch or through a physical link. Message length is fixed to 16 flits [**9, 1**]. Up to four messages can be simultaneously sent or received by each node.

Figure 3 shows the average message delay versus traffic for a uniform distribution for message destination. Figure 4 compares the same routing algorithms on the same network when messages are sent locally. In this case, message destinations are uniformly distributed inside a cube centered at the source node with side equal to four channels.

As can be seen, both adaptive algorithms considerably increase performance over the deterministic one. The partially adaptive algorithm increases throughput by 70% for the uniform distribution and it doubles throughput when messages are sent locally. If we consider the usable part of the curves, throughput is also doubled for the uniform distribution (0.18 versus 0.36 flits/cycle/node).

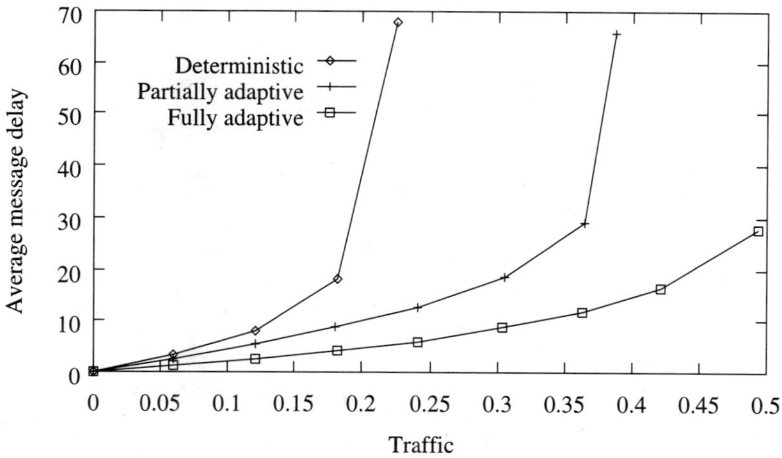

FIGURE 3. Average message delay versus traffic for a uniform distribution of message destinations

The explanation is very simple. Our algorithm allows most messages to choose between two virtual channels instead of one. The advantages of virtual-channel flow control were properly highlighted in [**10**]. These results are improved by the fully adaptive algorithm, which increases throughput over the deterministic algorithm by factors of 2.2 and 2.4 for the uniform and the locality distribution, respectively. The higher performance achieved by the fully adaptive algorithm is mainly due to the additional freedom to cross dimensions in any order. The adaptive algorithms also achieve a reduction in message delay with respect to the deterministic one for the full range of traffic.

Recently, it has been pointed out that virtual channels are expensive and that adaptive routing requires more complex switches, increasing node delay accordingly [**6**]. However, the implementation of our partially adaptive algorithm is almost as simple as the deterministic one. Both of them use the same number of virtual channels. Also, a very similar low-size crossbar [**7**] or multistage routing automata [**18**] can be used, because dimensions are crossed in order. The only additional cost is produced by considering the status bits (free or busy) of the two output virtual channels in the current dimension while computing a routing operation. This low extra implementation cost is unusual in adaptive algorithms [**6**], making our proposal very attractive from the practical point of view.

The fully adaptive routing algorithm requires three virtual channels per physical channel and a more complex switch. Thus, node delay will be higher. However, we have simulated all the routing algorithms assuming the same clock period. The higher node delay and the additional hardware complexity are not compensated by the higher performance achieved by the fully adaptive algorithm when traffic is low and uniform (with or without locality). However, as indicated in [**11**], adaptive routing is required for many non-uniform traffic pat-

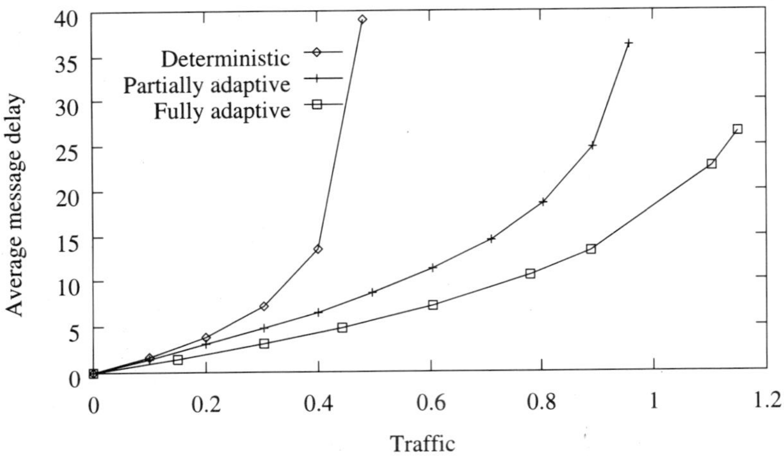

FIGURE 4. Average message delay versus traffic for local traffic

terns. Otherwise, performance is very poor. Currently, we are evaluating the above proposed routing algorithms under several non-uniform traffic patterns.

7. Conclusions

We have proposed a more relaxed sufficient condition for deadlock-free adaptive routing in wormhole networks. This condition is the key for the design of highly adaptive routing algorithms.

The theorem proposed in this paper is an extension of our previous theorem. Both theories allow the existence of cyclic dependencies between channels. To apply the theorems, routing is restricted. The restricted routing function (defined as routing subfunction) must be connected and free of cyclic dependencies to guarantee that the unrestricted routing function is deadlock-free.

In our previous theory, we classify channels as belonging or not belonging to the routing subfunction. Instead, we restrict routing in this paper in a more general way. A given channel may belong or not to the routing subfunction depending on message destination. This generalization of the definition of routing subfunction gives rise to two new kinds of channel dependency (direct cross dependency and indirect cross dependency). The additional complexity of this approach is required to obtain a more relaxed sufficient condition for deadlock-freedom.

Finally, we have proposed two routing algorithms for n-dimensional tori. The first algorithm is partially adaptive and it only requires two virtual channels per physical channel. The evaluation shows that, using the same resources, this algorithm significantly improves performance over the deterministic routing algorithm. The second algorithm is fully adaptive. It requires three virtual channels per physical channel and a more complex switch. It increases performance even further at the cost of a higher switch complexity.

REFERENCES

1. A. Agarwal, "Limits on interconnection network performance", *IEEE Trans. Parallel and Distributed Systems*, vol. 2, no. 4, pp. 398–412, Oct. 1991.
2. W.C. Athas and C.L. Seitz, "Multicomputers: Message-passing concurrent computers," *IEEE Comput. Mag.*, vol. 21, no. 8, pp. 9–24, Aug. 1988.
3. P.E. Berman, L. Gravano, G.D. Pifarré and J.L.C. Sanz, "Adaptive deadlock- and livelock-free routing with all minimal paths in torus networks," in *Proc. 4th ACM Symp. Parallel Algorithms Architectures*, June 1992.
4. S. Borkar et al., "iWarp: An integrated solution to high-speed parallel computing," in *Proc. Supercomputing'88*, Nov. 1988.
5. A.A. Chien and J.H. Kim, "Planar-adaptive routing: Low-cost adaptive networks for multiprocessors," in *Proc. 19th Annu. Int. Symp. Comput. Architecture*, May 1992.
6. A.A. Chien, "A cost and speed model for k-ary n-cube wormhole routers," in *Proc. Hot Interconnects'93*, Aug. 1993.
7. W.J. Dally and C.L. Seitz, "The torus routing chip," *Distributed Comput.*, vol. 1, no. 3, pp. 187–196, Oct. 1986.
8. W.J. Dally and C.L. Seitz, "Deadlock-free message routing in multiprocessor interconnection networks," *IEEE Trans. Comput.*, vol. C-36, no. 5, pp. 547–553, May 1987.
9. W.J. Dally, "Performance analysis of k-ary n-cube interconnection networks," *IEEE Trans. Comput.*, vol. C-39, no. 6, pp. 775–785, June 1990.
10. W.J. Dally, "Virtual-channel flow control," *IEEE Trans. Parallel Distributed Syst.*, vol. 3, no. 2, pp. 194–205, Mar. 1992.
11. W.J. Dally and H. Aoki, "Deadlock-free adaptive routing in multicomputer networks using virtual channels," *IEEE Trans. Parallel Distributed Syst.*, vol. 4, no. 4, pp. 466–475, April 1993.
12. J. Duato, "On the design of deadlock-free adaptive routing algorithms for multicomputers: design methodologies," in *Proc. Parallel Architectures Languages Europe 91*, June 1991.
13. J. Duato, "Deadlock-free adaptive routing algorithms for multicomputers: evaluation of a new algorithm," in *Proc. 3rd IEEE Int. Symp. Parallel Distributed Processing*, Dec. 1991.
14. J. Duato, "Improving the efficiency of virtual channels with time-dependent selection functions", in *Proc. Parallel Architectures Languages Europe 92*, June 1992.
15. J. Duato, "A theory to increase the effective redundancy in wormhole networks," in *Proc. Int. Conf. Decentralized Distributed Syst.*, Sept. 1993.
16. J. Duato, "A new theory of deadlock-free adaptive multicast routing in wormhole networks," in *Proc. 5th IEEE Int. Symp. Parallel Distributed Processing*, Dec. 1993.
17. J. Duato, "A new theory of deadlock-free adaptive routing in wormhole networks," *IEEE Trans. Parallel Distributed Syst.*, vol. 4, no. 12, pp. 1320–1331, Dec. 1993.
18. C. M. Flaig, "VLSI mesh routing systems," Master's thesis, California Institute of Technology, Department of Computer Science, May 1987.
19. P.T. Gaughan and S. Yalamanchili, "Adaptive routing protocols for hypercube interconnection networks," *IEEE Comput. Mag.*, vol. 26, no. 5, pp. 12–23, May 1993.
20. D. Gelernter, "A DAG-based algorithm for prevention of store-and-forward deadlock in packet networks," *IEEE Trans. Comput.*, vol. C-30, pp. 709–715, Oct. 1981.
21. C.J. Glass and L.M. Ni, "The turn model for adaptive routing," in *Proc. 19th Annu. Int. Symp. Comput. Architecture*, May 1992.
22. C.J. Glass and L.M. Ni, "Maximally fully adaptive routing in 2D meshes," in *Proc. Int. Conf. Parallel Processing*, Aug. 1992.
23. K.D. Gunther, "Prevention of deadlocks in packet-switched data transport systems," *IEEE Trans. Commun.*, vol. COM-29, pp. 512–524, Apr. 1981.
24. C.R. Jesshope, P.R. Miller, and J.T. Yantchev, "High performance communications in processor networks," in *Proc. 16th Annu. Int. Symp. Comput. Architecture*, May–June 1989.
25. J.H. Kim and A.A. Chien, "The impact of packetization in wormhole-routed networks," in *Proc. Parallel Architectures Languages Europe 93*, June 1993.
26. H.T. Kung, "Deadlock avoidance for systolic communication," in *Proc. 15th Annu. Int.

Symp. Comput. Architecture, May–June 1988.

27. X. Lin and L.M. Ni, "Deadlock-free multicast wormhole routing in multicomputer networks," in *Proc. 18th Annu. Int. Symp. Comput. Architecture*, May 1991.
28. X. Lin, P.K. McKinley and L.M. Ni, "The message flow model for routing in wormhole-routed networks," in *Proc. 1993 Int. Conf. Parallel Processing*, Aug. 1993.
29. D.H. Linder and J.C. Harden, "An adaptive and fault tolerant wormhole routing strategy for k-ary n-cubes," *IEEE Trans. Comput.*, vol. C-40, no. 1, pp. 2–12, Jan. 1991.
30. Z. Liu and H. Wu, "Performance evaluation of adaptive wormhole routing in 3D mesh networks," in *Proc. 26th Annu. Simulation Symp.*, 1993.
31. Z. Liu, J. Duato and L.-E. Thorelli, "Grouping virtual channels for deadlock-free adaptive wormhole routing," in *Proc. Parallel Architectures Languages Europe 93*, June 1993.
32. P. López and J. Duato, "Deadlock-free adaptive routing algorithms for the 3D-torus: limitations and solutions," in *Proc. Parallel Architectures Languages Europe 93*, June 1993.
33. P.M. Merlin and P.J. Schweitzer, "Deadlock avoidance in store-and-forward networks – I: Store-and-forward deadlock," *IEEE Trans. Commun.*, vol. COM-28, pp. 345–354, Mar. 1980.

Facultad de Informática, Universidad Politécnica de Valencia, Valencia, SPAIN
E-mail address: {jduato,plopez}@aii.upv.es

DIMACS Series in Discrete Mathematics
and Theoretical Computer Science
Volume **21**, 1995

Conflict-Free Access to Constant-Perimeter Rectangular Subarrays

DOREEN L. GALLI ERICKSON AND CHARLES J. COLBOURN

ABSTRACT. In a parallel computing environment, we consider conflict-free access to constant-perimeter rectangular subarrays, using a natural formulation in terms of latin squares. For parallel matrix computations, there are frequently portions of data, referred to as templates, that one wants stored and retrieved in so that one can operate on the data simultaneously with each processor working on one element of the template. If more than one processor attempts to retrieve data from a single memory module during the same memory cycle, a memory conflict arises. When the array is stored to allow the desired templates to be accessible from the memory modules without memory conflicts, the array is conflict-free for that set of templates. In this paper, we examine a set of structured templates where the number of template instances defined by a template grows with respect to the maximum size of a template. In particular, we examine the set of constant-perimeter rectangular subarrays.

A square is *perimeter rectangular conflict-free* (p_{rcf}) if it is conflict-free for all rectangular subarrays whose perimeter is less than or equal to $2p$. If p is even, the problem is to provide conflict-free access to all rectangular subarrays of size $(\frac{p}{2} - i) \times (\frac{p}{2} + i)$ for $-\frac{p}{2} < i < \frac{p}{2}$. We show that the necessary number of memory modules is $\frac{p^2}{2} - p + 1$, and there is a linear skewing scheme that realizes this bound. If p is odd, the problem is to provide conflict-free access to all rectangular subarrays of size $(\lfloor\frac{p}{2}\rfloor - i) \times (\lceil\frac{p}{2}\rceil + i)$ for $-\lfloor\frac{p}{2}\rfloor < i < \lfloor\frac{p}{2}\rfloor$. We show that the necessary number of memory modules is $\lfloor\frac{p}{2}\rfloor^2$, and there is a linear skewing scheme that realizes this bound. Furthermore, the linear results for the constant-perimeter rectangular subarrays hold when the subarrays are stretched by a factor of v, where v is relatively prime to the number of memory modules.

1. Introduction

In a parallel computing environment, consider the problem of storing an $n \times n$ array in n parallel memory modules. If the entire array is stored in the same

1991 *Mathematics Subject Classification.* Primary 68R05; Secondary 68P99, 05B15.

memory module, a memory bottle neck occurs when multiple processors are simultaneously operating on the array. More formally, a *memory conflict* occurs when more than one request is given to the same memory module during a single memory cycle. Hence, one may want to spread the information among the various memory modules. Ideally, one would optimize the distribution of information among the n memory modules. This may be accomplished by determining which parts of the array will be requested by the processors at the same time. Historically, to make this determination, one would assume that all the processors work under the same clock. This assumption would imply the use of the SIMD (single instruction stream, multiple data stream) model for parallel computation (see Figure 1). One could instead assume that the synchronization

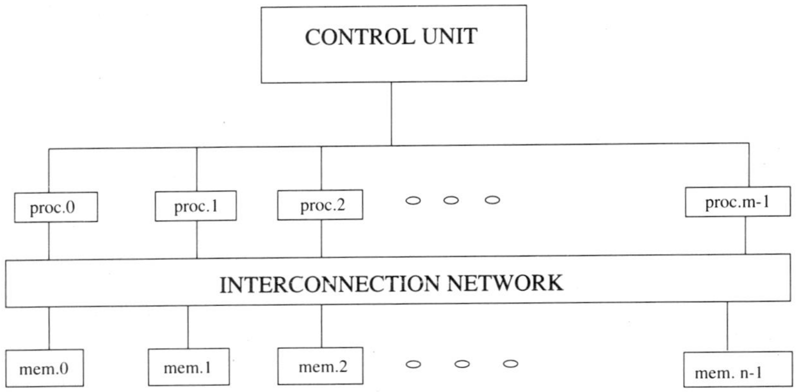

FIGURE 1. SIMD model of computation

is done at the software level, and thus generalize to any parallel computation model. Intuitively, a *template* specifies the shape of a subarray, all instances of which should be conflict-free. More formally, a template consists of a distinguished position called the handle and possibly other array positions defined by their relative locations with respect to this handle. If the actual position of the handle is not specified, it is assumed to be the leftmost element in the uppermost row. Common templates include the row template, the column template, and front and back diagonal templates. All *template instances* are translates of a template. Thus, a template instance is a specific instantiation of a template such as the first row. Only one template instance is accessed in a given memory cycle. An array element could refer to an elementary entity such as an integer, or it could refer to a composite entity such as a portion of a larger array. Without loss of generality, it is considered to be an elementary entity. The size of a template is the number of elements contained in the template. The function that maps the array elements into the memory modules is a *skewing scheme.* A skewing scheme is *linear* if it maps the array elements $a_{i,j} \equiv q(i-\alpha) + r(j-\beta) \pmod{n}$ for some fixed integers q, r, α and β [**Kuc68**]. Otherwise, the skewing scheme is *non-linear*. A square defined by a linear skewing scheme is a cyclic square. A skewing scheme may provide conflict-free access to a single template such as row template, or it may provide conflict-free access to a set of templates such as rows, columns and constant-perimeter rectangular subarrays.

In a more general framework, an efficient implementation is desired for a parallel algorithm. One aspect of the efficient implementation entails the storage and retrieval of data. The following are several factors or points that would permit this action to occur efficiently.

POINT 1: The set of templates accessible without memory conflicts should reflect the needs of algorithms.

POINT 2: The computation to generate the address of the appropriate memory module should require constant time.

POINT 3: There should exist an interconnection network that can efficiently realize the processor/memory connections needed by the skewing scheme.

POINT 4: The memory utilization should be high.

These points are all significant if a parallel program is to run efficiently. Suppose there are no skewing schemes that provide conflict-free access to the exact set of templates needed for a particular algorithm, or the existing skewing schemes are unacceptable according to any of the aforementioned points. This situation calls for a new skewing scheme. Thus, the next step would involve determining what is theoretically possible, to help narrow down the search for a skewing scheme. While limited theoretical results can be found in [**Sha78, Wij89**], there may be no pertinent theoretical results available, one would want to have some techniques to discover what is possible. When developing skewing schemes, one must predetermine what set of templates are to be conflict-free. One reasonable approach in choosing a set of templates is to accommodate those that are frequently found in parallel computing applications. In this paper, we focus on rectangular subarrays. These appear in algorithms for applications involving image processing [**SMJ92, DH91**], biological simulation [**RM85**] and scientific computing [**Fei82, Chu88**]. We present lower bounds that apply to both linear and non-linear skewing schemes. We also present constructions for linear skewing schemes.

In the literature, the set of templates that are accommodated by existing skewing schemes contain a fixed number of templates. In contrast, the number of templates in the set of templates accommodated by skewing schemes in this paper grow with respect to the maximum size of a template. Specifically, in Section 3, we examine the templates that provides conflict-free access to constant-perimeter rectangular subarrays. The number of templates is a function of the perimeter. In general, for the perimeter, $\gamma = 2p$, there are $p-1$ templates for a given constant perimeter. Furthermore, if the rectangular subarrays are to be stretched, each stretch produces a unique template, and the number of templates in the set is once again increased. Preliminary results were provided in [**EC92**]. Other results relating to rectangular templates can be found in [**Eri93a**].

2. Background for Rectangular Templates

A scheme proposed by Budnik and Kuck [**BK71**] provided conflict-free access to rows, columns, diagonals and $\sqrt{n-1} \times \sqrt{n-1}$ subarrays. This scheme uses a prime, P, larger than the dimension of the array as the number of memory modules. In addition, a skewing factor, s, is used. Specifically, $a_{i,j} \equiv is + j \pmod P$. A similar set of templates is accessible by the scheme proposed by Lawrie [**Law75**]. The set of templates here includes the row template, the column template, diagonals and $\sqrt{n} \times \sqrt{n}$ rectangular subarrays. This scheme is not conflict-free for all of these templates, but guarantees that each template instance can be accessed in no more than two memory cycles.

In 1989, Kim and Kumar [**KK89**] expressed the problem of skewing schemes in terms of *latin squares.* A latin square of order n is composed of the symbols 0 to $(n-1)$ such that no symbol appears more than once in any row or column [**DK74**]. Each array location is mapped to the memory module identified by the integer in the center of the square. Table 1 shows a latin square of order 3.

0	1	2
1	2	0
2	0	1

TABLE 1. Latin square of order 3

A *diagonal latin square* is a latin square such that no symbol appears more than once in either one of its two main diagonals [**DK74**].

A new type of latin square, a *perfect latin square*, was proposed by Kim and Kumar [**KK89, HKK92**] in 1989. A perfect latin square of order n is a diagonal latin square such that no symbol appears more than once in any *main subsquare.* The order, n, must be a perfect square. A subsquare of order $\sqrt{n}$, $S_{i,j}$, is a main subsquare if its top left cell is (i,j) where $i \equiv 0 \bmod \sqrt{n}$ and $j \equiv 0 \bmod \sqrt{n}$ [**HKK92**].

Kim and Kumar [**KK89**] also show that all subsquares of size $\sqrt{n}$ are accessible in two memory cycles when using perfect latin squares. When the perfect latin squares are of order 2^{2k} for k an integer, the address generation may be accomplished in constant time. In addition, they present skewing schemes for $3D$ arrays $(n \times n \times n)$. Additional correlations to combinatorics and the conflict-free access problem can be found in [**EC93a**].

Colbourn and Heinrich [**CH92**] used latin squares to access an $x \times y$ rectangular subarray. Specifically, there is an *(x, y) conflict-free latin square** for all $n = xy + p$ for $1 \leq p \leq y$. An (x, y) conflict-free latin square is a latin square in which no element appears more than once in any $x \times y$ subarray nor more then once in any row and any column. Since p may be equal to one, only one spare memory is required for conflict-free storage. They have also shown that if $x > y + 1$, there is an *$\{x, y\}$ conflict-free latin square** of order n for $n \geq xy + y$

[**CH92**]. An $\{x, y\}$ conflict-free latin square is a latin square in which no element appears more than once in any $x \times y$ **or** $y \times x$ subarrays nor more then once in any row and any column. If $2 \leq y < x < 2y$, they showed that a $\{x, y\}$ conflict-free latin square of order n exists if $n \geq xy + (2y - x)(x - y) + 1$. This implies that, if $x > y$, then a $\{x, y\}$ conflict-free latin square of order n exists if $n \geq xy + y$ [**CH92**]. They improved the lower bounds for the situation in which $x \not\equiv 1 \pmod y$, $x \geq 2y - 1$, and $x > y > 3$. In this situation, a $\{x, y\}$ conflict-free latin square of order n exists if $n > xy + y$. If $x \equiv 1 \pmod y$, then a $\{x, y\}$ conflict-free latin square of order n exists if $n = xy + y$.

Additional results can be found in the monograph by Wijshoff [**Wij89**]. One result of particular interest concerns block templates. For a $(p_1, v_1) \times (p_2, v_2) \times \ldots \times (p_d, v_d)$-block template B on a d-dimensional array, the nth tuple relates to the nth dimension. The p value of this tuple determines the number of elements chosen, and the v value indicates every v^{th} element in that dimension is included in the template [**Wij89**]. They showed that one cannot store two different block templates that have equal stretch and size n in n memory modules [**Wij89**]. The stretch of a block defined above is $(v_1, v_2, \ldots v_d)$ [**Wij89**].

3. All Constant-Perimeter Subarrays

In this section, we examine the templates that represent all constant-perimeter rectangular subarrays. As the perimeter desired increases, the skewing scheme has a larger set of templates whose access must be conflict-free.

Let the perimeter of the rectangular subarray desired to be conflict-free be defined as $\gamma = 2p$, where $p = x + y$, $x = \lfloor \frac{p}{2} \rfloor, y = \lceil \frac{p}{2} \rceil$. A square is said to be a *perimeter rectangular conflict-free square with stretch* v ($p_{\mathrm{rcf},v}$) if it is conflict-free for all rectangular subarrays whose perimeter is less than or equal to γ. If the $p_{\mathrm{rcf},v}$ square with stretch v is also latin, it is a *perimeter latin rectangular conflict-free square with stretch* v ($p_{\mathrm{lrcf},v}$).

To prove our bounds, we use a mega-template that encapsulates the essence of the interaction of the set of templates we want to be conflict-free [**EC93b**].

3.1. Preliminary Definitions and Theorems.

The following definitions are needed to define a p_v-diamond that we use as a mega-template. In addition, we proceed with some fundamental theorems needed for the proofs of $p_{\mathrm{rcf},v}$ and $p_{\mathrm{lrcf},v}$ squares.

- Let S be an $l \times l$ subarray. The rectangular subarray $S(a, b : c, d)$, refers to the subarray S with the rows a to b and columns c to d inclusive, a $(b-a+1) \times (d-c+1)$ rectangle. Furthermore, $S(a : c, d)$ is a partial row. Specifically, row a from column c to d inclusive. Similarly, $S(a, b : c)$ is a partial column of column c from row a to b inclusive. A similar notation is used in [**CH92, EC92**].
- The *base* is $S_v(0, x - 1 : 0, y - 1)$.

- A *stair* at level L on the *North* side is $S_v(-L : L, y-1-L) : 0 < L \leq \lfloor \frac{y-1}{2} \rfloor$.
- A *stair* at level L on the *South* side is $S_v(x-1+L : L, y-1-L) : 0 < L \leq \lfloor \frac{y-1}{2} \rfloor$.
- A *stair* at level L on the *West* side is $S_v(L, x-1-L : -L) : 0 < L \leq \lfloor \frac{x-1}{2} \rfloor$.
- A *stair* at level L on the *East* side is $S_v(L, x-1-L : y-1+L) : 0 < L \leq \lfloor \frac{x-1}{2} \rfloor$.
- A p_v*-diamond* is a template that is the union of the base, the North, South, West and East stairs (see Figure 2) or any translate in an $n \times n$ square with wrapping.

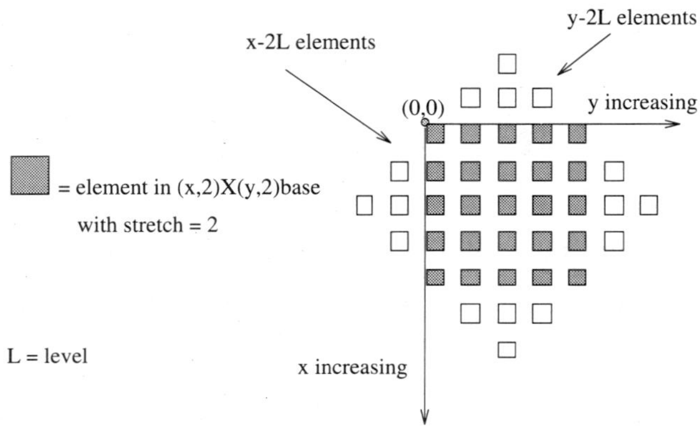

FIGURE 2. The p_2-diamond

THEOREM 3.1. *Let M be a latin square. The following two conditions are equivalent.*

(i) *Each p_v-diamond is conflict-free in M.*

(ii) *For each i satisfying $0 < i < p$, each $(p-i, v) \times (i, v)$ template instance is conflict-free in M.*

Proof: Since M is latin all $(p-1, v) \times (1, v)$ and $(1, v) \times (p-1, v)$, rectangular subarrays are conflict-free.

(1) $\Rightarrow$ (2): Assume that (2) is false, so that some $(p-i, v) \times (i, v)$ subarray does not contain distinct symbols. The definition of the p_v-diamond ensures that every $(p-i, v) \times (i, v)$ subarray is contained in some p_v-diamond or is a $(p-1, v) \times (1, v)$ or $(1, v) \times (p-1, v)$ subarray. Thus, (1) does not hold if (2) does not, and hence (1) $\Rightarrow$ (2).

(2) $\Rightarrow$ (1): Assume first that some p_v-diamond does not contain distinct elements. Let the two equal entities be at horizontal distance vh and vertical distance vk from one another. Then $vk \leq v(\lfloor \frac{p}{2} \rfloor - 1)$, since the diamond has

height $v\lfloor\frac{p}{2}\rfloor$, and $v(k+h) \leq v(\lfloor\frac{p}{2}\rfloor + \lceil\frac{p}{2}\rceil - 2)$ for any two entries of the diamond. Now since $v(k+h) \leq v(\lfloor\frac{p}{2}\rfloor + \lceil\frac{p}{2}\rceil - 2)$ and $vk \leq v(\lfloor\frac{p}{2}\rfloor - 1)$, and some $(k+1, v) \times (h-1, v)$ rectangular subarray is not conflict-free, and thus, some $(p-i, v) \times (i, v)$ template instance is not conflict-free and (2) does not hold. Thus (2) $\Rightarrow$ (1). □

COROLLARY 3.2. *Let M be a cyclic latin square. The following two conditions are equivalent.*

(i) *Some p_v-diamond is conflict-free in M.*

(ii) *For each i satisfying $0 < i < p$, some $(p-i, v) \times (i, v)$ template instance is conflict-free in M.*

Proof: One instance of a template is conflict-free if and only if all instances in M are conflict-free. This follows immediately by the cyclic construction. Thus, the proof is complete by Theorem 3.1. □

We now proceed to count the elements contained in the p-diamond. We use a convenient notation suggested by Knuth [**Knu92**], denoting by $[x \text{ odd}]$, the function whose value is one if x is odd, and zero otherwise. The stairs can be examined in two separate cases. Let z be the length of the side of the base upon which the stairs are built. Let $n(z)$ be the number of elements in the set of stairs on side z.

LEMMA 3.3. *$n(z) = \frac{1}{4}z^2 - \frac{1}{2}z + \frac{1}{4}$ if z is odd.*

Proof: For odd z, the number of elements in the set of stairs is:

$$\begin{aligned} n(z) &= \sum_{i=1}^{\frac{z-1}{2}} (z - 2i) \\ &= \frac{z-1}{2} + 2\sum_{i=0}^{\frac{z-3}{2}} i \\ &= \frac{(z-1)^2}{4} \\ &= \frac{1}{4}z^2 - \frac{1}{2}z + \frac{1}{4}. \end{aligned}$$

□

LEMMA 3.4. *$n(z) = \frac{1}{4}z^2 - \frac{1}{2}z$ if z is even.*

Proof: If z is even, then the number of elements in the set of stairs is:

$$\begin{aligned} n(z) &= \sum_{i=1}^{\frac{z}{2}}(z-2i) \\ &= 2\sum_{i=0}^{\frac{z-2}{2}}(i) \\ &= (\frac{z-2}{2})(\frac{z}{2}) \\ &= \frac{1}{4}z^2 - \frac{1}{2}z. \end{aligned}$$

□

LEMMA 3.5. *The p_v-diamond contains $xy + \frac{1}{2}x^2 + \frac{1}{2}y^2 - x - y + \frac{1}{2}[x \text{ odd}] + \frac{1}{2}[y \text{ odd}]$ elements.*

Proof: The number of elements in a p_v-diamond is the sum of the base and the four sets of stairs. Let z be the length of the side of the base upon which the stairs are built. Let $n(z)$ be the number of elements in the set of stairs on side z. By Lemma 3.3, $n(z) = \frac{1}{4}z^2 - \frac{1}{2}z + \frac{1}{4}$ if z is odd. By Lemma 3.4, $n(z) = \frac{1}{4}z^2 - \frac{1}{2}z$ if z is even. Thus, the number of elements in the p_v-diamond is $xy + 2n(x) + 2n(y) = xy + \frac{1}{2}x^2 + \frac{1}{2}y^2 - x - y + \frac{1}{2}[x \text{ odd}] + \frac{1}{2}[y \text{ odd}]$. □

Define a (t, s, r, v) *shift-box* by $S_v(1, t : 1, s) \cup S_v(t + 1 : 1, r)$.

Let $n = st + r$ and define an $n \times n$ cylic square $c = (a_{i,j}), a_{i,j} \equiv (i-1)s + j \pmod{n}$.

THEOREM 3.6. *If W is a (t, s, r, v) shift-box, then W contains distinct elements in c where v is relatively prime to n.*

Proof: Recall the skewing function for a cyclic square. Successive array locations in a row are mapped to successive integers modulo n. If v equals one, the first row in W is mapped to the first s integers. The first element in the second row is mapped to the integer $s + 1$. Thus, the first t rows are mapped to the first st integers. Thus, there are r integers remaining that are mapped into row $t + 1$. Since $n = st + r$, all the array locations in the shift-box are mapped to distinct integers that represent the memory modules. If v is greater than one but relatively prime to n, the array locations in the first row of the (t, s, r, v) shift-box are mapped to $1, v + 1, 2v + 1, \ldots, s(v - 1) + 1$. The second row is mapped to $v(s) + 1$. If v is relatively prime to n, the mapping will cycle through all the integers, since the mapping is modulo n. Thus all the elements in the shift-box are mapped to distinct integers. □

LEMMA 3.7. *If* $a_{i,j} \equiv a_{i+b,j+c} \pmod{n}$, *then* $a_{i,j} \equiv a_{i+vb,j+vc} \pmod{n}$ *in a cyclic square.*

Proof: If $a_{i,j} \equiv a_{i+b,j+c} \pmod{n}$, then according to the cyclic construction, $bq + cr \equiv 0 \pmod{n}$. Thus, $v(bq+cr) \equiv 0 \pmod{n} \equiv vbq + vcr \pmod{n}$. This implies that $a_{i,j} \equiv a_{i+vb,j+vc} \pmod{n}$. □

OBSERVATION 3.8. *If a skewing scheme is conflict-free for a template, then the transpose of that skewing scheme is conflict-free for the transpose of that template. Furthermore, observe that the transpose of a cyclic skewing scheme is cyclic.*

3.2. Bounds and Constructions. We now proceed with the proofs for $p_{\text{rcf},v}$ and $p_{\text{lrcf},v}$ squares. We first treat the case when p is odd. Let $x = \lfloor \frac{p}{2} \rfloor, y = \lceil \frac{p}{2} \rceil$ (recall by Observation 3.8 that the results apply if $x = \lceil \frac{p}{2} \rceil, y = \lfloor \frac{p}{2} \rfloor$). We want conflict-free access to all $(x-i, v) \times (y+i, v)$ rectangular subarrays of an $n \times n$ array, where $-x < i < x$. Observe that these templates are exactly the same as each $(p-i, v) \times (i, v)$ rectangular subarrays for $0 < i < p$, or all rectangular subarrays of perimeter γ be conflict-free where $\gamma = 2p$.

THEOREM 3.9. *If an* $n \times n$ $p_{\text{rcf},v}$ *square exists for odd* p, *then* $n \geq 2x^2$.

Proof: By Theorem 3.1, all the elements in the p_v-diamond must be distinct. It follows by Lemma 3.5 that $n \geq xy + \frac{1}{2}x^2 + \frac{1}{2}y^2 - x - y + \frac{1}{2}[x \text{ odd}] + \frac{1}{2}[y \text{ odd}]$ where $y = x + 1$.
$n \geq x(x+1) + \frac{1}{2}x^2 + \frac{1}{2}(x+1)^2 - x - (x+1) + \frac{1}{2}[x \text{ odd}] + \frac{1}{2}[(x+1) \text{ odd}] =$

$$
\begin{aligned}
& x(x+1) + \frac{1}{2}x^2 - x + \frac{1}{2} + \frac{1}{2}(x+1)^2 - (x+1) \\
= \;& x^2 + x + \frac{1}{2}x^2 - x + \frac{1}{2} + \frac{1}{2}(x^2 + 2x + 1) - (x+1) \\
= \;& \frac{3}{2}x^2 + \frac{1}{2} + \frac{1}{2}x^2 + x + \frac{1}{2} - x - 1 \\
= \;& 2x^2.
\end{aligned}
$$

□

Let $n = 2x^2$ and define an $n \times n$ square $A = (a_{i,j}), a_{i,j} \equiv (i-1)(2x-1) + j \pmod{n}$.

THEOREM 3.10. *A is a $p_{rcf,v}$ and a $p_{lrcf,v}$ square for odd p when v is relatively prime to n.*

The general strategy is to exhibit a mega-template that is easily seen to be conflict-free, and for which the absence of conflicts implies the whole square is conflict-free for the set of templates. Rather than directly show that the p_v-diamond is conflict-free within the square, we frequently insert an intermediate step involving a box, usually a shift-box. The primary remaining step is to show the relationship between the shift-box and the p_v-diamond. Thus, the proof of Theorem 3.10 is based on the following lemma.

LEMMA 3.11. *If an $(x-1, 2x-1, 3x-1, v)$ shift-box contains distinct elements in A, then the p_v-diamond contains distinct elements for all x, odd p and v relatively prime to n.*

Proof: First we examine the p_1-diamond. The upper left-hand corner of the shift-box is placed over the North-West corner of the North-most stair of the p_1-diamond as displayed in Figure 3. $a_{i,j} \equiv a_{i-1,j+2x-1} \pmod{n}$, since

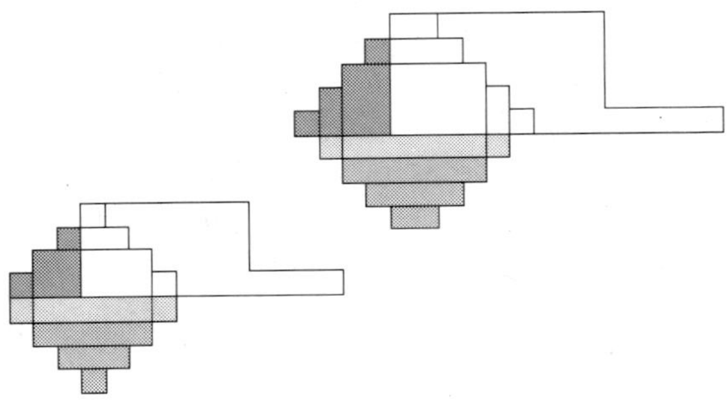

FIGURE 3. Shift-box placed over the p_1-diamond

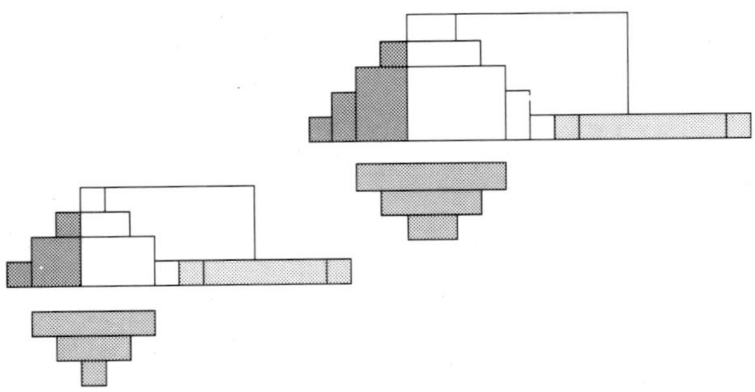

FIGURE 4. Shifting row $x+1$ to shift-box

$(-1)(2x-1)+(2x-1) \equiv 0 \pmod{n}$, and the symbol is the same due to the construction of the square. Thus, row $(x+1)$ is shifted up one row and over $(2x-1)$ columns as displayed in Figure 4. Now, $a_{i,j} \equiv a_{i-x-1,j+x-1} \pmod{n}$, since $(-x-1)(2x-1)+(x-1) \equiv -2x^2 \equiv 0 \pmod{n}$, and the symbol is the same

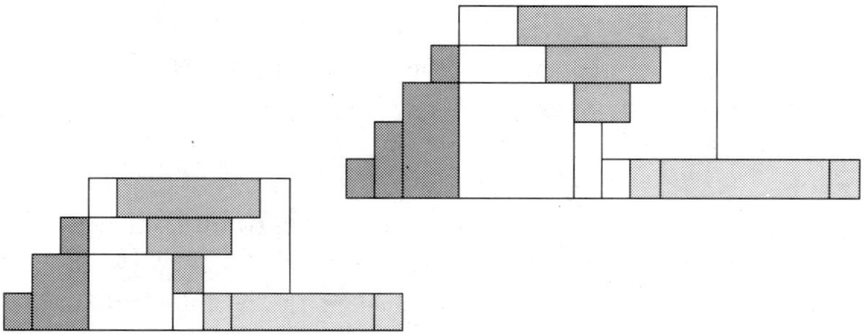

FIGURE 5. Shifting rows $x+2$ through $2x-1[x \text{ odd}]$ to shift-box

due to the construction of the square. Thus, rows $(x+2)$ through $(2x-1[x \text{ odd}])$ are shifted up $(x+1)$ rows and over $(x-1)$ columns as displayed in Figure 5. Finally, $a_{i,j} \equiv a_{i-1,j+2x-1} \pmod n$, since $(-1)(2x-1)+(2x-1) \equiv 0 \pmod n$, and the symbol is the same due to the construction of the square. Thus, all the elements in rows 2 through x not contained in the shift-box are shifted up 1 row and over $(2x-1)$ columns. This final shift is displayed in Figure 6. Finally, the above shifts are valid for the p_v-diamond by Lemma 3.7. □

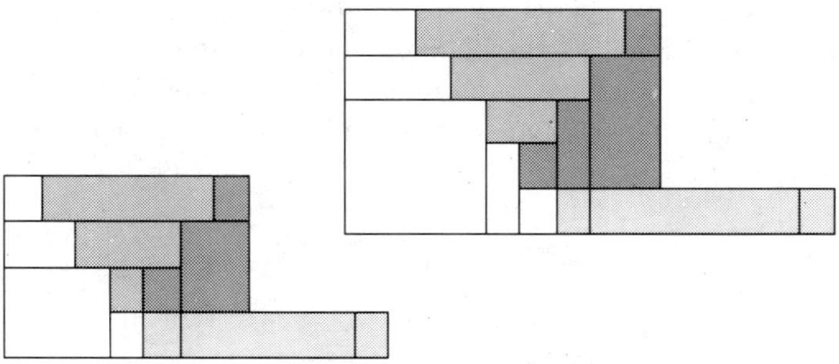

FIGURE 6. Shifting rows 1 through x to shift-box

Proof of Theorem 3.10: Since $n = 2x^2$ is relatively prime to $s = 2x-1$, the square is latin. Thus, by Corollary 3.2, if A is conflict-free for some p_v-diamond, then it is a $p_{\mathrm{rcf},v}$ square. By Lemma 3.11, a p_v-diamond is conflict-free if the shift-box is conflict-free. Finally, by Theorem 3.6, all of the elements in the shift box are distinct, and A is a $p_{\mathrm{lrcf},v}$ square. □

Let σ be the maximum size of any template in the set of constant-perimeter templates. For odd p, $\sigma = x^2 + x$ and for even p, $\sigma = x^2$.

COROLLARY 3.12. *The $p_{lrcf,v}$ square for odd p can be accessed in two memory cycles when stored in σ memory modules if v is relatively prime to n.*

Proof: Since $n = 2x^2 \leq 2x(x+1) = 2\sigma$ for all $x \geq 0$, any memory modules from σ to $2\sigma - 1$ can be mapped onto the memory modules 0 to $\sigma - 1$, and can be accessed in the second memory cycle. □

Thus, if one wants to increase memory utilization to satisfy **Point 3**, one can store the templates in σ memory modules.

Now, we turn to the second main case, when the perimeter $\gamma = 2p$ and p is even. We write $p = 2x$, and thus $y = x = \frac{p}{2}$. We want conflict-free access to all $(x - i, v) \times (x + i, v)$ rectangular subarrays of an $n \times n$ array for $-x < i < x$.

THEOREM 3.13. *If a $p_{\mathrm{rcf},v}$ square of order n exists for even p,* $n \geq 2x^2 - 2x + 1$.

Proof: By Theorem 3.1, all of the elements in the p_v-diamond must be distinct. It follows by Lemma 3.5 that $n \geq 2x^2 - 2x + [x \text{ odd}]$. Thus, when x is odd, we achieve the bound in the statement of the theorem. When x is even, we obtain $n \geq 2x^2 - 2x$ by Lemma 3.5. Thus, we only need to show that $n \neq 2x^2 - 2x$. Shapiro shows in general that a template tiles a plane if and only if a skewing scheme exists [**Sha78**]. Thus, any $p_{\mathrm{rcf},v}$ square of order $2x^2 - 2x$ must correspond to a tiling of the plane with a p_v-diamond. Thus, one can choose an arbitrary external element of the p_v-diamond and attempt to place it next to every possible external element in a replica copy of a p_v-diamond. This type of case analysis reveals that there are only two possible tilings that are represented in Figure 7 and Figure 8. In the first tiling, a $(1, v) \times (x + x - 1, v)$ template

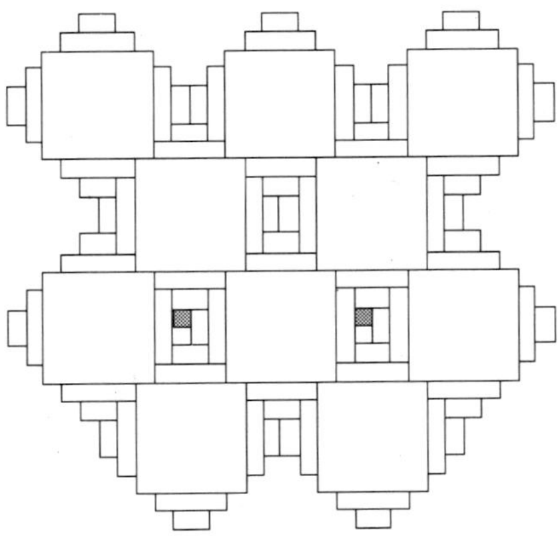

FIGURE 7. First possible tiling of the p_v-diamond for even p and even x

is not conflict-free. Thus, while the p_v-diamond tiles the plane, the tiling does not provides a $p_{\mathrm{rcf},v}$ square. In the second tiling, an $(x + x - 1, v) \times (1, v)$ template is not conflict-free. Thus, there is no possible tiling of the plane with the p_v-diamond and $n \geq 2x^2 - 2x + 1$. □

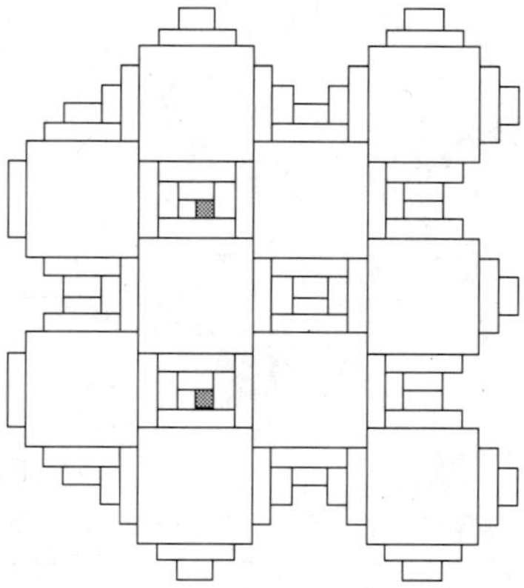

FIGURE 8. Second possible tiling of the p_v-diamond for even p and even x

Let $n = 2x^2 - 2x + 1$, and define an $n \times n$ square $B = (a_{i,j})$, $a_{i,j} \equiv (i-1)(2x - 1) + j \pmod{n}$.

A p_v-*zircon* is the p_v-diamond for even p and even x together with the cell $S_v(\frac{x}{2} - 1 : x + \frac{x}{2} - 1)$, which is an element added onto the North-East side of the East stairs as depicted in Figure 9.

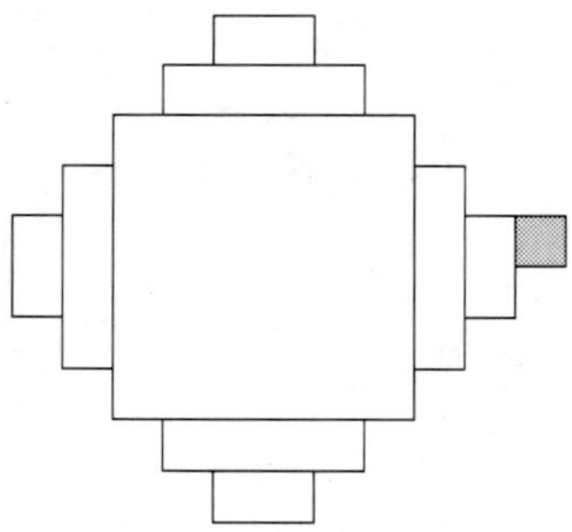

FIGURE 9. The p_v-zircon

THEOREM 3.14. *B is a $p_{rcf,v}$ and a $p_{lrcf,v}$ square for even p when v is relatively prime to n.*

The proof is based on the following lemma.

LEMMA 3.15. *If an $(x-1, 2x-1, x, v)$ shift-box contains distinct elements in B, then the p_v-diamond for odd x and the p_v-zircon for even x contain distinct elements for even p when v is relatively prime to n.*

Proof: First, examine the p_1-diamond and the p_1-zircon. The upper left-hand corner of the shift-box is placed over the North-West corner of the North-most stair of the p_1-diamond and the p_1-zircon as displayed in Figure 10. Next,

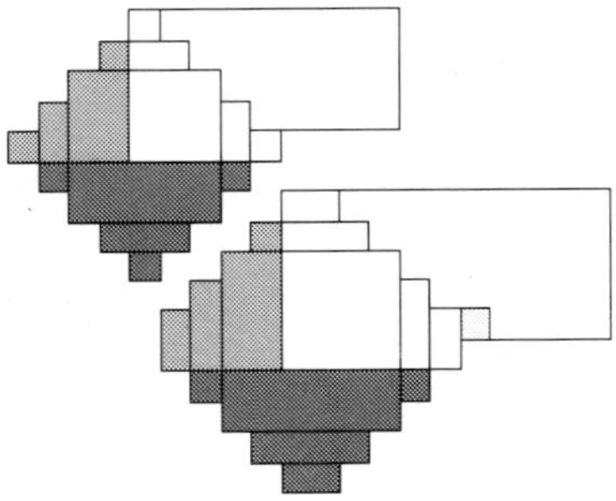

FIGURE 10. Shift-box placed over the p_1-diamond and p_1-zircon

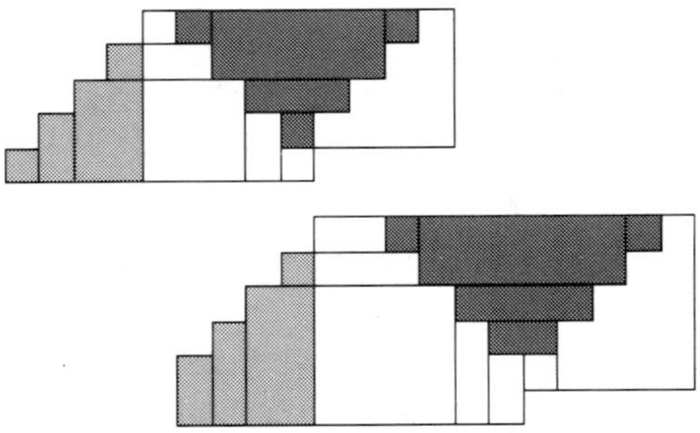

FIGURE 11. Shifting rows $x+1$ through $2x-1$ to shift-box

$a_{i,j} \equiv a_{i-x,j+x-1} \pmod n$, since $(-x)(2x-1)+(x-1) \equiv 0 \pmod n$. Thus, according to the cyclic construction, this element is the same element, and rows $(x+1)$ through $(2x-1)$ are shifted up x rows and over $(x-1)$ columns as displayed in Figure 11. Now, $a_{i,j} \equiv a_{i-1,j+2x-1} \pmod n$, since $(-1)(2x-1)+2x-1 \equiv 0 \pmod n$, and the symbol is is the same due to the construction of the square. Thus, all of the elements in rows 2 through x not contained in the shift-box are shifted up one row and over $(2x-1)$ columns. This shift is displayed in Figure 12. Finally, by Lemma 3.7, the above shifts are valid for p_v-diamonds and p_v-zircons. □

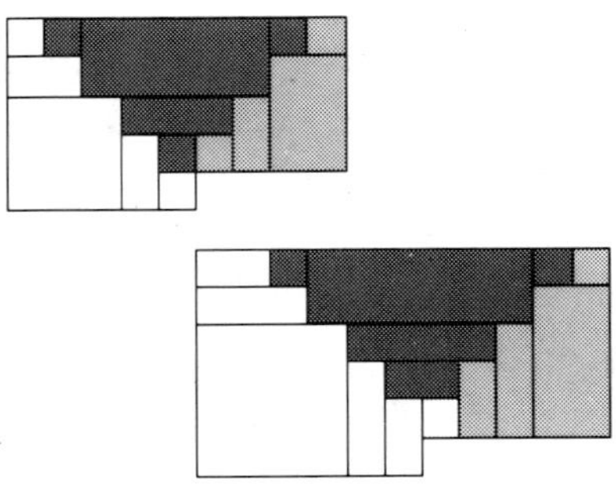

FIGURE 12. Shifting rows 1 through x to shift-box

Proof of Theorem 3.14. Since $(2x-1)$ is relatively prime to $(2x^2-2x+1) =$

$(2x-1)(x-1)+x$, B is latin. Thus, by Corollary 3.2, if B is conflict-free for some p_v-diamond, then it is a $p_{\mathrm{rcf},v}$ square. Since the p_v-zircon contains the p_v-diamond, this also holds for the p_v-zircon. By Lemma 3.15, the p_v-diamond and the p_v-zircon are conflict-free if the shift-box is conflict-free. Finally, by Theorem 3.6, all of the elements in the shift box are distinct and B is a $p_{\mathrm{lrcf},v}$ square. □

Recall that σ is the maximum size of any template in the set of constant-perimeter templates.

COROLLARY 3.16. *The $p_{lrcf,v}$ square for even p can be accessed in two memory cycles when stored in $\sigma = x^2$ memory modules if v is relatively prime to n.*

Proof: Since $n = 2x^2 - 2x + 1 < 2x^2 = 2\sigma$ for all $x > 0$, any memory modules from σ to $2\sigma - 1$ can be mapped onto the memory modules 0 to $(\sigma - 1)$ and be accessed in the second memory cycle. □

Once again, storing the templates in σ memory modules can increase memory utilization. This would lead to a higher memory utilization and accommodate **Point 3**. Observe that, if $vu = n$, that is to say, v is not relatively prime to n, there are v symbols in the p-diamond that are repeated u times. Each set of v symbols requires two memory cycles. Thus, for $vu = n$, $2u$ memory cycles are required for conflict-free access to the templates when the array is stored in σ memory modules.

Since the skewing scheme should reflect the needs of algorithms (**Point 1**) and maximize memory utilization (**Point 3**), one may desire more flexibility for the situation when v is not relatively prime to n. If the number of memory modules used in the first scheme is relatively prime to the number of memory modules used in the additional scheme, then the only stretches that cannot be accommodated are those that have a common prime factor with each scheme.

Table 2 indicates computational results for additional linear skewing schemes for $p_{\mathrm{lrcf},v}$ squares for even p. They are displayed by their x value and by the number of memory modules utilized above the minimum linear skewing scheme. The bold-face numbers represent the families of skewing schemes proved in [**Eri93a, Eri93b**]. Table 3 indicates computational results for additional linear skewing schemes for $p_{\mathrm{lrcf},v}$ squares for odd p. They are displayed by their x value and by the number of memory modules utilized above the minimum linear skewing scheme. The bold-face numbers indicate those schemes that are equivalent to using the skewing scheme for $p+1$.

In general, to find a family of skewing schemes, we attempt to find a pattern in the number of memory modules. If a pattern is located, we then examine all the related skew factors and try to find a pattern. If a pattern is found, we then attempt to prove that this pattern of skews represents a family of skewing

x	Difference above minimum n																
2	**2**	3	4		6	7	8	9		11	12	13	14	15	16	17	18
3	2		**4**	5	6		8	9	10	11	12	13	14	15	16	17	18
4			**4**		6	7	8	9			12	13	14	15	16		18
5	**2**		4		6		8	9	10		12		14		16	17	18
6			**4**		6		8		10	11	12	13	14	15	16	17	18
7			**4**		6				10		12	13	14		16		18
8	**2**		4				8		10		12		14	15	16	17	18
9			**4**		6		8				12		14		16	17	18
10			**4**		6				10		12				16		18
11	**2**				6		8		10		12				16		18
12			**4**		6						12		14		16		18
13			**4**		6				10		12				16		18
14	**2**		4				8				12		14		16		18
15			**4**						10		12		14		16		18
16			**4**		6				10		12				16		18
17	**2**		4		6		8				12		14		16		18
18			**4**		6				10		12				16		18
19			**4**		6						12				16		18
20	**2**				6		8		10		12				16		
21			**4**						10		12				16		18
22			**4**								12				16		18

TABLE 2. Additional skewing schemes found for even p

x	Difference above minimum n																
x	3	4	5	6	7	8	9	10	11	12	13	14	15	16	17	18	19
2	**3**		5		7		9		11		13		15		17		19
3			**5**		7				11		13				17		19
4			5		**7**		9		11		13		15		17		19
5							**9**		11						17		19
6					7				**11**		13				17		19
7							9		11		**13**		15		17		19
8							9		11				**15**		17		
9									11						**17**		19
10									11		13				17		**19**
11													15		17		

TABLE 3. Additional skewing schemes found for odd p

schemes by proving that the elements in the p-diamond are distinct. It would be of substantial interest to find a mechanical technique for finding such patterns. By examining Tables 2 and 3, one realizes that one cannot arbitrarily choose a value for n above the minimum when implementing a skewing scheme. All n values, even if larger than the upper bound, do not realize a construction because of constraints on the relationship of the skew value, the n value and the set of templates. Furthermore, the aforementioned techniques do not seem to be helpful in identifying additional linear skewing schemes for $p_{\mathrm{lrcf},v}$ squares.

4. Additional Implications

In addition to the above results, there are additional implications when one is dealing with $(x,v)\times(y,v)$ rectangular subarrays because of the theorem regarding shift-boxes that allows them to be stretched, in particular Theorem 3.6 and because of the lemma that allows the shifts to hold, in particular, Lemma 3.7. Wijshoff states that two blocks of equal size n and equal stretch cannot be stored in n memory modules, but no precise bounds are presented [**Wij89**].

Define a $(x,y)_v$ *conflict-free latin square* to be a (x,y) conflict-free latin square with stretch v.

LEMMA 4.1. *There is a $(x,y)_v$ conflict-free latin square of order n with $n = xy + p$ for $1 \le p \le y$.*

Proof: By Theorem 3.6, the shift-box equivalent may be stretched. By Lemma 3.7, shifts are valid if the blocks are stretched. The non-stretch proofs are presented in [**CH92**]. □

Define a $\{x,y\}_v$ *conflict-free latin square* to be a latin square in which no element appears more than once in any $(x,v)\times(y,v)$ **or** $(y,v)\times(x,v)$ subarray nor more than once in any row and any column. Table 4 displays extensions to the results from [**CH92**], which follow immediately from Theorem 3.6 and Lemma 3.7 for v relatively prime to n.

relations for x and y	n value
$x > y+1$	$n \ge xy + y$
$2 \le y < x < 2y$	$n \ge xy + (2y-x)(x-y) + 1$
$x > y$	$n \ge xy + y$
$x > y \ge 2$ and $x \equiv 1 \pmod{y}$	$n = xy + y$
$x > y \ge 2$ and $gcd(x,y) = 1$	$n = y(x+y-1)$

TABLE 4. Results for blocks of equal size and stretch

LEMMA 4.2. *When $x > 4$ is even, there is no $\{x,2\}_v$ conflict-free latin square of order $2x+2$ or $2x+3$.*

Proof: As before, by Theorem 3.7, Lemma 3.6 and by [**CH92**]. □

LEMMA 4.3. *If* $x \not\equiv 1 \pmod{y}, x \geq 2y-1, x > y \geq 3$, *there is no* $\{x,y\}_v$ *conflict-free latin square of order* $xy+y$.

Proof: As before, by Theorem 3.7, Lemma 3.6 and by [**CH92**]. □

5. Conclusions

To provide conflict-free access to all constant-perimeter rectangular subarrays, one has to accommodate the entire set of templates. The size of this set is a function of the perimeter. In addition, the templates are all of different shapes. The set of templates further expands when the rectangular subarrays are stretched. In this paper, we employ the p-diamond as a mega-template that encapsulates the interactions of the templates contained in the set of constant-perimeter templates and, thus, the essence of the constant-perimeter problem. This mega-template is employed not only to find lower bounds that apply to any type of skewing scheme, but is also employed to find a construction. This set of templates appears to be useful for various applications, satisfying **Point 2**, which requires that the templates accessible without memory conflicts reflect the needs of algorithms. In addition, the linear skewing schemes are efficient, satisfying **Point 1**.

In 1989, Kim and Kumar [**KK89, HKK92**] showed that two memory cycles are necessary for the retrieval of $\sqrt{n} \times \sqrt{n}$ subsquares of an $n \times n$ square when stored in n memory modules. Thus, one of the more surprising aspects of the results presented in this paper is that all constant-perimeter rectangular subarrays can be accessed in two memory cycles when stored in σ memory modules, where σ is the maximum size of a template in the set of constant-perimeter templates. This satisfies **Point 3**, which requires high memory utilization. Table 5 presents a summary of all the upper and lower bounds found for constant-perimeter rectangular subarrays with stretch v when v is relatively prime to n. While experimental data reveals that in most cases the upper bound for $(x,y)_{\mathrm{lrcf},v}$ squares is minimal for a linear skewing scheme, there are instances that can beat this upper bound. For example, the upper bound and the related construction provided for a $(2,3)_{\mathrm{lrcf},v}$ square uses 8 memory modules when one could store it in 7 using a skew factor of 3. Table 6 gives the constructions for the linear skewing schemes presented in this paper. The order of the square is the upper bound given in Table 5.

	p	x cond.	y cond.	lower bound	upper bound
$p_{\mathrm{lrcf},v}$	even		$y=x$	$2x^2-2x+1$	$2x^2-2x+1$
	odd		$y=x+1$	$2x^2$	$2x^2$

TABLE 5. Table of upper and lower bounds

Thus, we have presented a technique to find bounds for sets of templates. In the problems we examined, the number of templates is a function of the maxi-

	p	x cond.	y cond.	skew factor
$p_{\mathrm{lrcf},v}$	even		$y = x$	$2x - 1$
	odd		$y = x + 1$	$2x - 1$

TABLE 6. Table of constructions

mum size of a template. For $p_{\mathrm{lrcf},v}$ squares, these techniques have found tight bounds. The lower bounds apply to any type of implementation one may choose to employ. In addition, we have found linear skewing schemes that are efficient and satisfy **Point 2**. When linear skewing schemes are deemed acceptable, our constructions can be applied to solve the conflict-free access problem. Future work includes determining which interconnection networks can efficiently realize the skewing schemes presented and, thus, how these skewing schemes satisfy **Point 4**.

REFERENCES

[BK71] P. Budnik and D.J. Kuck, *The organization and use of parallel memories*, IEEE Trans. Computers **C-20** (1971), 1566–1569.

[CH92] C.J. Colbourn and K. Heinrich, *Conflict Free Access to Parallel Memories*, Journal of Parallel and Distributed Computing **14** (1992), 193–200.

[Chu88] E. Chu, *Orthogonal Decomposition of Dense and Sparse Matrices on Multiprocessors*, Ph.D. thesis, University of Waterloo, 1988, Department of Computer Science.

[DH91] F. Dehne and S.E. Hambrusch, *Parallel Algorithms for Determining k - width connectivity in Binary Images*, Journal of Parallel and Distributed Computing **12** (1991), 12–23.

[DK74] J. Dénes and D. Keedwell, *Latin Squares and Their Applications*, Academic Press, 1974.

[EC92] D.L. Erickson and C.J. Colbourn, *Conflict-Free Access to Rectangular Subarrays*, Congressus Numerantium **90** (1992), 239–253.

[EC93a] D.L. Erickson and C.J. Colbourn, *Combinatorics and the Conflict-Free Access Problem*, Congressus Numerantium (1993), 115–121.

[EC93b] D.L. Erickson and C.J. Colbourn, *Conflict-Free Access for Collections of Templates*, Proceedings of the Sixth SIAM Conference on Parallel Processing for Scientific Computing: Vol II, March 1993, Norfolk, Virginia, pp. 949–952.

[Eri93a] D.L. Erickson, *Conflict-Free Access to Rectangular Subarrays in Parallel Memory Modules*, Ph.D. thesis, University of Waterloo, 1993, Department of Computer Science.

[Eri93b] D.L. Erickson, *Conflict-Free Access to Rectangular Subarrays in Parallel Memory Modules*, Tech. Report CS-93-24, University of Waterloo, 1993, FTP only.

[Fei82] M. Feilmeier, *Parallel numerical algorithms*, Parallel Processing Systems (D.J. Evans, ed.), Cambridge University Press, 1982, pp. 285–338.

[HKK92] K. Heinrich, K. Kim, and V.K. Kumar, *Perfect Latin Squares*, Discrete Applied Mathematics **37/38** (1992), 281–286.

[KK89] K. Kim and V.K.P. Kumar, *Perfect latin squares and parallel memory access*, Proc. Sixteenth Int. Symp. Computer Architecture, 1989, pp. 372–379.

[Knu92] D.E. Knuth, *Two Notes on Notations*, The American Mathematical Monthly **99** (1992), no. 5, 403–422.

[Kuc68] D.J. Kuck, *ILLIAC IV software and application programming*, IEEE Trans. Computers **C-17** (1968), 758–770.

[Law75] D.H. Lawrie, *Access and alignment of data in an array processor*, IEEE Trans. Computers **C-24** (1975), 1145–1155.

[RM85] R. Ransom and R.J. Matela, *Computers in Biology: An Introduction*, Open University Press, 1985.

[Sha78] H.D. Shapiro, *Theoretical limitations on the effective use of parallel memories*, IEEE Trans. Computers **C-27** (1978), 421–428.

[SMJ92] W. Sung, S.K. Mitra, and B. Jeren, *Multiprocessor Implementation of Digital Filtering Algorithms Using a Parallel Block Processing Model*, IEEE Trans. Parallel and Distributed Systems **3** (1992), no. 1, 110–120.

[Wij89] H.A.G. Wijshoff, *Data Organization in Parallel Computers*, Kluwer Academic Publishers, 1989.

Associate Professor of Computer Science; Southern College of Technology; Marietta, Georgia, 30060-2896 USA, erickson@sct.edu

Chair, Dept. of Combinatorics and Optimization, University of Waterloo; Waterloo, Ontario, N2L 3G1 Canada, cjcolbourn@math.uwaterloo.ca

DIMACS Series in Discrete Mathematics
and Theoretical Computer Science
Volume **21**, 1995

MAKESPAN MINIMIZATION OF TASK GRAPHS WITH RANDOM TASK RUNNING TIMES

LUCIAN FINTA AND ZHEN LIU

First draft in April 1994, revised in November 1994

ABSTRACT. The problem of scheduling a set of tasks on two parallel and identical processors is considered. The executions of tasks are constrained by precedence relations. The running times of the tasks are independent random variables with a common exponential distribution. The goal of scheduling is to minimize the makespan, i.e. the maximum task completion time. A simple optimal preemptive policy is proven to stochastically minimize the makespan when the precedence graph belongs to a class of forest-cut graphs.

1. INTRODUCTION

Parallel programs are usually represented by task graphs which are directed acyclic graphs where vertices represent tasks and arcs represent precedence relations between tasks. The executions of these tasks have to satisfy these precedence constraints in such a way that a task can start execution only when all its predecessor tasks have completed execution. For any given task graph, the scheduling problem consists in assigning tasks to a set of processors in such a way that the makespan, i.e. the maximum task completion time, is minimized. Due to the partial order relation defined on the set of tasks, such a scheduling problem is NP-hard in general (see Ullman [15]), even when the task running times are equal. The reader is referred to Liu and Sanlaville [10] for a survey of complexity results and optimal polynomial algorithms for special cases.

Most scheduling literature is concerned with the deterministic scheduling problem, where task running times are assumed to be known constants. However, in practice, task running times are difficult to obtain in advance. For example, when a task contains loops, the number of iterations can depend on the input data. Even if a task contains a fixed number of executions of several instructions, the number of memory cycles per instruction can depend on the contentions on shared memory.

In this paper, we consider the scheduling problem where task running times are random variables. More specifically, we assume that task running times are independent and identically distributed (i.i.d.) random variables with a common

1991 *Mathematics Subject Classification.* Primary 90B35, 68M20; Secondary 90A80.

Key words and phrases. stochastic scheduling, makespan, parallel processors, precedence constraint, forest-cut graph, in-forest, out-forest.

Research of both authors was supported in part by the CEC DG XIII under the ESPRIT BRA grant QMIPS.

exponential distribution. The goal of scheduling is to find optimal preemptive schedule that stochastically minimizes the makespan.

When the task graph is an in-forest, and there are two (parallel and identical) processors, Chandy and Reynolds [2] proved that the Highest Level first (HL) policy minimizes the expected makespan. Here, the level of a task is simply the distance from it to the root of the tree it appears. They also have given one counterexample for in-trees under HL policy with three processors. Bruno [1] subsequently showed that HL stochastically minimizes the makespan when the system has two identical parallel processors. Pinedo and Weiss [13] extended this last result to the case where tasks at different levels may have different expected task running times. Frostig [7] further generalized the result of Pinedo and Weiss to include increasing likelihood ratio distributions for the task running times. When the number of identical parallel processors in the system is arbitrarily fixed, and the task running times have a common exponential distribution, Papadimitriou and Tsitsiklis [11] proved that HL is asymptotically optimal as the number of tasks tends to infinity.

Coffman and Liu [3] investigated the stochastic scheduling of out-forest with exponential task running times. For the uniform out-forests where all the subtrees are ordered by an embedding relation, they showed that an intuitive priority scheduling policy induced by the embedding relation, referred to as the Largest Tree (LT) policy in this paper, stochastically minimizes the makespan when there are two processors. For a more restrictive graph, the r-uniform out-forests, they showed the optimality of LT policy for arbitrary number of processors.

Liu and Sanlaville [9] proved the optimality of Most Successor (MS) policy when the task graph is an interval-order graph and when the task running times have a common exponential distribution. MS stochastically minimizes the makespan on any fixed number of processors.

In this paper, we investigate scheduling of fork-join graphs obtained from combination of in-forests and out-forests. We show that when there are two processors, the policy Highest Level Most Successors (HLMS) stochastically minimizes the makespan on two processors. This result generalizes known results on stochastic scheduling of forests on two processors.

In the next section, we define the notation used in the paper. In Section 3 we prove that the policy HLMS stochastically minimizes the makespan of task graphs belonging to the class of forest-cut graphs. Concluding remarks are provided in Section 4.

2. Notation

A *task graph* $G = (V, E)$ is a directed acyclic graph, where $V = \{1, 2, \cdots, |V|\}$ is the set of vertices representing tasks, $E \subset V \times V$ is the set of arcs representing *precedence constraints*: $(i, j) \in E$ if and only if task i must be completed before task j can start. Denote by $p(i)$ and $s(i)$ the sets of immediate *predecessors* and *successors* of $i \in V$, respectively, i.e.,

$$p(i) = \{j : (j, i) \in E\}, \qquad s(i) = \{j : (i, j) \in E\}.$$

Graph G is an *in-forest*, illustrated in Figure 1 (resp. *out-forest*, illustrated in Figure 2) if $|s(i)| \leq 1$ (resp. $|p(i)| \leq 1$) for all $i \in G$. Let $S(i)$ be the set of (not

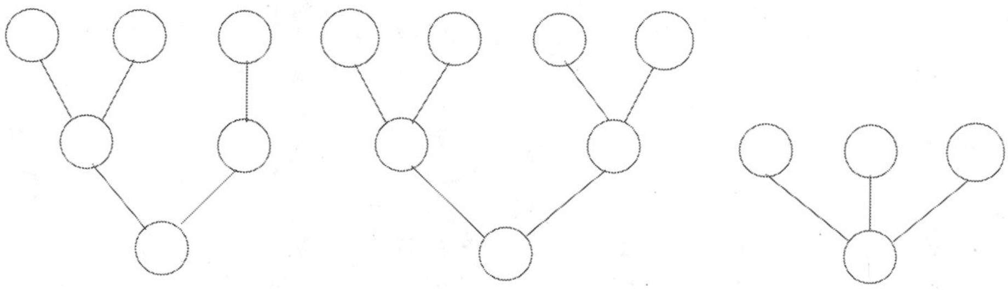

FIGURE 1. An example of in-forest

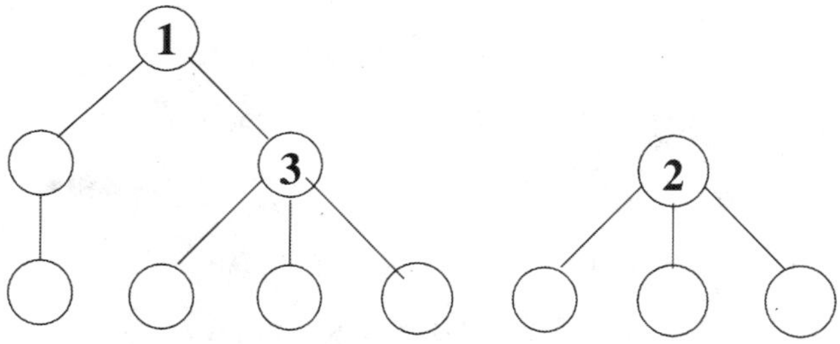

FIGURE 2. An example of uniforme out-forest

necessarily immediate) successors of $i \in V$, i.e.,

$$\forall i: \quad \text{if} \quad s(i) = \emptyset \quad \text{then} \quad S(i) = \emptyset \quad \text{else} \quad S(i) = s(i) \bigcup \left(\bigcup_{j \in s(i)} S(j) \right).$$

The set of successors $S(i)$ and the vertex i form a subgraph of G denoted by $S_i(G)$.

All vertices $i \in V$ that have no predecessor (resp. successor) are *initial* (resp. *final*) vertices. The level of the final vertices is zero by convention. The level of a vertex $i \in V$, denoted by $L(i)$, is the length of the longest path from it to final vertices. The level of graph G, denoted by $L(G)$, is the maximum level of its vertices. Let $N_k(G)$ be the number of tasks of G at level k, $k \geq 0$. Clearly $N_k(G) = 0$ for all $k > L(G)$.

For any task graph $G = (V, E)$ and a level l, let $T_l(G) = (V(T_l(G)), E(T_l(G))$ and $D_l(G) = (V(D_l(G)), E(D_l(G)))$ be two subgraphs of G consisting of top and down vertices separated by level l. More precisely,

$$V(T_l(G)) = \{v | v \in V, \ L(v) \geq l\}, \quad V(D_l(G)) = \{v | v \in V, \ L(v) \leq l\},$$

and $E(T_l(G))$ and $E(D_l(G))$ are restrictions of E on $V(T_l(G))$ and $V(D_l(G))$. In order to simplify the notations we will denote only by $T(G)$ and $D(G)$ the top and the down of some graph separated by level l, when l is some fixed level for all graphs.

We now introduce majorization notions for comparing graphs. Two graphs F and G are identical, $F = G$, if they have the same sets of vertices and arcs. We say that F is *flatter than* G, denoted by $F \prec_f G$, if

$$\forall i, \ i \geq 0: \qquad \sum_{k \geq i} N_k(F) \leq \sum_{k \geq i} N_k(G).$$

Let G_1, G_2 be two out-trees. Out-tree G_1 is said to *embed* out-tree G_2, or G_2 is *embedded* in G_1, denoted by $G_1 \succ_e G_2$ or $G_2 \prec_e G_1$, if G_2 is isomorphic to a subgraph of G_1. Formally, G_1 embeds G_2 if there exists an injective function f from G_2 into G_1 such that $\forall u, v \in G_2$, $v \in s(u)$ implies $f(v) \in s(f(u))$. Function f is called an embedding function. An out-forest G is said to be *uniform* if all its subtrees $\{S_v(G),\ v \in G\}$ can be ordered by the embedding relation. For example, the out-forest in Figure 2 is a uniform out-forest.

The embedding relation is extended to uniform out-forests: Let $G_1 = (V_1, E_1)$ and $G_2 = (V_2, E_2)$ be two uniform out-forests. Assume that the vertices of G_1 and G_2 are indexed in such a way that

$$S_1(G_1) \succ_e S_2(G_1) \succ_e \cdots \succ_e S_{|V_1|}(G_1),$$

$$S_1(G_2) \succ_e S_2(G_2) \succ_e \cdots \succ_e S_{|V_2|}(G_2).$$

Out-forest G_1 is embedded in G_2, referred to as $G_1 \prec^e G_2$, if and only if

$$|V_1| \leq |V_2|, \quad \text{and} \quad \forall i,\ 1 \leq i \leq |V_1|: \quad S_i(G_1) \prec_e S_i(G_2).$$

Note that if G is a uniform out-forest, then for any $v \in G$ with $p(v) = \emptyset$, $G - \{v\} \prec^e G$, where $G - \{v\}$ denotes the graph obtained by removing v from G.

We shall study a class $\mathcal{G}$ of task graphs, referred to as *forest-cut* graphs, which can be cut into an in-forest and a uniform out-forest. More precisely, $\mathcal{G}$ is the class of graphs satisfying the following conditions.

Definition 1. *For any graph $G = (V, E) \in \mathcal{G}$, there exists a level l such that*

(1) *$D_l(G)$ is an in-forest;*
(2) *$T_l(G)$ is a uniform out-forest, whose vertices can be labeled in such a way that*

$$S_1(T_l(G)) \succ_e S_2(T_l(G)) \succ_e \cdots \succ_e S_{|V(T_l(G))|}(T_l(G)),$$

(3) *and that the subgraphs are in the flatness relation*

$$S_1(G) \succ_f S_2(G) \succ_f \cdots \succ_f S_{|V(T_l(G))|}(G).$$

The level l will be called the cut level.

As an example, by combining the inforest of Figure 1 and the out-forest of Figure 2, we obtain the forest-cut graph in Figure 3.

Another example of forest-cut graphs are fork-join task graphs, cf. Figure 4.

Note that the cut level of a graph is not unique. Note also that this class of graphs has the following closure property: if $G = (V, E) \in \mathcal{G}$ then $G - \{v\} \in \mathcal{G}$ for all $v \in V$ such that $p(v) = \emptyset$, where $G - \{v\}$ is the graph obtained by deleting vertex v and its adjacent edges. Such a closure property (by deletion) will be used in establishing our results.

We extend the notion of flatness as follows. Let $F, G \in \mathcal{G}$ be two forest-cut graphs with the same cut level l, the relation of *extended flatness*, denoted by $F \prec^f G$, means that

$$|V(T_l(F))| \leq |V(T_l(G))|, \quad \text{and} \quad \forall i, \quad 1 \leq i \leq |V(T_l(F))|: \quad S_i(F) \prec_f S_i(G).$$

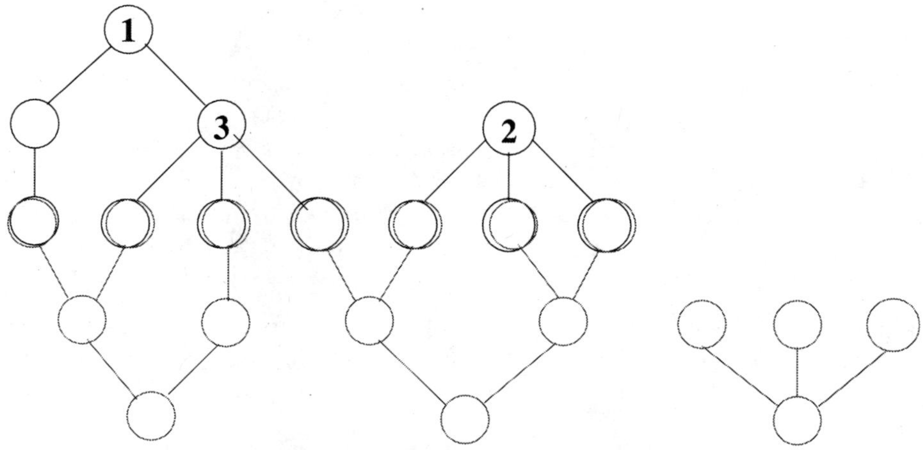

FIGURE 3. An example of forest-cut graph

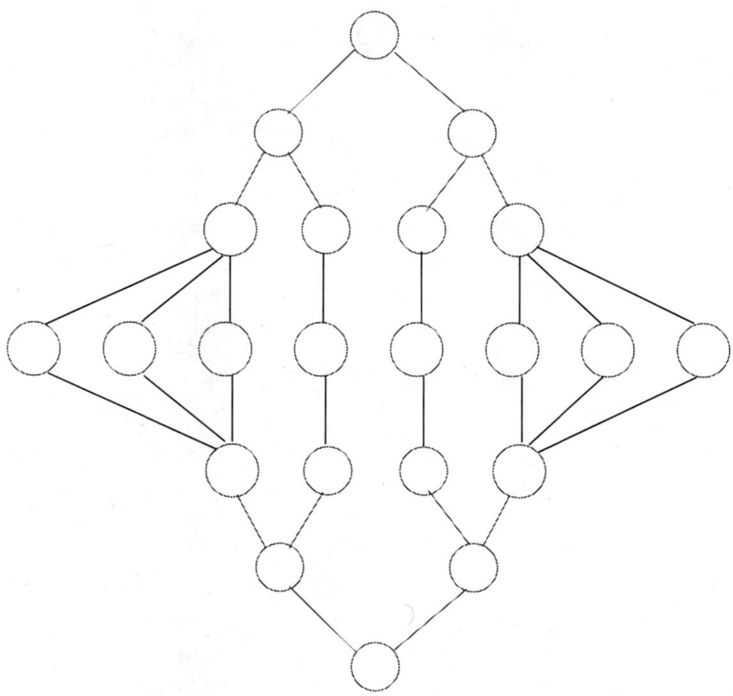

FIGURE 4. An example of fork-join graph

Definition 2. *Two forest-cut graphs $F, G \in \mathcal{G}$ with the same cut level l are in the relation $F \prec G$ if and only if they satisfy*

$$F \prec_f G, \qquad T(F) \prec^e T(G), \qquad F \prec^f G.$$

Note that if F and G are two in-forests, then $F \prec_f G$ implies $F \prec G$. If F and G are two uniform out-forests, then $F \prec^e G$ implies $F \prec G$.

Finally, we define the notion of stochastic comparison $\leq_{st}$. Random variable $X \in \mathbf{R}$ is stochastically smaller than $Y \in \mathbf{R}$, denoted by $X \leq_{st} Y$, if for all $x \in \mathbf{R}$, the tail distributions of X is smaller than Y: $P(X > x) \leq P(Y > x)$.

The proofs of our main result will use a coupling argument based on the following well-known result due to Strassen [14], where $=_{st}$ denotes equality in distribution.

Lemma 1 (Strassen). *Two random variables $X, Y \in \mathbf{R}$ satisfy $X \leq_{st} Y$ if and only if there exist two random variables $\hat{X}$ and $\hat{Y}$ defined on a common probability space such that $X =_{st} \hat{X}$, $Y =_{st} \hat{Y}$, and $\hat{X} \leq \hat{Y}$ almost surely (a.s.).*

3. Stochastic Minimization of Makespan

We define the Highest Level Most Successors (HLMS) policy as follows. The tasks of $G \in \mathcal{G}$ are labeled by $1, 2, \cdots, |V|$ in such a way that for any $1 \leq i \leq |V| - 1$, $L(i) \geq L(i+1)$, and if $L(i) = L(i+1)$ then $|S(i)| \geq |S(i+1)|$. At any time, only the enabled tasks (those without any unfinished predecessors, i.e. the initial vertices in the remaining task graph) with the smallest labels are executed on the parallel processors. In other words, the HLMS assigns the tasks at the highest level among the enabled tasks to available processors. If several enabled tasks are at the same level, the priority is given to the task with most successors. By convention, the available task with the lowest label is assigned to processor 1. We will consider *preemptive* policies where executions of tasks can be preempted and resumed without penalty.

Let Ψ denote the class of preemptive policies which use no information on task running times. For any policy π, let $\pi(G)$ denote the makespan of G, i.e. the completion time of all tasks of G under policy π. Let γ denote policy HLMS. We will show that for two processors, a policy is optimal for the class $\mathcal{G}$ of task graphs if and only if the policy is HLMS.

Theorem 1. *Assume there are two parallel processors and task running times are i.i.d. with a common exponential distribution. Then for any graph $G \in \mathcal{G}$, and any $\pi \in \Psi$,*

$$\gamma(G) \leq_{st} \pi(G). \tag{1}$$

Moreover, HLMS is the only policy that satisfies relation (1).

In order to simplify the proofs of this result, we make some restrictions on the class of policies Ψ. Observe first that due to the memorylessness property of exponential distributions, the distribution of the remaining running time of a task running on a processor is still exponential with the same parameter. If we represent the state of the system by the remaining task graph and the distributions of the remaining running times of the tasks, then the state does not change between the instants of task completions. Therefore, we can, without loss of generality, confine ourselves to the class of policies where preemptions and new task assignments occur only at the instants of task completions. These instants are referred to as the *decision epochs*. Hence, we assume that all the policies in Ψ make their scheduling decisions at these time instants only.

A second consequence of the memorylessness property is that any idling scheduling policy (A policy is idling if it leaves some processors idle whereas there are enabled tasks waiting for execution.) is never optimal. Therefore, we assume without loss of generality that Ψ contains only nonidling preemptive policies.

The proof of Theorem 1 makes use of the following lemmas.

Lemma 2. *Let F and G be two subgraphs of a uniform out-forest H, obtained by successively deleting initial vertices of H or of previously obtained subgraphs. Let F' (resp. G') be the remaining subgraph when the second highest initial vertex of F (resp. G), if any, is deleted. If $F \prec^e G$, then $F' \prec^e G'$.*

Proof. The embedding relation of the uniform out-forest F and G means that $|V(F)| \leq |V(G)|$ and also gives the embedding relations for the uniform out-trees of successors (provided that vertex 1 have the highest number of successors in F and respectively in G):

$$\forall i,\ 1 \leq i \leq |V(F)| : \quad S_i(F) \prec_e S_i(G).$$

We analyze the embedding of the remaining uniform out-forests after removing the second highest initial vertex from both graphs, if any. If out-forests F and G are empty, the embedding relation holds trivially (both remaining out-forests are empty). It is simple to see that if F is empty the embedding relation is preserved (if a vertex is deleted from G). If F and G are not empty, then we consider three cases:

Case 1: In F there is only one initial vertex (the highest one) and in G vertex $s \in G$ is the second highest initial vertex. It means that only vertex $s \in G$ is deleted.

The embedding relation of the trees of successors of the highest vertex in F, respectively in G, guarantees the extended embedding relation

$$F = S_1(F) \prec_e S_1(G) \prec^e G - \{s\}.$$

Case 2: In G there is only one initial vertex (the highest one) and $r \in F$ is the second highest initial vertex of F. It is trivial that the removal of r from F implies

$$F - \{r\} \prec^e F \prec^e G,$$

which means that the extended embedding relation is preserved.

Case 3: Both F and G have at least two initial vertices. Let r (resp. s) be the second highest initial vertex of F (resp. G). There are two subcases:

Case 3.1: $r > s$. Then relations

$$\forall i \in \{1, ..., s-1\} : \qquad S_i(F) \prec_e S_i(G)$$

are the same as before the deletion of vertices r and s. It is simple to see that relations

$$\forall i \in \{r+1, ..., |V(F)|\} : \qquad S_i(F) \prec_e S_i(G)$$

still hold (provided there is no renumbering of the vertices).

Since

$$S_1(F) \prec_e S_1(G)$$

and since $S_i(F)$ is a subtree of $S_1(F)$ for all $i \in \{1, 2, ..., r-1\}$, we have

$$S_i(F) \prec_e S_{f(i)}(G),$$

where f is the embedding function from $S_1(F)$ to $S_1(G)$. Let $f(i) = h_i$, $1 \leq i \leq r-1$. It is easy to see that $h_1 < h_2 < ... < h_{r-1}$. Since $h_i \in S_1(G)$, $1 \leq i \leq r-1$, and since $S_1(G)$ and $S_s(G)$ are disjoint trees, we obtain that $s \notin \{h_1, h_2, ..., h_{r-1}\}$ so that

$$h_i \geq i+1\ , \quad s \leq i \leq r-1.$$

The last two relations imply

$$S_i(F) \prec_e S_{i+1}(G)\ , \quad s \leq i \leq r-1.$$

Thus, the extended embedding relation is preserved after deletion of r and s. *Case 3.2:* $r \leq s$. Then relations

$$\begin{aligned} \forall i \in \{1, ..., r-1\}: &\quad S_i(F) \prec_e S_i(G), \\ \forall i \in \{s+1, ..., |V(F)|\}: &\quad S_i(F) \prec_e S_i(G) \end{aligned}$$

are preserved. If $r < s$ then we have also

$$\forall i \in \{r, ..., s-1\}: \qquad S_{i+1}(F) \prec_e S_i(F) \prec_e S_i(G).$$

This allows us to conclude that the extended embedding relation is preserved by the deletion of the second highest initial vertices of F and G, if any. □

Lemma 3. *Let F and G be two subgraphs of forest-cut graph $H \in \mathcal{G}$ obtained by successively deleting initial vertices of H or of previously obtained subgraphs. If $T(F) \prec^e T(G)$ then*

$$F \prec^f G.$$

Proof. Since $H \in \mathcal{G}$ is a forest-cut graph, we obtain a cut level l for which $T(H) = (V(T(H)), E(T(H)))$ is a uniform out-forest, and that for $i \in \{1, \ldots, |V(T(H))|\}$,

$$S_1(T(H)) \succ_e S_2(T(H)) \succ_e \cdots \succ_e S_{|V(T(H))|}(T(H)), \tag{2}$$

$$S_1(H) \succ_f S_2(H) \succ_f \cdots \succ_f S_{|V(T(H))|}(H). \tag{3}$$

Owing to the transitivity of relations $\succ_e$ and $\succ_f$, the embedding relation (2) and the flatness relation (3) implies that for any two subgraphs F, G of H, if $T(F) \prec^e T(G)$ then $F \prec^f G$. □

Lemma 4. *Let $F = (V(F), E(F))$ and $G = (V(G), E(G))$ be two subgraphs of $H \in \mathcal{G}$ obtained by successively deleting initial vertices of H or of previously obtained subgraphs. If $F \prec G$, then*

$$\gamma(F) \leq_{st} \gamma(G) \tag{4}$$

Proof. The lemma is shown by combining the ideas of the proofs of Theorem 1 in [3] and Lemma 4.2 in [9].

We consider the coupled processing model where both processors $1, 2$ are continuously executing tasks. When a task is assigned to a processor, it is assigned a running time equal to the remainder of the running time already underway at that processor. Thus, if tasks $u \in F$ and $v \in G$ are assigned to the same processor at the same time, they have the same (remaining) running time. We will show that in such a probability space

$$\gamma(F) \leq \gamma(G) \quad a.s. \tag{5}$$

Denote by $\{c_n\}_{n=1}^{\infty}$ the (increasing) sequence of completion times of the tasks in F and G at both processors under the HLMS policy. Let $c_0 = 0$ be the starting time. Let $F_n^\gamma = (V(F_n^\gamma), E(F_n^\gamma))$ and $G_n^\gamma = (V(G_n^\gamma), E(G_n^\gamma))$ be the remaining graphs of F, respectively G, at time c_n under HLMS in the coupled model. We show by induction that for all $n \geq 0$,

$$F_n^\gamma \prec G_n^\gamma, \tag{6}$$

which immediately implies (5).

For $n = 0$, it is trivial that

$$F_0^\gamma = F \prec G = G_0^\gamma.$$

Assume that (6) holds for some $n \geq 0$.

Let $u_1, \cdots, u_{m_1}$ (resp. $v_1, \cdots, v_{m_2}$) be the tasks of F_n^γ (resp. G_n^γ) that are assigned to the processors $1, \cdots, m_1$ (resp. $1, \cdots, m_2$) under the *HLMS* policy. By definition, $\max(m_1, m_2) \leq 2$. Let

$$L(u_1) \geq \cdots \geq L(u_{m_1}), \qquad L(v_1) \geq \cdots \geq L(v_{m_2}).$$

If F_n^γ is empty, then (6) trivially holds for $n+1$. Assume that F_n^γ is not empty, then c_{n+1} corresponds to a completion at some processor, say processor h, $1 \leq h \leq \max(m_1, m_2)$. Let u_h, if any, be at level l_1, and v_h, if any, be at level l_2. Tasks u_h and/or v_h complete at time c_{n+1}. If $m_1 = 1$ (resp. $m_2 = 1$) then there is only one enabled task in F_n^γ (resp. G_n^γ). Moreover, the decision epoch c_{n+1} corresponds to a completion either at processor 1, i.e. $h = 1$ and tasks u_1 and v_1 are completed, or at processor 2, i.e. $h = 2$ and tasks u_2 and v_2 (if any) are completed. Therefore, we consider these two cases.

Case 1: $h = 1$.

In this case, both tasks u_1 of F and v_1 of G complete. Under the inductive assumption we have

$$L(F_n^\gamma) \leq L(G_n^\gamma).$$

Since tasks u_1 and v_1 are at the highest level in both graphs, we obtain that

$$\forall i,\ L(F_n^\gamma) + 1 \leq i \leq L(G_n^\gamma):$$
$$\sum_{k \geq i} N_k(F_{n+1}^\gamma) = \sum_{k \geq i} N_k(F_n^\gamma) = 0 \leq \sum_{k \geq i} N_k(G_n^\gamma) - 1 = \sum_{k \geq i} N_k(G_{n+1}^\gamma),$$
$$\forall i,\ 0 \leq i \leq L(F_n^\gamma):$$
$$\sum_{k \geq i} N_k(F_{n+1}^\gamma) = \sum_{k \geq i} N_k(F_n^\gamma) - 1 \leq \sum_{k \geq i} N_k(G_n^\gamma) - 1 = \sum_{k \geq i} N_k(F_{n+1}^\gamma),$$

which implies:

$$F_{n+1}^{\gamma} \prec_f G_{n+1}^{\gamma}.$$

Owing to the facts that $T(F_n^{\gamma}) \prec^e T(G_n^{\gamma})$ and that tasks u_1 and v_1 are at the highest level in both graphs, we obtain

$$T(F_{n+1}^{\gamma}) = T(F_n^{\gamma}) - \{u_1\} \prec^e T(G_n^{\gamma}) - \{v_1\} = T(G_{n+1}^{\gamma}).$$

Moreover, according to Lemma 3, the extended embedding relation of uniform out-forests implies the extended flatness relation for the subgraphs

$$F_{n+1}^{\gamma} \prec^f G_{n+1}^{\gamma}.$$

Therefore, relation (6) holds for $n+1$ in Case 1.

Case 2: $h = 2$ (completion at processor 2).

Using Lemma 2, we conclude that the completion of the second highest enabled task of F_n^{γ} and G_n^{γ} preserves the extended embedding relation of uniform out-forests

$$T(F_{n+1}^{\gamma}) \prec^e T(G_{n+1}^{\gamma}).$$

Applying further Lemma 3 implies the extended flatness relation

$$F_{n+1}^{\gamma} \prec^f G_{n+1}^{\gamma}.$$

Thus, we have only to verify that the ordinary flatness relation for graphs F_{n+1}^{γ} and G_{n+1}^{γ} holds at time c_{n+1}. We consider three subcases:

Case 2.1: $m_1 = 2$ and $m_2 = 1$ $(h = 2)$.

Since $m_2 = 1$, there is only one enabled task in G at time c_n. Thus, only task $v_2 \in F$ completes at time c_{n+1}, so that

$$F_{n+1}^{\gamma} \prec_f F_n^{\gamma} \prec_f G_n^{\gamma} = G_{n+1}^{\gamma}.$$

Case 2.2: $m_1 = 1$ and $m_2 = 2$ $(h = 2)$.

In this case only task $v_2 \in G_n^{\gamma}$ completes.

Thus, $S_{u_1}(F_n^{\gamma})$ is identical to F_n^{γ}, i.e. all the vertices (but u_1) of F_n^{γ}, are successors of task u_1. It follows, that

$$F_{n+1}^{\gamma} = F_n^{\gamma} = S_{u_1}(F_n^{\gamma}) \prec_f S_{v_1}(G_n^{\gamma}) = S_{v_1}(G_{n+1}^{\gamma}) \prec_f G_{n+1}^{\gamma}.$$

Case 2.3: $m_1 = m_2 = 2$ $(h = 2)$.

In this case, both tasks $u_2 \in F_n^{\gamma}$ and $v_2 \in G_n^{\gamma}$ complete. Let $L(u_2) = l_1$ and $L(v_2) = l_2$. There are two subcases to be considered.

Case 2.3.1: $l_1 > l_2$.

It is simple to see that

$$\forall i,\ i \geq l_2 + 1: \quad \sum_{k \geq i} N_k(F_{n+1}^{\gamma}) \leq \sum_{k \geq i} N_k(F_n^{\gamma}) \leq \sum_{k \geq i} N_k(G_n^{\gamma}) = \sum_{k \geq i} N_k(G_{n+1}^{\gamma}),$$

$$\forall i,\ 0 \leq i \leq l_2: \quad \sum_{k \geq i} N_k(F_{n+1}^{\gamma}) = \sum_{k \geq i} N_k(F_n^{\gamma}) - 1 \leq \sum_{k \geq i} N_k(G_n^{\gamma}) - 1 = \sum_{k \geq i} N_k(G_{n+1}^{\gamma}).$$

So the flatness relation

$$F_{n+1}^{\gamma} \prec_f G_{n+1}^{\gamma}$$

holds for $n+1$.

Case 2.3.2: $l_1 \le l_2$.

Since in both graphs there is a completed task, we obtain

$$\forall i,\ 0 \le i \le l_1 :$$
$$\sum_{k \ge i} N_k(F_{n+1}^\gamma) = \sum_{k \ge i} N_k(F_n^\gamma) - 1 \le \sum_{k \ge i} N_k(G_n^\gamma) - 1 = \sum_{k \ge i} N_k(G_{n+1}^\gamma).$$

Moreover, since tasks $u_1 \in F_n^\gamma$ and $v_1 \in G_n^\gamma$ are not completed, we have

$$L(F_{n+1}^\gamma) = L(F_n^\gamma) \le L(G_n^\gamma) = L(G_{n+1}^\gamma).$$

There are two possibilities: either $L(F_n^\gamma) < l$ or $L(F_n^\gamma) \ge l$, where l is the cut level of graph H.

Case 2.3.2.a: $L(F_n^\gamma) < l$.

Task graph F_n^γ is an in-forest with the second highest enabled task at level l_1. Thus, there is at most one task at levels $i \ge l_1 + 1$, so that

$$\forall i \ge l_1 + 1 :$$
$$\sum_{k \ge i} N_k(F_{n+1}^\gamma) = L(F_{n+1}^\gamma) - i + 1 \le L(G_{n+1}^\gamma) - i + 1 \le \sum_{k \ge i} N_k(G_{n+1}^\gamma).$$

Case 2.3.2.b: $L(F_n^\gamma) \ge l$.

Using the extended flatness relation on graphs $F_{n+1}^\gamma \prec^f G_{n+1}^\gamma$, we obtain

$$\forall i \ge l_1 + 1 :$$
$$\sum_{k \ge i} N_k(F_{n+1}^\gamma) = \sum_{k \ge i} N_k(S_1(F_{n+1}^\gamma)) \le \sum_{k \ge i} N_k(S_1(G_{n+1}^\gamma)) \le \sum_{k \ge i} N_k(G_{n+1}^\gamma),$$

which implies

$$F_{n+1}^\gamma \prec_f G_{n+1}^\gamma.$$

Therefore, the relation (6) holds for step $n+1$ in Case 2.

Thus, by induction, relation (6) holds for all $n \ge 1$. Consequently, $\gamma(F) \le \gamma(G)$ *a.s.* in that probability space. Applying further Strassen's Theorem allows us to conclude. □

Lemma 5. *Let $H \in \mathcal{G}$ be a task graph. Let $\pi, \rho \in \Psi$ be two policies which follow the HLMS rule all the time except at the first decision epoch. At the first decision epoch, policy π assigns tasks v_1, v_2 to processors $1, 2$, respectively, whereas ρ assigns tasks v_1, v to processors $1, 2$, respectively. If $S_{v_2}(H) \prec_f S_v(H)$, then*

$$\rho(H) \le_{st} \pi(H).$$

Proof. We couple the running times on processors 1 and 2 in such a way that under both policies π and ρ, the running time on processor n starting from time 0 is τ_n, $1 \le n \le 2$. In such a coupled model, the second decision epoch occur at the same time under both policies. Let this time epoch be fixed, referred to as $c_1 = \min(\tau_1, \tau_2)$.

Then, for any increasing function $f: \mathbf{R} \to \mathbf{R}$,

$$E\left[f\left(\pi(H)-c_1\right)\right] = \frac{1}{2}E\left[f\left(\gamma(H-\{v_1\})\right)\right]+\frac{1}{2}E\left[f\left(\gamma(H-\{v_2\})\right)\right], \tag{7}$$

$$E\left[f\left(\rho(H)-c_1\right)\right] = \frac{1}{2}E\left[f\left(\gamma(H-\{v_1\})\right)\right]+\frac{1}{2}E\left[f\left(\gamma(H-\{v\})\right)\right]. \tag{8}$$

According to Lemma 4, $\gamma(H-\{v\}) \leq_{st} \gamma(H-\{v_2\})$, so that

$$E\left[f\left(\gamma(H-\{v\})\right)\right] \leq E\left[f\left(\gamma(H-\{v_2\})\right)\right].$$

Thus, cf. (7) and (8),

$$E\left[f\left(\rho(H)-c_1\right)\right] \leq E\left[f\left(\pi(H)-c_1\right)\right].$$

Now, uncondition c_1 implies

$$\rho(H) \leq_{st} \pi(H).$$

□

Lemma 6. *Let $H \in \mathcal{G}$ be a task graph. Let $\pi \in \Psi$ be a policy which follows the HLMS rule all the time except at the first decision epoch. Then,*

$$\gamma(G) \leq_{st} \pi(G).$$

Proof. The assertion of the lemma is an easy consequence of Lemma 5. Indeed, due to Lemma 5, in order to stochastically minimize the makespan, the processors should be assigned the enabled tasks with the largest sets of successors. □

Proof of Theorem 1. In order to prove the theorem, we construct a finite sequence of policies $\{\pi_n\}_{n=0}^N$, $|V|-1 \leq N \leq 2|V|$. For $1 \leq n \leq N$, policy π_n takes exactly the same scheduling decisions as π at times $c_0, c_1, \cdots, c_{n-1}$, i.e., the same tasks are assigned to the same processors at these time instants under policies π and π_n. From time c_n, policy π_n follows the HLMS rule. By definition, π_0 is an HLMS policy.

We use a backward induction on n to show that for all n, $0 \leq n \leq N$,

$$\pi_n(H) \leq_{st} \pi(H). \tag{9}$$

For $1 \leq n \leq N$, let H_n be the remaining task graph of H at time c_n under π. It is clear that $H_1 = H$ and H_N has a single vertex. Applying Lemma 6 to the task graph H_N implies that

$$\pi_N(H) = c_N + \gamma(H_N) \leq_{st} c_N + \pi(H_N) = \pi(H).$$

Therefore, (9) holds for $n = N$. Assume that relation (9) holds for some $0 \leq n \leq N$. Then

$$\pi_{n-1}(H) = c_{n-1} + \gamma(H_{n-1}) \leq_{st} c_{n-1} + \pi_n(H_{n-1}) = \pi_n(H),$$

where we used Lemma 6 to obtain the inequality. Thus, by induction, relation (9) holds for all $0 \leq n \leq N$. Taking $n = 0$ immediately implies relation (1).

Moreover, it is not difficult to see that HLMS is the only policy that satisfies relation (1). Indeed, from the proof of Lemma 5, one can show that if a policy π violates the HLMS rule, then, with positive probability $\gamma(H) < \pi(H)$. □

4. Concluding Remarks

In this paper, we have analyzed the stochastic scheduling problem for the minimization of makespan of a set of partially ordered tasks. When the task running times are i.i.d. random variables with a common exponential distribution, and the task graph is forest-cut, we have shown that HLMS policy stochastically minimizes the makespan on two processors.

Notice that the optimality result of HLMS policy does not hold when there are at least three processors. In fact, even in the special cases of in-forests and out-forests, counterexamples have been provided in the literature for the minimization of the average makespan, see [2] for in-forests and [3] for out-forests.

The proof techniques that we have used are based on majorization relations (for comparing task graphs) and coupling arguments (for sample path analysis). These techniques turn out to be very useful to tackle complex scheduling problems like the one considered in the paper.

Our result can be generalized to the case of two uniform processors (processors with different speeds) with variable profile (where the set of available processors varies in time). This generalization can be carried out using an analogous approach of Liu and Sanlaville [9].

Throughout this paper, we assumed that the task running times are exponentially distributed random variables. We conjecture that the optimality result extends to the case when task running times have an increasing-likelihood-ratio distribution.

References

1. J. Bruno, *On Scheduling Tasks with Exponential Service Times and In-Tree Precedence Constraints*, Acta Informatica, **22** (1985), pp. 139–148.
2. K. M. Chandy and P. F. Reynolds, *Scheduling Partially Ordered Tasks with Probabilistic Execution Times*, Operating System Review, **9** (1975), pp. 169–177.
3. E. G. Coffman and Z. Liu, *On the Optimal Stochastic Scheduling of Out-Forests*, Rapport de Recherche INRIA No. 1156, 1990, Operations Research, **40** (1992), pp. S67–S75.
4. D. Dolev and M. K. Warmuth, *Scheduling Precedence Graphs of Bounded Height*, J. of Algorithms, **5** (1984), pp. 48–59.
5. D. Dolev and M. K. Warmuth, *Scheduling Flat Graphs*, SIAM J. on Comput., **14** (1985), pp. 638–657.
6. D. Dolev and M. K. Warmuth, *Profile Scheduling of Opposing Forests and Level Orders*, SIAM J. Alg. Disc. Meth., **6** (1985), pp. 665–687.
7. E. Frostig, *A Stochastic Scheduling Problem with Intree Precedence Constraints*, Operations Research, **36** (1988), pp. 937–943.
8. M. R. Garey, D. S. Johnson, R. E. Tarjan and M. Yanakakis, *Scheduling Opposite Forests*, SIAM J. Alg. Disc. Meth., **4** (1983), pp. 72–93.
9. Z. Liu, E. Sanlaville, *Stochastic Scheduling with Variable Profile and Precedence Constraints*, Rapport de Recherche INRIA, No. 1525, September 1991.
10. Z. Liu, E. Sanlaville, *Profile Scheduling by List Algorithms.* In Scheduling Theory and Its Applications, P. Chretienne et al. (Eds.), J. Wiley, to appear, 1995.
11. C. H. Papadimitriou, J. N. Tsitsiklis, *On Stochastic Scheduling with In-Tree Precedence Constraints*, SIAM J. Comput., **16** (1987), pp. 1–6.
12. C. H. Papadimitriou, M. Yannakakis, *Scheduling Interval-Ordered Tasks*, SIAM J. Comput., **8** (1979), pp. 405–409.
13. M. Pinedo, G. Weiss, *Scheduling Jobs with Exponentially Distributed Processing Times and Intree Precedence Constraints on Two Parallel Machines*, Operations Research, **33** (1985), pp. 1381–1388.

14. V. Strassen, *The Existence of Probability Measures with Given Marginals*, Ann. Math. Sta., **36** (1965), pp. 423–439.
15. J. D. Ullman, *NP-Complete Scheduling Problems*, J. Comput. System Sci., **10** (1975), pp. 384–393.

(Lucian FINTA) I.N.R.I.A. CENTRE SOPHIA-ANTIPOLIS, 2004 ROUTE DES LUCIOLES, 06902, SOPHIA-ANTIPOLIS CEDEX, FRANCE
E-mail address, Lucian FINTA: Lucian.Finta@sophia.inria.fr

(Zhen LIU) I.N.R.I.A. CENTRE SOPHIA-ANTIPOLIS, 2004 ROUTE DES LUCIOLES, 06902, SOPHIA-ANTIPOLIS CEDEX, FRANCE
E-mail address, Zhen LIU: Zhen.Liu@sophia.inria.fr

DIMACS Series in Discrete Mathematics
and Theoretical Computer Science
Volume **21**, 1995

SCHEDULING OF STRUCTURED AND UNSTRUCTURED COMPUTATION

APOSTOLOS GERASOULIS[1], JIA JIAO[1] AND TAO YANG[2]

ABSTRACT. In this paper, we give an overview of scheduling techniques and investigate the practicality of automatic task scheduling on scalable parallel architectures where communication cost is significant. We analyze the run-time performance of the PYRROS scheduling and code generation system and report our experience with this system in several regular and irregular scientific applications including the dense Gaussian and Gauss Jordan Elimination algorithms, sparse matrix computation, fast fourier transforms and fast hierarchical algorithms for 2D N-body simulations. The preliminary experiments on nCUBE-2 have produced very promising results with regard to the practicality of scheduling for scalable architectures.

1. INTRODUCTION

Given a sequential program and a parallel architecture, one of the most important problems in parallel processing today is the development of programming tools that will automatically map and execute this program on the given parallel architecture. The lack of such tools and compilers has had a negative impact on the widespread use of massively parallel scalable architectures. The parallelization process is shown in Fig. 1. There are two fundamental problems that must be solved:

(i) *Parallelism detection* at the the *granularity level* of the parallel machine.

The work presented here was in part supported by ARPA contract DABT-63-93-C-0064, by the Office of Naval research under grant N000149310114, and by NSF Research Initiation Award CCR-9409695. The content of the information herein does not necessarily reflect the position of the Government and official endorsement should not be inferred.

AMS(MOS) subject classifications: 68M20; 65Y05, 65F05, 65F50, 70F10, 90B35.

(ii) *Mapping* parallelism onto the given architecture.

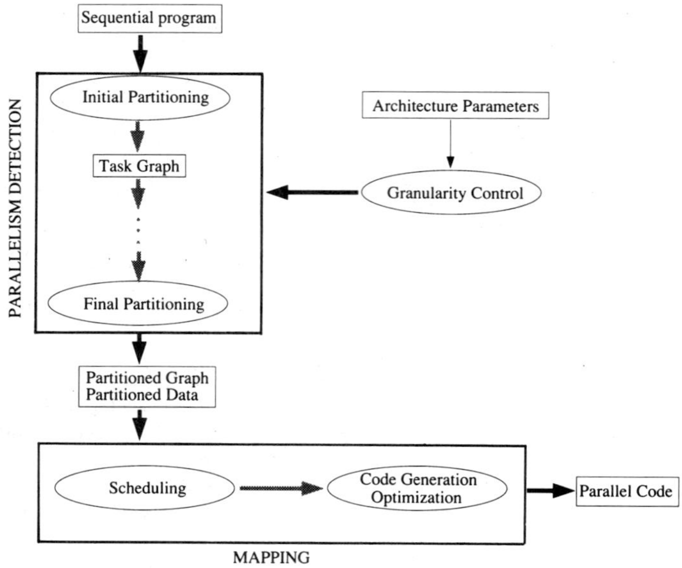

FIGURE 1. Parallelization steps.

In this paper, we will discuss issues related to the mapping of a partitioned program onto parallel architectures. Task scheduling theory and algorithms have been of a long-standing research interest because of the promise that could better utilize the power of multicomputers. Finding optimal scheduling solutions is difficult, especially in the presence of communication. Instead many scheduling heuristic algorithms have been proposed in the literature [**6, 12, 16, 22, 31, 33**]. Also many attempts have been made to build scheduling systems that predict the performance of task graph computation so that better parallel programs could be written, [**12, 33**]. Few work has been conducted to demonstrate the practicality of scheduling in a large class of applications on real machines.

The following are the issues addressed in this paper.

- How could a scheduling algorithm be used to execute programs on parallel machines. We will discuss an automatic scheduling and code generation system for message-passing machines.
- What is the discrepancy between the static performance predicted by the scheduler at compile time and the actual run-time performance. Since there is always a discrepancy between the weight estimation and the actual run-time performance, it is of interest to know that small estimation

errors will not dramatically alter the run-time performance behavior.

- A practical scheduling system should not only perform well but also solve large size real-world problems. Thus, we need to evaluate the run-time behavior of automatic scheduling algorithms in numerous applications and analyze the factors that could affect their performance.

The paper is organized as follows. Section 2 discusses the macro-dataflow task graph model used to represent parallel computation. Section 3 describes the organization of PYRROS. Section 4 discusses run-time performance. Section 5 presents experiments for Gaussian and Gauss Jordan elimination, sparse matrix computation and FFT. Section 6 presents our latest experiments on scheduling for the N-body problem. Section 7 discusses related work.

2. SCHEDULING STRUCTURED AND UNSTRUCTURED GRAPHS

Given a program, we partition this program into basic units of computation that could contain statement(s) and/or procedure(s). Each basic unit of computation is called a *task* and the dependence[1] relation between tasks constitutes their communication edges. Such a graph is called a dataflow task graph. We can assign weights to nodes and edges. A node weight is the time for executing this computation unit. An edge (n_i, n_j) is associated with a data item that is produced in n_i and used in n_j. Its edge weight is the shortest time for transferring this data item between two processors if n_i and n_j are in separate processors. The edge weight is assumed to be zero if both nodes are mapped onto the same processor, because fetching data from local memory costs significantly less than that from remote off-processor sites. The method and parameters for estimating such weights on INTEL and nCUBE hypercubes can be found in Dunigan [**11**]. The compiler algorithms for deriving the average weight values are discussed in Sarkar [**32**].

Formally a task graph is a directed acyclic task graph (DAG) defined by a tuple $G = (V, E, \mathcal{C}, \mathcal{T})$ where V is the set of task nodes and $v = |V|$ is the number of nodes, E is the set of communication edges and $e = |E|$ is the number of edges, $\mathcal{C}$ is the set of edge communication costs and $\mathcal{T}$ is the set of node

[1]The true dependencies dominate the communication. The anti-dependence and output dependence are not true communication edges. They could be incorporated in scheduling algorithm for the enforcement of ordering but they can be eliminated by a renaming strategy.

computation costs. The value $c_{i,j} \in \mathcal{C}$ is the communication cost incurred along the edge $e_{i,j} = (n_i, n_j) \in E$, which is zero if both nodes are mapped in the same processor. The value $\tau_i \in \mathcal{T}$ is the execution time of node $n_i \in V$. $PRED(n_x)$ is the set of immediate predecessors of n_x and $SUCC(n_x)$ is the set of immediate successors of n_x. We use the *macro-dataflow task model* of execution. A task receives all input in parallel before starting execution, executes to completion without interruption, and immediately sends the output to all successor tasks in parallel.

A task graph is called structured if task dependencies follow certain regularities, otherwise it is called unstructured. For example, a program that sums an array of numbers results in a tree structure if the array is dense, otherwise in an unstructured graph if the array is sparse. A schedule derived for the tree DAG is usually optimized for this structure. When the input array of numbers is sparse, then the tree schedule could perform poorly. On the other hand, an automatic scheduler can take advantage of the special structure of the graph to produce an optimized graph specific schedule. As long as the scheduling cost is small compared to the total parallel time then scheduling is beneficial.

2.1. AN EXAMPLE OF SCHEDULING. We demonstrate program parallelization and the importance of program scheduling using the following simple program.

```
x(1)=1;
for i = 2 to 4
   x(i)=x(1);
end
for i = 1 to 2
   y(i+1)=2*x(i);
end
```

We assume each assignment statement to be a task and the partitioned program is shown below:

```
S1: x(1)=1;
   for i = 2 to 4
S2:        x(i)=x(1);
   end
   for i = 1 to 2
S3:        y(i+1)=2*x(i);
   end
```

The dependence task graph is shown in Figure 2(a) where task node S_2^i stands for the i-th loop iteration in executing statement S_2. We assume that the computation time of S_1 and S_2 is 1 time unit and S_3 is 2 time units. The transmission delay between two processors for one data element takes 4 time units. We mark the computation weight in the nodes and the communication delay in the edges in Figure 2(a).

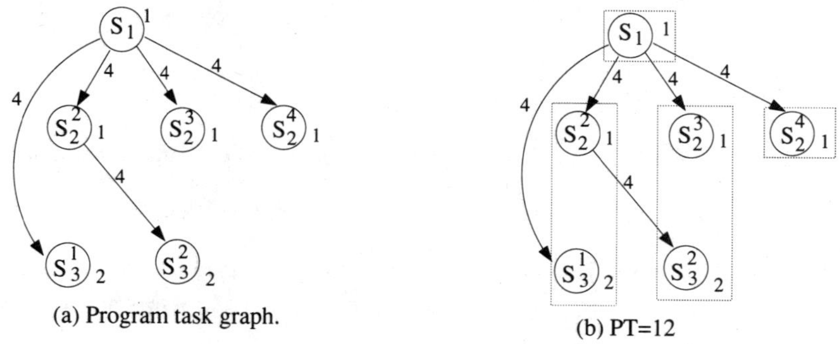

FIGURE 2. (a) An example of task graph. (b) An unoptimized mapping.

We list three scheduling solutions. The parallel time is computed assuming that processors are fully connected. A nonoptimal solution is depicted in Figure 2(b). The parallel time(PT) of executing this task graph is equal to 12. This processor assignment can be produced by using the so called "owner computes rule" [**3**]. Owner computes rule is a simple but powerful mapping solution such that a user can specify the distribution of data items and the system will assign program statements to processors based on whether the data items are modified by these statements. A processor executes a computation unit if this unit modifies the data that the processor owns. Thus the optimization issue is how a user could distribute the data items such that the parallel time can be minimized.

Assume that a user distributes array x and y according to the element indices, i.e. processor i owns $x(i)$ and $y(i)$. Then the owner computes rule implies the following processor assignment of tasks: $PA(S_1) = 0$ and $PA(S_2^i) = i - 1$ and $PA(S_3^i) = i$. This results in the assignment shown in Figure 2(b).

Another solution is shown in Figure 3(a) the PT is 8 which is better than solution 1. A user can improve the previous processor mapping by analyzing the relationship between array x and y in the statement S_3. By changing the distribution of array y so that processor i owns $x(i)$ and $y(i+1)$ then by the owner computes rule the new processor assignment is shown in Figure 3(a). The weights of communication edge (S_1, S_3^1) and (S_2^2, S_3^2) become zero and the parallel time is reduced to 8.

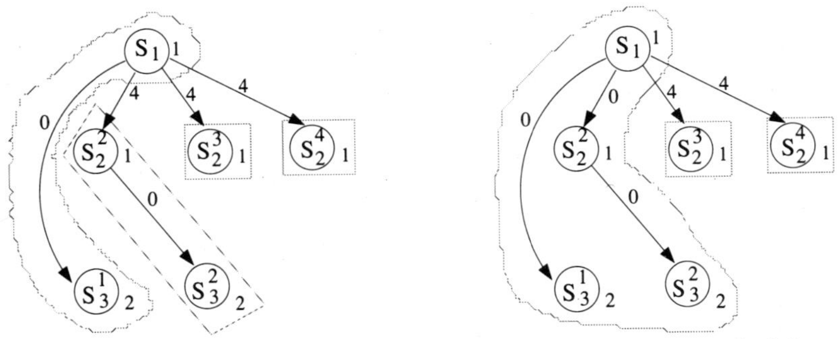

(a) PT=8 after alignment. (b) PT=6 by program scheduling.

FIGURE 3. (a) An improved schedule. (b) The best solution.

A third and better solution can be obtained by a program scheduling algorithm that utilizes the computation and communication weight information to optimize the mapping of data items and tasks so that the parallel time is minimized. Figure 3(b) shows a processor assignment where 3 processors are used and S_1, S_2^2 and S_3^1 and S_3^2 are mapped into one processor. The optimized parallel time is 6.

2.2. IMPACT OF COMMUNICATION AND GRANULARITY. From the previous example, we can see that by placing two tasks in different processors *parallelizes* the computation process but also induces higher communication cost. Placing tasks in one processor eliminates communication cost but also *sequentializes* the computation process. The current scalable architectures such as iPSC/860 and NCUBE-2, require coarser grain task partitioning. In the above example, the task partitioning is at the statement level: a simple assignment statement is a task and the message communicated between tasks is a scalar data item. Such partitioning is too fine grain because of a large overhead in the startup time for message transmission. The following table gives the communication and single-precision computation speeds of two machines in microseconds(μs) taken from Dunigan [**11**]. The time delay for transmitting N-byte data items between two adjacent processors is $\alpha + N\beta$.

	α	β ($\mu s/byte$)	add or mult
iPSC/860	136	0.4	0.1
NCUBE-2	200	0.6	0.8

For example, using the parameters of iPSC/860, the communication weights of the task graph in Figure 2(a) will be 137.6μs for a single precision floating point number. The computation time of statement S_3 is 0.1μs and the time of statement S_1 and S_2 is assumed to be $0.05\mu s$. By letting processor i own both $x(i)$ and $y(i+1)$ and using the owner computes rule, the parallel time of executing this graph on iPSC/860 will be $137.8\mu s$, as shown in Figure 4(a). The best solution is to assign all tasks into one processor and sequentialize their execution, which results in PT=$0.6\mu s$.

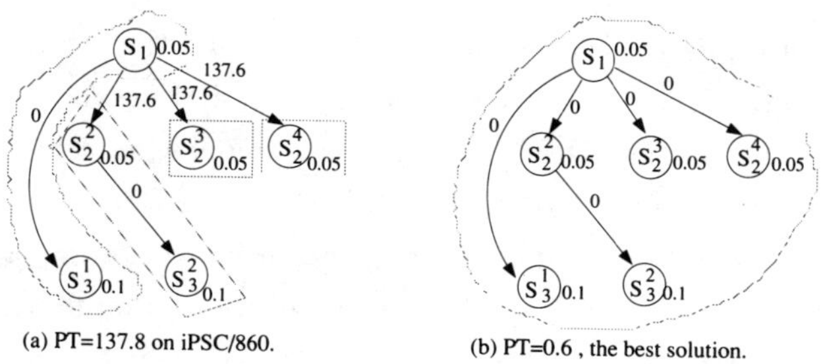

FIGURE 4. (a)Preserve the parallelism on iPSC/860. (b) Sequentialize parallel tasks.

Finding the optimal partitioning, corresponding to the optimum tradeoff point between parallelization and sequentialization, that leads to the shortest parallel time is NP-complete [**6, 25, 31**]. But we can find a tradeoff point based on the granularity information when we assume that the processors are fully connected and there are sufficient number of processors.

We define the grain of a task T_x as

$$g_x = min(\frac{\min_{T_k \in PRED(T_x)} \tau_k}{\max_{T_k \in PRED(T_x)} c_{k,x}}, \frac{\min_{T_k \in SUCC(T_x)} \tau_k}{\max_{T_k \in SUCC(T_x)} c_{x,k}})$$

and the *granularity* of a DAG G as

$$g(G) = \min_{x=1:v} \{g_x\}.$$

Formally we can show that

PROPERTY 1. *For a DAG with $g(G) \geq 1$, any schedule that sequentializes independent tasks can be always transformed into a schedule that executes these tasks in parallel with shorter parallel time. In addition, any schedule that does not sequentialize independent tasks is always within 50% of the optimum.*

The conclusion is that a coarse grain partitioning with sufficient parallelism not only can reduce the complexity of scheduling but also can achieve scalable performance. On the other hand, for fine grain graphs a schedule that preserves fine-grain parallelism may result in poor performance. Thus, a good scheduling algorithm should sequentialize the fine grain parts of a task graph and keep the coarse grain parts parallel. We present such an algorithm in the next section.

3. THE PYRROS SYSTEM AND SCHEDULING ALGORITHMS

The input of PYRROS is a weighted task graph and the associated sequential C or Fortran code. The output is a static schedule and parallel code for a given architecture. The current PYRROS tool has the following components: a task graph language with an interface to C and Fortran, allowing users to define partitioned programs and data; a scheduling system for clustering the graph, load balancing and physical mapping, and communication/computation ordering; a graphic displayer for displaying task graphs and scheduling results; a code generator that inserts synchronization primitives and performs code optimization for NCUBE-1, NCUBE-2 and INTEL iPSC/860 hypercube machines.

3.1. SCHEDULING ALGORITHMS. Scheduling parallel tasks with precedence relations over distributed memory multiprocessors has been found to be much more difficult than the classical scheduling problem [**7, 23**]. This is because data transferring between processors requires substantial transmission delays. Without imposing a limitation on the number of processors, the scheduling problem that ignores communication overhead is solvable in polynomial time. When communication overhead is present, however, the problem becomes NP-complete [**31, 6, 25**]. Several scheduling algorithms have been proposed in the literature, e.g. [**12, 24, 22, 31, 33**]. The main focus of our research is to develop low complexity scheduling algorithms without compromising performance.

There are two classes of scheduling strategies. One is to travel through the given graph once, and produce a solution in one stage. Another is a multi-stage approach which divides the scheduling process into several stages [**22, 31**]. Kim and Browne [**22**] have performed a comparison of the one-stage and the multistage method, and found that the multi-stage method produces good solutions. We have used the following multi-stage approach to design low-complexity scheduling algorithms: 1) Assign tasks to a set of clusters. Tasks in the same cluster will be executed in the same processor. 2) Merge the u clusters into p completely connected virtual processors if $u > p$ and map the p virtual processors into p physical processors. 3) Order the execution of tasks within each processor. We discuss the details below. The optimization in this multi-stage approach is to eliminate unnecessary communication, balance the computation load, and overlap computation with communication.

Clustering: The goal of clustering is to identify "useful" parallelism and eliminate unnecessary communication. Sarkar [**31**] was one of the first to recognize the importance of clustering, which he calls internalization pre-pass. This problem is equivalent to scheduling on an unbounded number of processors. A detailed comparison of clustering techniques is given in [**17**].

PYRROS uses the Dominant Sequence Algorithm (DSC) with a complexity of $O((v+e)\log v)$ to automatically determine the clustering for task graphs. It performs a sequence of clustering refinement steps and at each refinement step, it tries to zero an edge to reduce the parallel time (PT). The parallel time is determined by the longest path in the scheduled graph. Chretienne [**6**] shows that the problems of scheduling a join, fork DAG or coarse grain tree DAG are solvable in polynomial time, but the complexity jumps to NP-complete for scheduling fine-grain tree DAGs and a DAG structure obtained by concatenating a fork and a join together. In [**37**], we show that DSC performs well for general DAGs by examining a set of randomly generated DAGs and also produces the optimal solutions for fork, join and coarse grain tree DAGs.

Mapping clusters to processors: PYRROS uses a variation of *work profiling method* suggested by George et. al. [**13**] for cluster merging in order to balance the load of computation among processors. This method is simple and has been shown to work well in practice, e.g. Saad [**29**], Gerasoulis and Nelken [**14**]. The complexity of this algorithm is $O(v\log v)$.

In this stage, we have p virtual processors (or clusters) and p physical processors. Since physical processors are not completely connected, we have to take the processor distance into account. We use a heuristic algorithm due to Bokhari [**2**]. His algorithm starts from an initial assignment, then performs a series of pairwise interchanges so that the cost function (the overall communication volume) is reduced monotonically.

Task ordering: After the physical mapping has been fixed, the interprocessor communication delay between tasks can be determined. We further order the execution of tasks within each processor with a goal of overlapping the communication with computation to hide communication latency so that the total parallel time is minimized. Finding a task ordering that minimizes the parallel time is NP-hard even for chains of tasks [**20**]. We have proposed a heuristic algorithm RCP based on the critical-task-first principle and compared its performance with others for randomly generated graphs, and we have shown that RCP is optimal for fork and join DAGs (see [**36**]).

4. THE RUN-TIME METHOD FOR EXECUTING A SCHEDULE

Once a schedule is determined it must be executed on the real architecture. Let us assume that the target is a message-passing parallel machine, each processor has its own private memory and a communication buffer to hold the outgoing and incoming messages. The basic model of the run-time schedule executor for processor i is described in figure 5.

Tasklist = ordered tasks assigned to processor i based on the static schedule.
While Tasklist is not empty **Do**
    Remove the first task T_x in the list,
    Examine the incoming edges of this task T_x. If data items it needs are not in the local memory, issue a message-receiving to fetch desired data items from the communication buffer.
    Perform the task computation.
    Send out the produced data items to the processors that need these data items.
ENDWHILE

FIGURE 5. The run-time execution of a static schedule.

Parallel machines are complex systems and accurate estimation of communication and computation weight at compile time is very difficult. Therefore, it is important to know the sensitivity of the schedule executor on inaccurate weight estimation and also other run time weight variations. For example, in solving the N-body problem as described in Section 6, the scheduling tool can be used to map a tree DAG onto parallel processors. Particles gradually move during each iteration, and the weights of computation and communication change between iterations. Because there is a cost of rescheduling at each iteration it is of interest to know how the weight variations will affect the run-time scheduler so that rescheduling is done only after a sufficiently large number of iterations.

4.1. AN ANALYSIS OF RUN-TIME PERFORMANCE. We first address the sensitivity of the scheduling performance on the variation of task communication and computation weights. Let G be the graph for which the scheduling algorithm produces a schedule S. At run-time, the weights of this graph change, we call the new graph as G^r. We examine the performance of schedule S when used to execute G^r.

We assume that at run-time, each communication weight $c_{i,j}$ is changed to $c^r_{i,j}$ and the computation weight to τ^r_i. They satisfy

$$(1-f^c_1)c_{i,j} \le c^r_{i,j} \le (1+f^c_2)c_{i,j}, \qquad (1-f^\tau_1)\tau_i \le \tau^r_i \le (1+f^\tau_2)\tau_i$$

where $f_1 = max(f^c_1, f^\tau_1)$ and $f_2 = max(f^c_2, f^\tau_2)$. The lower and upper bounds in the inequalities above are chosen to be sharp.

Let the length of the schedule S for G be $PT(G)$. The run-time method in Fig. 5 uses the mapping and task ordering of S to execute graph G^r, producing a parallel time $PT(G^r)$.

Let us assume that $PT(G) \le B \times PT_{opt}(G)$ where $PT_{opt}(G)$ is the optimal schedule length for G where the minimum is taken over all possible schedules S. Similarly, let $PT_{opt}(G^r)$ is the optimal schedule length for G^r. If B is close to 1 then the scheduler, say for example the PYRROS static scheduler, produces close to the optimum schedules. What happens to the actual run-time execution of this schedule is described by the following property:

PROPERTY 2. *Assume that $PT(G,S) \le B \times PT_{opt}(G)$, then*

$$\frac{PT(G^r)}{PT_{opt}(G^r)} \le min\ \left(\frac{1+f_2}{1-f_1},\ \frac{1+f^\tau_2}{1-f^\tau_1}(1+\frac{1}{g(G^r)})(1+\frac{1}{g(G)}) \right) \times B$$

Proof: Given a schedule S of G which defines the mapping and ordering of task to processors. By adding an edge from T_x to T_y if T_x is executed immediately before T_y in the same processor and T_x and T_y are independent, then we obtain the execution graph G_s for G under schedule S. The graph structure of G_s is identical to G^r_s, the graphs differing only in the values for communication and computation weights. We set $PT(G^r)$, the length of the run-time schedule for the method in Figure 5 when using schedule S. We have that

$$PT(G^r) = \sum_{CP(G^r_s)} (\tau^r_i + c^r_{i,i+1}) \le (1+f_2) \sum_{P(G_s)} (\tau_i + c_{i,i+1}) \le (1+f_2)PT(G).$$

where $CP(G_s^r)$ is the critical path in G_s^r and $P(G_s)$ is the same path in G_s which might not be a critical path in G_s because of the change in the weights. The summations are taken over all nodes in the corresponding paths in the graph.

Now, let $PT_{opt}(G^r)$ be the length of the optimal schedule $S = ropt$ for G^r. Let $PT^*(G)$, be the parallel time of this optimum schedule applied to G. We have that

$$(1-f_1)PT_{opt}(G) \leq (1-f_1)PT^*(G) = \sum_{CP(G_{ropt})} ((1-f_1)\tau_i + (1-f_1)c_{i,i+1}) \leq$$

$$\leq \sum_{P(G^r_{ropt})} (\tau_i^r + c^r_{i,i+1}) \leq \sum_{CP(G^r_{ropt})} (\tau_i^r + c^r_{i,i+1}) = PT_{opt}(G^r_{ropt}).$$

The proof of the first bound is obvious.

To prove the second bound we use the definition of granularity to derive another upper bound for run-time communication with respect to computation: $c^r_{i,i+1} \leq \frac{\tau_i^r}{g(G^r)}$. We have that

$$PT(G^r) = \sum_{CP(G_s^r)} (\tau_i^r + c^r_{i,i+1}) \leq (1 + \frac{1}{g(G^r)}) \sum_{CP(G_s^r)} \tau_i^r \leq$$

$$\leq (1+f_2^r)(1+\frac{1}{g(G^r)}) \sum_{P(G_s)} (\tau_i + c_{i,i+1}) \leq$$

$$\leq (1+f_2^r)(1+\frac{1}{g(G^r)}) \sum_{CP(G_s)} (\tau_i + c_{i,i+1}) = (1+f_2^r)(1+\frac{1}{g(G^r)})PT(G).$$

For the lower bound assume again that the schedule $S = ropt$ is optimum for G^r and $PT^*(G)$ is the corresponding parallel time when $ropt$ is applied to G. Then we use the facts that $PT_{opt}(G) \leq PT^*(G)$ and $c_{i,i+1} \leq \frac{\tau_i}{g(G)}$ to show that

$$(1-f_1^r)\frac{PT_{opt}(G)}{1+\frac{1}{g(G)}} \leq (1-f_1^r) \sum_{CP(G_{ropt})} \frac{\tau_i + c_{i,i+1}}{1+\frac{1}{g(G)}} \leq \sum_{P(G^r_{ropt})} \tau_i^r$$

$$\leq \sum_{P(G^r_{ropt})} \tau_i^r + c^r_{i,i+1} \leq \sum_{CP(G^r_{ropt})} \tau_i^r + c^r_{i,i+1} = PT_{opt}(G^r).$$

Combining the lower and upper bounds we prove the property[2]. ∎

This property implies that the run-time performance depends on how good the compile time schedule is and how small the variations of the weights at run-time are. It should be mentioned that for coarse grain task graphs it is

[2] This result is still true if we define the grain of each task as $\frac{\tau_x}{\max_{T_k \in SUCC(T_x)} c_{k,x}}$. The granularity $g(G)$ is the minimum of grain values of tasks.

easier to produce good schedules. As a matter of fact for a completely connected architecture with unbounded number of processors we can always produce a schedule, see [**15**], such that

$$\frac{PT(G)}{PT_{opt}(G)} \leq 1 + \frac{1}{g(G)}.$$

This implies that $B = 1 + 1/g(G)$ tends to 1 as the grain of graph becomes coarser. Also for coarse grain task graphs we can estimate the computation and communication weights much more accurately as long as there is no resource contention.

5. EXPERIMENTS

The purpose of this section is to experimentally investigate the usefulness of scheduling in practice. We consider several different scientific computing applications and study the performance of PYRROS run-time execution and the compile time scheduler. The performance of the scheduled program depends on the input partitioning and its granularity as well as the degree of the parallelism in the graph. We will give the results in the following format:

Graph Structure: Regular(structured) or irregular(unstructured)
Weights: We estimate the weights using ω for the computation cost of $a(i) = b(i) + c(i) * d(i)$, and $\alpha + v\beta$ for the cost of communicating a vector of size v. For the NCUBE-2, these parameters are approximately for single precision arithmetic,

$$\omega = 1.6\mu s, \quad \alpha = 200\mu s, \quad \beta = 2.4\mu s/word.$$

Bounds on speedup: An analytic expression of an upper bound for the speedup is

$$\min(\text{Degree of parallelism}, \frac{T_1}{CPL})$$

where T_1 is the sequential execution time and CPL is the length of critical path without including communication and degree of parallelism is the width of the graph. This bound is loose since communication overhead has not been included.
Granularity: If computable we will give the granularity of the graph.
Performance summary: We will provide performance summary for PYRROS compile and run-time execution for each graph.

5.1. THE BLOCK GJ AND LU FACTORIZATION.

5.1.1. *LU FACTORIZATION.* The first example is the column block partitioning version of the LU factorization algorithm.

```
for k = 1 to N
  T_k^k : { Factorize A_{k,k} as L_k * U_k
          Compute L_k^{-1} and U_k^{-1}
          for i = k+1 to N
              A_{i,k} = A_{i,k} * U_k^{-1}
          end}
  for j = k+1 to N
      T_k^j : { A_{k,j} = L_k^{-1} * A_{k,j}
              For i = k+1 to N
                  A_{i,j} = A_{i,j} - A_{i,k} * A_{k,j}
              end}
  end
end
```

FIGURE 6. Block LU factorization and its task partitioning.

Graph Structure: The block LU program without pivoting is shown in Figure 6. A $n \times n$ matrix is divided into N block columns, each block column consists of N submatrices with size $r \times r$ where $r = n/N$ assumed to be an integer. A dependence structure in which all transitive edges have been removed is shown Figure 7. When partial pivoting is introduced, the task T_k^k performs an additional operation of finding the pivot element in each column which is local to that task. The row exchange information due to pivoting needs to be used in updating operation of T_k^j. And the dependence structure remains the same as in Figure 7.

Weights: From Figure 6, task T_k^k computes the L_kU_k at a cost of $r^3\omega/3$, finds the L_k^{-1} and U_k^{-1} at a cost of $r^2\omega$ and then performs $N-k$ matrix multiplications with a lower triangular matrix at cost of $(N-k)r^3\omega/2$. Thus the total cost, by ignoring constant factors, for T_k^k is about $\tau_k^k = (N-k)r^3\omega/2$. Similarly for the T_k^j task the cost is about $\tau_k^j = (N-k)r^3\omega_1$ since the matrix operation $A_{i,j} = A_{i,j} - A_{i,k} * A_{k,j}$ contains an additional subtraction which costs about $.5\omega$, so that $\omega_1 = \omega(1+.5) = 2.4\mu s$. The communication weights are $c_{k,k+1}^k = \alpha + (N-k+1)r^2\beta$ from T_k^k to T_{k+1}^k since L_k^{-1} and $A_{i,k}$ are communicated and from T_k^j to T_{k+1}^j it is $c_{k,k+1}^j = \alpha + (N-k)r^2\beta$.

Column partial pivoting cost is of lower order $O((N-k)r)$ and adds only a small amount to the T_k^k cost which we ignore here.

Bounds on speedup: The critical path contains the tasks T_k^k and T_k^{k+1} for $k = 1 : N-1$ and the task T_N^N. Its length without including communication

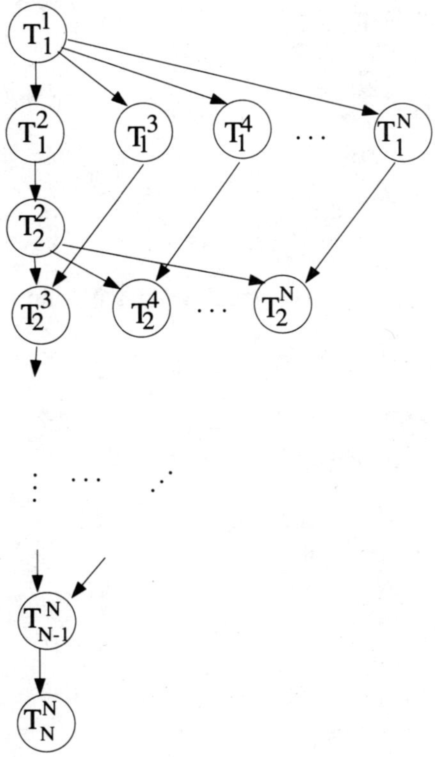

FIGURE 7. The task graph of block LU factorization with or without pivoting.

edges is approximately equal to

$$CPL \approx \sum_{k=1}^{N-1}(\tau_k^k + \tau_k^{k+1}) + \tau_n^n \approx \sum_{k=1}^{N-1} 2.0(N-k)r^3\omega \approx N^2r^3\omega.$$

Since $T_1 = n^3\omega_1/3$ then an upper bound to the speedup is

$$min(N-1, \frac{T_1}{CPL}) = \frac{N\omega_1}{3\omega} = .5N.$$

Granularity: Since $1 \leq k \leq N$ the graph is coarse grain when

$$(N-k)r^3\omega/2 \geq \alpha + (N-k+1)r^2\beta$$

while it becomes fine grain after some value of the k. If N is large then the major part of the graph is coarse grain and a small part is fine grain.

Summary for LU graph
Graph Structure: Regular.
Weights: $\tau_k^k = (N-k)r^3\omega/2$, $\tau_k^j = (N-k)r^3\omega_1$ $c_{k,k+1}^{k+1} = \alpha + (N-k+1)r^2\beta$, $c_{k,k+1}^j = \alpha + (N-k)r^2\beta$. **Bounds on speedup:** $Speedup \leq N/2$
Granularity: Partially coarse(top part) and fine(bottom part). 226z **Performance results:** PYRROS performs well provided that the data partition gives sufficient parallelism, say for example $N = n/r \geq 2p$.

Performance results: In Table 1, we show the compile time predicted performance of PYRROS(PY-Compile), the run-time execution of PYRROS parallel program(PY-Run), the speedup (Speedup) using an estimated value of sequential time ($n = 1024$ is $T_1 = 858.6$ seconds with single precision) and the performance of an optimized hand written (Hand) program based on column block wrap mapping and level scheduling, Saad [**29**]. The table contains the results for two different block partitions, $r = 8$ and $r = 16$ which result in parallelism degree of $N = 128$ and $N = 64$ respectively when $n = 1024$.

	$r=8$ $N=128$				$r=16$ $N=64$		
#Proc.	PY-Compile	PY-Run	Speedup	Hand	PY-Run	Speedup	Hand
p=4	225.6	225.5	3.8	225.3	230.2	3.7	232.4
p=8	120.5	120.8	7.1	120.7	128.3	6.7	129.1
p=16	68.2	69.1	12.5	68.6	77.0	11.2	77.3
p=32	42.5	44.2	19.4	42.7	39.8	21.2	51.0
p=64	24.1	24.0	35.7	29.7	36.4	23.6	36.9
	Speedup Bound ≤ 64				Speedup Bound ≤ 32		

TABLE 1. The block LU with pivoting performance of PYRROS for $n = 1024$.

#Proc.	PY-Compile	PY-Run	Hand	Speedup
p=4	299.0	283.4	288.7	4.0
p=8	155.7	150.5	150.4	7.6
p=16	84.5	81.0	81.4	14.1
p=32	50.0	48.7	47.4	24.3
p=64	33.0	31.8	32.8	35.1
				Speedup Bound ≤ 50

TABLE 2. The performance of GJ for $n = 1000$ and $r = 10$ on the NCUBE-2.

The PYRROS performance is affected by the granularity of the graph and the existence of sufficient parallelism. Notice that as long as the width of the graph $N \geq 2p$ the performance of PYRROS is similar for both $r = 8$ and $r = 16$. But when $p = 64$ the performance for $r = 8$ is much better because of the existence of twice as much parallelism than for $r = 16$. Thus, data partitions that result in graphs with sufficient parallelism and with high enough granularity are necessary to optimize PYRROS performance. In conclusion, PYRROS has performed very well both in terms of upper bound performance and in comparison to an optimized hand written program.

5.1.2. *GJ ALGORITHM.* The second part of this example is the block Gauss-Jordan (GJ) algorithm without pivoting. The program is similar to Figure 6 except that T_k^k is no longer present since it has been incorporated into the i loop. The T_k^j task can be derived by replacing L_k^{-1} with $A_{k,k}^{-1}$ and the inner i loop with $i = 1 : N$ and $i \neq k$. The graph structure is similar except that T_k^k are removed and T_{k+1}^j, $j = k+1 : N$ depend on T_k^{k+1} instead. All computation weights are equal to $\tau = Nr^3\omega_1$ and all communication weights are equal to $c = \alpha + Nr^2\beta$. A summary is given below.

Summary for GJ graph
Graph Structure: Regular.
Weights: $\tau = Nr^3\omega_1 \quad c = \alpha + Nr^2\beta$.
Bounds on speedup: $Speedup \leq \omega_1 N/2\omega_1 = N/2$
Granularity: $g(GJ) = \tau/c \approx r\omega_1/\beta$ for sufficiently large N. For the NCUBE-2 parameters the graph is coarse grain for all $r \geq 2$.
Performance results: Similar to LU.

Performance results: For this example, we choose $r = 10$ and $n = 1000$. The sequential time is about 1150. The results are similar to LU. Notice that the compile-time predicted performance is close to the run-time performance and we provide a justification below.

5.1.3. *PERFORMANCE SENSITIVITY ON GRAPH WEIGHTS FOR GJ.*. In the third part of this example we investigate the sensitivity of PYRROS on inaccurate weight inputs using the GJ example. Let T_p be the run-time parallel time using the normal weights $\tau = Nr^3\omega_1$ and $c = \alpha + Nr^2\beta$ in PYRROS. Let T_{wXcY} be the run-time of PYRROS using X as computation weight and Y as communication weight. In Figure 8 we give the difference ratios $(1 - T_p/T_{wXcY})$ for a variety of input weights. The normal graph is coarse grain. The experiments show that as long as the input weights result in coarse grain graph the Ratio is within 10%. Otherwise the errors in the weight input could have a significant impact in performance.

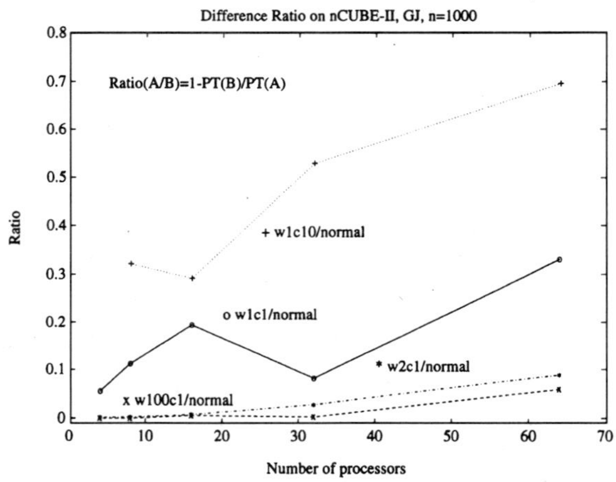

FIGURE 8. The impact of graph weight assignments on PYRROS performance for a GJ DAG. $wXcY$ stands for using X as task weight and Y as an edge weight.

We provide an explanation for this result using Property 2:

$$\frac{PT(G^r)}{PT_{opt}(G^r)} \leq \frac{1+f_2^{\tau}}{1-f_1^{\tau}}(1+\frac{1}{g(G^r)})(1+\frac{1}{g(G)}) \times B$$

where G^r is the run-time task graph of the GJ program using the normal task weights and $PT(G^r) = T_{wXcY}$.

The previous experimental results show near optimal performance of PYRROS when sufficient parallelism is present, e.g. see the performance for $p = 4$ in Table 2, so we assume that $PT_{opt}(G^r) \approx T_p$. B is the performance bound of PYRROS compile time scheduling which we conjecture that it is near optimal when the graph is coarse grain, e.g. see the performance for the normal weights and $p = 64$

the speedup is about 35 for $p = 64$ and an upper loose bound is 50 (the parallelism is $N = 100$ and even better than 35 speedup is derived when p=128 is chosen). Thus we assume here that $B \approx 1$.

We have that $g(G^r) = r\omega_1/\beta \approx 10$ and $g(G) = X/Y$. Thus when $X/Y \geq 1$, G is coarse grained. Also the normal computation weights of a GJ graph are all equal, so by using a uniform weight X for all tasks, $(1+f_2^T)/(1-f_1^T)$ is close to 1. Thus we have

$$super\frac{T_{wXcY}}{T_p} \approx 1.1 \times (1 + \frac{Y}{X}).$$

Let $V = super\{1 - \frac{T_p}{T_{wXcY}}\}$, then

$$V \approx 1 - \frac{1}{1.1 \times (1 + \frac{Y}{X})}.$$

In Figure 8, we can see that the run-time performance difference is within 10% for cases $w100c1$ and $w2c1$, is within 35% for case $w1c1$. That is consistent with our estimation, e.g $V = 0.10$ for case $w100c1$ and $V = 0.39$ for case $w2c1$, and $V = 0.55$ for case $w1c1$. For case $w1c10$, $g(G) = 0.1$ and $V = 0.91$. In Fig. 8, the performance difference varies from 30% to 70%. Thus our theoretical sensitivity analysis does provide a good explanation on the variation of the run-time performance.

5.2. THE RECURSIVE FAST FOURIER TRANSFORMATION.

The next example is the classical FFT transform of an array of complex numbers A with length $n = 2^m$ into a array B:

$$B_j = \sum_{k=1}^{n} A_k e^{-\frac{2\pi i}{n}(k-1)j} = \sum_{k\ even} A_k e^{-\frac{2\pi i}{n}(k-1)j} + \sum_{k\ odd} A_k e^{-\frac{2\pi i}{n}(k-1)j}.$$

We consider the recursive FFT algorithm, which computes the even and odd elements of A and then combining them together recursively.

Graph Structure: The dependence task graph can be considered as the concatenation of two trees: the MERGE tree and SPLIT tree. Each task in the SPLIT tree splits a vector into even and odd parts as shown in the FFT algorithm. Each task in the MERGE tree uses the result of two incoming data vectors Y and Z of size $h/2$ into a vector of size h. Figure 9 gives an example when $n = 8$.

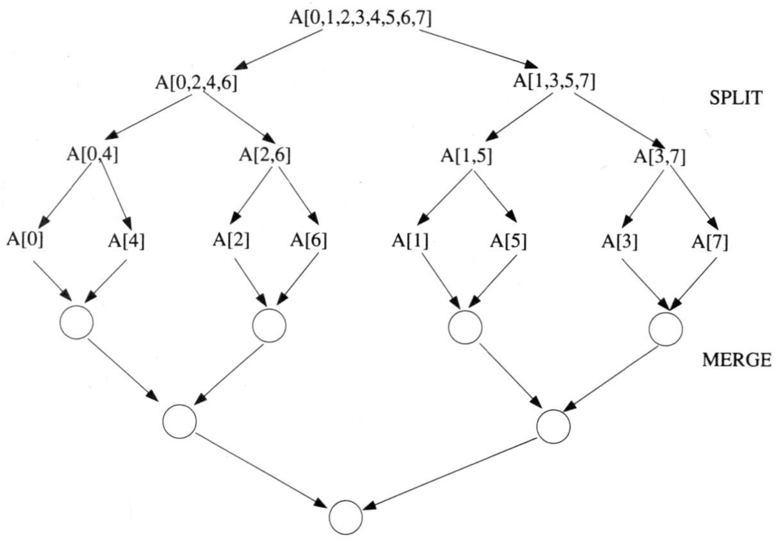

FIGURE 9. The task graph of FFT when $n = 8$.

Weights: A SPLIT task, at levels $h = m : 1$, divides a 2^h vector into two 2^{h-1} vectors, has a computation weight of $\tau_h = 2^h\omega/4$ and communication weight for the outgoing edges $c_h = \alpha + 2^{h-1}\beta$.

The MERGE task, at levels $h = 0 : m - 1$, combines two 2^h vectors into a 2^{h+1} vector, has a weight of $\tau_h = 2^h(2+0.5)\omega$, where ω is the time for processing addition and multiplication of complex numbers. The communication edge cost is $c_h = \alpha + 2^h\beta$.

Bounds on speedup: Then the length of critical path of this DAG without considering communication delay is:

$$CPL = (2^m + 2^{m-1} + \cdots + 2)\omega/4 + (1 + \cdots 2^{m-1})\frac{5\omega}{2} \approx \frac{6}{4}2^{m+1}\omega.$$

The total sequential time is:

$$T_1 = m2^m\omega/4 + m2^{m-1}\frac{5\omega}{2} = \frac{6}{4}m2^m\omega.$$

The upper bound of optimal speedup is $\frac{T_1}{CPL} = \frac{m}{2}$.

Granularity: The grain for each level for the split graph $h = m : 1$ is given by $2^h w/(4(\alpha + 2^{h-1}\beta)$. For the merge part, it is about $2^h 2.5w/(\alpha + 2^h\beta)$, $h = 0 : m - 1$. Therefore the graph is fine grain for the majority of the levels with few exceptions near the roots of the merge part.

Summary for FFT graph

Graph structure: Regular. A concatenation of an out-tree and an in-tree.

Weights: $\tau_h^{split} = 2^h w/4$ and $c_h^{split} = \alpha + 2^{h-1}\beta$ at each level $h = m : 1$.

$\tau_h^{merge} = 2^h 2.5w$ and $c_h^{merge} = \alpha + 2^h \beta$ at each level $h = 0 : m-1$.

Bounds on speedup: $Speedup \leq m/2$

Granularity: The middle part of the graph, where most of the parallelism is present, is fine grain.

Performance Results: Performance of PYRROS speedup is limited by $m/2$ for a 2^m vector.

Performance Results: We execute this DAG on NCUBE-2 when $n = 2^{14}$. Figure 10 shows the result of PYRROS code and the speedup does not scale up because the speedup bound is 7. This upper bound is still too loose since communication is not included. The reason is that the task partitioning does not provide sufficient parallelism and also does not distribute arithmetic load evenly among tasks. The degree of parallelism for this case is $m = 14$. The root of the MERGE tree which combine large vectors has large arithmetic computation while the leaf of tree has a small operation.

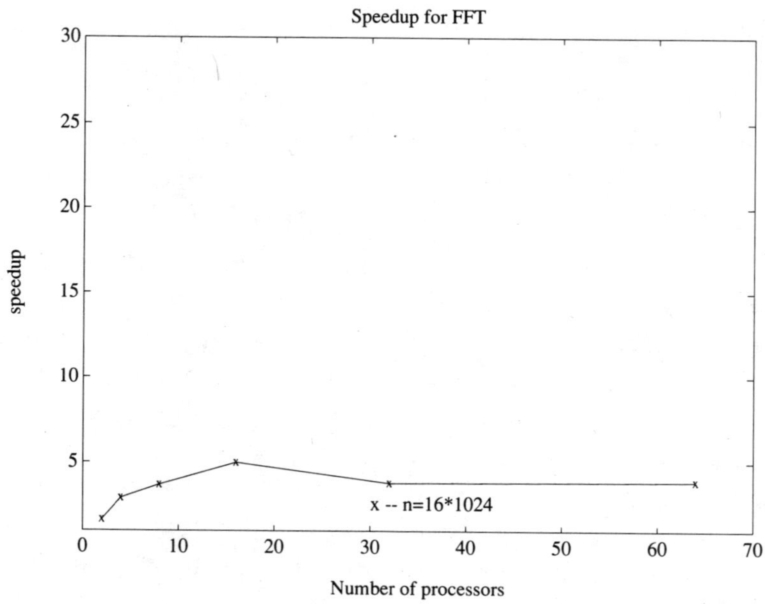

FIGURE 10. The speedup of FFT on NCUBE-2.

5.3. SPARSE MATRIX COMPUTATION. Since PYRROS can generate codes for irregular graphs, it can be used in applications where regular parallel programming is difficult or suffers from high overhead. An example is the solution of sparse matrix systems where dependence graphs are unstructured.

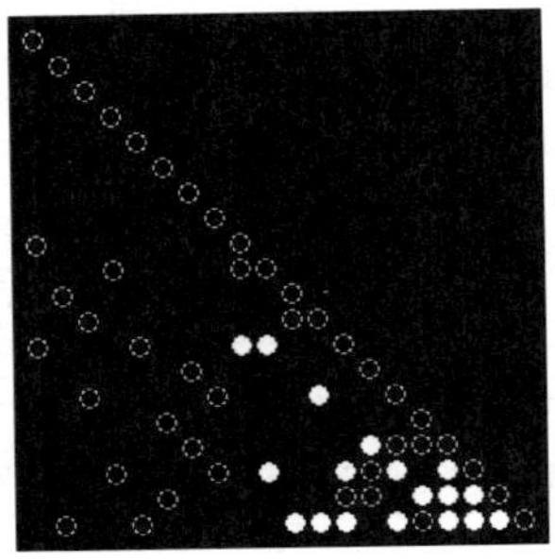

FIGURE 11. A factorized sparse matrix with filled in elements.

Finding an efficient scheduling solution for such kind of problem is important. An example is the area of circuit simulation and testing where the solution of a set of differential equations is often used. These problems are solved numerically by discretizing in time and space to reduce the problem to a large set of nonlinear equations. The solution is then derived by an iterative method such as Newton-Raphson which iterates over the same dataflow graph in solving a sparse matrix system. The topology of the iteration matrix remains the same but the data change at each step [**21**].

Graph Structure for Sparse Cholesky factorization: The dense Cholesky factorization has a graph structure similar to LU decomposition graph Figure 7. For sparse matrices, the graph is irregular and can be derived using symbolic factorization algorithms. For a positive definite sparse matrix system $Ax = b$, we first perform a symbolic factorization to find the fill-in element in the resulting sparse matrix L where $A = L \times L^t$. Figure 11 illustrates a 20×20 factorized sparse matrix. Then we derive a task graph using this symbolic factorization. Figure 12 is the task graph for this 20×20 sparse matrix. This is equivalent to deleting all zero nodes of the regular dense Colesky task graph. The complexity of this step is proportional to the number of the non-zero nodes and edges of the the irregular sparse graph.

Performance results: Figure 13 shows the speedup of PYRROS for test data matrix from Alvarado's sparse matrix manipulation system [**1**]. The PY-compile predicted performance is close to PY-Run time performance. PYRROS achieves a maximum speedup of about 7 and the speedup upper bound is less than 11 for this matrix. Since the speedup upper bound is loose, it is not clear if PYRROS performance can be improved further unless the communication cost is reduced. This example shows that PYRROS can be used for irregular sparse graph problems. The results are preliminary and more extensive experiments are needed to determine when scheduling for sparse matrices is beneficial.

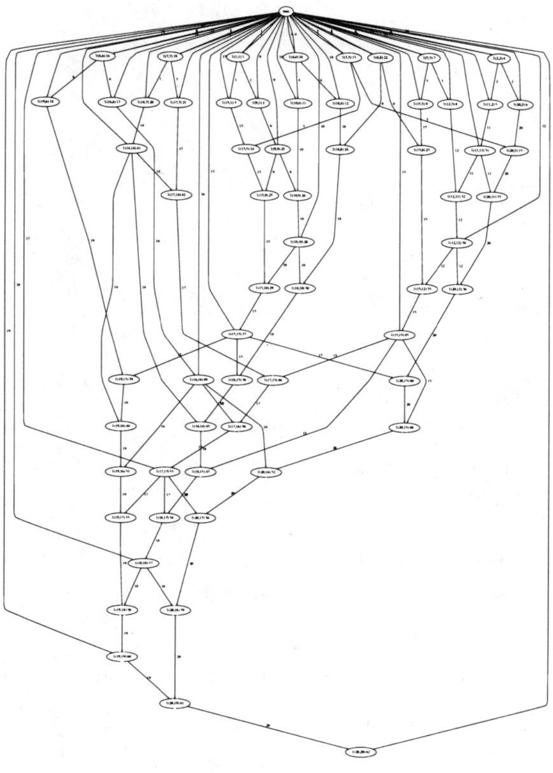

FIGURE 12. A task graph for factorizing a 20×20 sparse matrix.

Other researchers have used the PYRROS scheduling algorithms for sparse matrix computations. For example, in [**5**] the authors considered the iterative solution of linear systems for sparse matrix equations. They reported the highest speedups in the literature for these problems using the DSC algorithm, a clustering algorithm for PYRROS, and the CM5 scalable architecture. DSC achieved a speedup of 35 on 128 CM5 processors for small size sparse matrices, e.g. 5k size, as opposed to no higher than 4 speedup reported previously for similar size matrices [**5**]. The authors also report up to 75% improvements of DSC scheduling over several well known data driven mappings that use the "owners compute

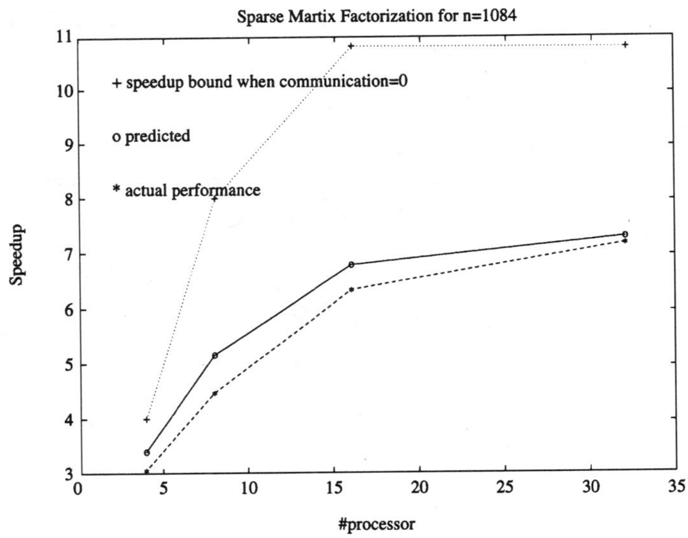

FIGURE 13. The sparse graph performance for PYRROS.

rule" for scheduling. The authors have used low level communication libraries in addition to scheduling to reduce communication and achieve what it appears to be the maximum possible performance for their matrices.

6. EXPERIMENTS ON SCHEDULING FOR THE N-BODY PROBLEM

The N-body problem is a common problem encountered in scientific computation applications. Given a distribution of n particles in space, and a rule governing their interactions, e.g. gravitational force, the problem is to compute the movement of the particles. This movement can be discretized in time steps, during each of which we compute the force applied on each particle, and update the particle positions. For example, in 2-D space force applied to a particle located at position z_j (note that we represent points in 2D using complex numbers) is :

$$F_j = \sum_{i=1, i\neq j}^{n} \frac{A_i}{z_i - z_j}$$

where A_i are the charge of particle i located at z_i.

Here we use PYRROS to parallelize the 2D Fast Multipole Method (FMM). The basic idea of the algorithm is to subdivide the space into boxes. The evaluation of particle positions (interactions) between far-away boxes can be approximated by series of expansions. The coeficients of the series are computed once and used by all "far-away" boxes. Depending on the input distribution, the algorithm comes in two flavors: (1) Non-adaptive: all boxes are of the same size. This is applied to uniform distributions. (2) Adaptive: boxes are subdivided further in the regions of higher particle density. This is used for highly-non-uniform distributions. For details see Greengard's thesis ([**19**]. Here is some rough idea. The force on a particle located at z_j is estimated for an entire box of particles. Assume that the 2D space is recursively divided into 4^l boxes where l is the level of subdivision and that each box contains s particles(uniform distribution). We have $4^l s = n$ and the evaluation in the summation is re-written as follows:

$$F_j = \sum_{i=1, i\neq j}^{n} \frac{A_i}{z_i - z_j} = \sum_{box=1}^{4^l} \sum_{z_i \in ibox} \frac{A_i}{z_i - z_j}.$$

The inner summation is approximated by a multipole series expansion with p terms. Assume z_i in a box (call it b) centered at position z_0 is evaluated for particles z_j which is far away, i.e. $|z_i - z_0| \leq c \times |z_j - z_0|$ where $c < 1$ is a separation factor. Then

$$\sum_{z_i \in b} \frac{A_i}{z_i - z_j} = -\sum_{z_i \in b} \frac{A_i}{(z_j - z_0) - (z_i - z_0)} = -\frac{1}{(z_j - z_0)} \sum_{z_i \in b} \frac{A_i}{1 - \frac{z_i - z_0}{z_j - z_0}}$$

$$\approx -\frac{1}{(z_j - z_0)} \sum_{k=0}^{p} \frac{a_k}{(z_j - z_0)^k}$$

where $a_k = \sum_{z_i \in b} A_i (z_i - z_0)^k$ are the terms of the expansion. They only depend on the particles within the box so such terms can be computed once and used for ALL far away particles. In this way we reduce complexity from s^2 to $O(ps)$ when we evaluate the interaction between s far away particles. To reduce the total complexity further the computations must be performed hierarchical using the tree structure. By evaluating the interactions at a constant number of boxes at each level then the complexity reduces and the entire space evaluation is covered.

6.1. SCHEDULING FOR THE NON-ADAPTIVE FMM ALGORITHM. We consider the non-adaptive algorithm first.

Graph Struture: Figure 14 is a 2-level ($l = 2$) regular subdivision and its corresponding quadtree.

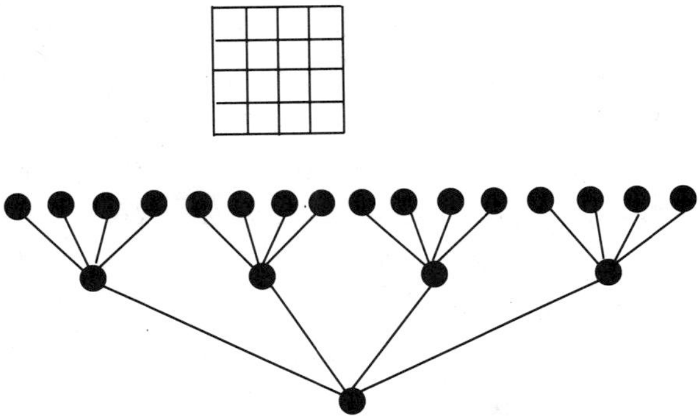

FIGURE 14. A 2-level non-adaptive subdivision

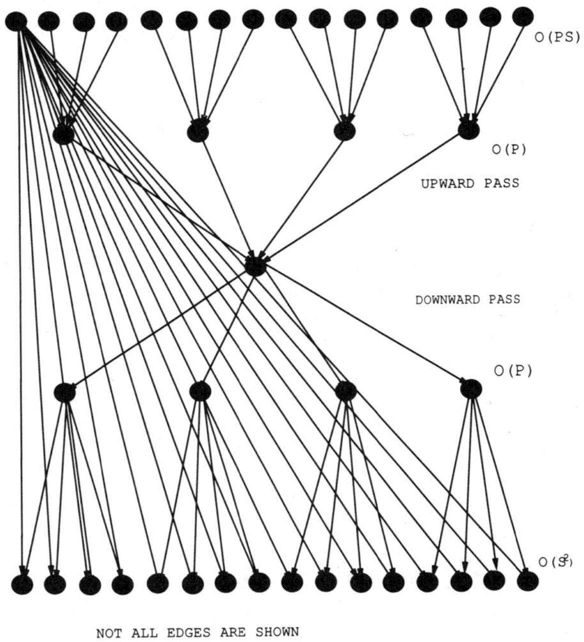

FIGURE 15. Task graph for a 2-level subdivision. The order of task weights are also shown. Only communication from one leave of the upward pass tree to the leaves of the downward pass tree is shown. The others are deleted for simplicity of presentation.

The basic structure of the task graph is a concatenation of this in-tree with an out-tree which is its reverse. Because of the regular subdivisions, the tree is a complete quad-tree. The leaves of the tree correspond to the boxes at the finest level, and the internal nodes correspond to the boxes at higher levels. Following the thesis of Greengard, we call the in-tree "upward pass", and the out-tree "downward pass." There are additional edges between the two parts of the graph, which represent interactions between boxes. Figure 15 is an example of the task graph for a level 2 division. For clarity we only draw the communication edges for one leaf node.

Note that although in this graph, every leaf communicates with all the other leaves, this is not true for divisions of higher levels. A leaf communicates with it's neighbors and "interaction list" (which is essentially the list of all second near neighbors, See figure 16). The number of boxes in each of the lists is bounded by constant, so the graph has $O(4^l)$ edges. It's not easy to draw these graphs clearly, but from the following description it should be easy to understand their structure.

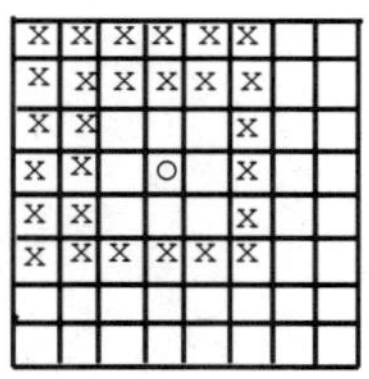

FIGURE 16. A 3-level non-adaptive subdivision. X are the boxes in the interaction list of box O. Any box has most 27 boxes in its interaction list, and at most 8 neighbors

Now we give some explanations of the functions of the nodes in the graph. These task definitions are derived directly from the algorithm description in Greengard's thesis, pp. 17-19. Although the graph has a regular structure, the task weights depends on the input and is irregular, so we only describe the order of the weights in terms of number of particles in a leaf box s and number of terms in the expansion p. (For simplicity we use s as upper bound to the number of particles in each leaf box. In a uniform distribution the actual number would vary from box to box but not too much.)

1. A leaf node in the upward pass :

a. Generates multipole expansions from the particles in the box.

b. Sends the mutipole expansion coefficients to its parent box in the upward

pass. and also to those boxes in the interaction list corresponding to the leaf level of the downward pass.
c. Sends the particle positions to those neighboring leaf boxes (including itself) in the downward pass.
Such node has computational weight $O(p * s)$.

2. A non-leaf node in the upward pass:
a. Receives the multipole expansions from each child, and shifts the expansion to its own box center. Then sums them up to obtain the multipole expansion for the whole box.
b. Sends the multipole expansion to it's parent (if it's not the root), and also, to the boxes in its interaction list in the downward pass.

3. A non-leaf node in the downward pass :
a. Receives the multipole expansions from the boxes in its interaction list in the upward pass. Then converts these mutipole expansions into local expansions. (Local expansion is a derived form of multipole expansion. It's used to reduce complexity, see Greengard thesis [**19**] for the definition of local expansion.)
b. Receives the local expansion from its parent, shifts it to its own center, then adds it to the local expansion computed in (a) above to obtain the total local expansion at this box.
c. Sends the local expansion to each of its children.

A non-leaf node of either kind has weights $O(p)$.

4. A leaf node in the downward pass:
It does the same as a non-leaf node, (Except that it does not send local expansion since it has no children) plus:
a. Receives the particle positions from neighboring boxes in the upward pass, and evaluates nearby interaction using direct, pairwise summation.
b. Computes the far away interaction using the local expansion, by evaluating the expansion at each particle in its box.
c. Sums up the "nearby" and "far away" interactions, obtaining the total interaction and updates particle positions.
Such node has weight $O(s^2)$.

Granularity: The graph is fine grain near the root but coarse grain at the leaf level where high degree of parallelism exists. So the speedup is expected to

proc	Time(ms)	PY-Run speedup	PY-compile speedup
p=4	320017	3.96	3.98
p=8	160320	7.90	7.95
p=16	81258	15.6	15.7
p=32	41638	30.4	31.0
p=64	21463	59.0	61.0

TABLE 3. The performance of the non-adaptive FMM on NCUBE-2s.

be very high.

Performance results: We use an example of 80000 randomly distribution particles, with a 5-level (32*32) subdivision. This graph has about 2700 nodes and 30000 edges. PYRROS scheduler can handle graph with such size quite rapidly. Our result in the case is very encouraging. The speed up is almost linear. See table 4. (Note this is the speedup for the first iteration. We will discuss the issue of the iterative execution later.)

The largest problem we were able to run on the NCUBE-2s machine is a problem with $8*10^5$ particles, with 128 processors, each having 16MB of memory. This shows that our system is capable to handle large real world problems.

Comparing FFT and FMM task graphs: It can be observed that although the task graph is somewhat similar to the FFT graph (if we don't consider transitive edges in the n-body graph.) the potential for parallelism is very high since the heavy weight tasks are the leaf tasks. From the scheduling result of the last subsection we can see this clearly. Whereas in the FFT graph the heavy weight tasks are near the root and have to be sequentialized, resulting in very limited speedup.

6.2. SCHEDULING FOR THE ADAPTIVE FMM ALGORITHM.

Although the non-adaptive algorithm has very good performance for uniform distributions of particles, it can not handle highly non-uniform distributions encountered in some applications, such as galaxy simulations. Adaptive subdivision must be used for such distributions. Figure 17 illustrates an example of an adaptive subdivision and the corresponding unbalanced tree for this partition. The real task graph consists of the concatenation of this in-tree with the symmetric out-tree and edges between the two parts.

Graph structure: The task graph structure is similar to that of the non-adaptive case in the sense that it still has an upward pass (in-tree) and a down-

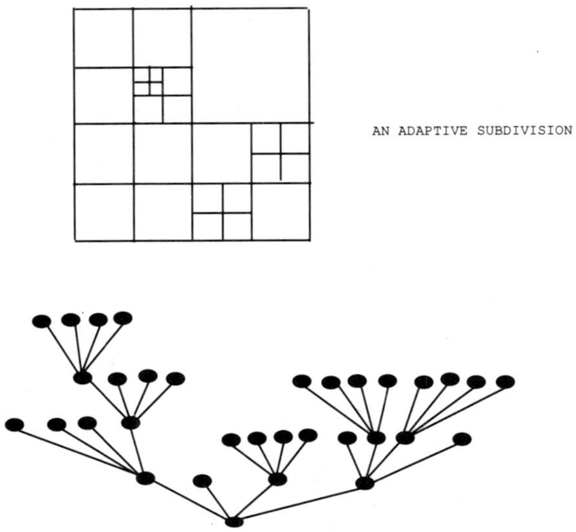

FIGURE 17. An adaptive subdivision

ward pass (out-tree). But the tree is unbalanced, and the height is larger for the places with higher particle density. So the tree structure varies, depending on the input distribution. For galaxy distributions, the tree is highly unstructured.

The function of the nodes are basically the same as in the non-adaptive algorithm, with the following complications: (see pp.28-38 in Greengard's thesis for detailed explanations.)

a. Neighboring leaf boxes can be of different size because they may not be in the same level of the tree.
b. Because Multipole expansions cannot be translated between boxes of different sizes, we must deal with "far-away" interactions between boxes in different levels of the tree in some special way.

With these extra cases in mind, a task graph can be generated by first creating different lists for each box. (These lists are classified as neighbors, far-away interaction lists consisting of boxes of the same size, smaller size, and bigger size.) then generating tasks and communication edges using these lists. The functions of the tasks are similar to those in the non-adaptive task graph, with the special cases of different sized boxes taken into account.

#proc	Time(ms)	PY-Run speedup	PY-compile speedup
p=4	105542	3.94	3.97
p=8	53663	7.76	7.83
p=16	28270	14.7	15.1
p=32	16648	25.0	28.0
p=64	10458	40.0	45.4

TABLE 4. The performance of the adaptive FMM on NCUBE-2s. Time is given in miliseconds(ms).

Performance results: Due to the irregular structure, it's necessary to use automatic scheduling tool to schedule such task graphs. PYRROS scheduling tool is well-suited for this purpose. Our example is a highly non-uniform galaxy distribution with 40000 particles. This distribution results in a highly unbalanced tree, with the maximum depth of the leaves being as large as 9, and the minimum depth of the leaves being only 1. Such kind of distribution is very difficult to parallelize. Nevertheless, PYRROS is able to achieve good performance, with speed up of 40 using 64 processors. See table 4.

6.3. STATIC SCHEDULING PERFORMANCE FOR ITERATIVE EXECUTIONS. The N-body problem is a dynamic problem, in the sense that particles are moving, their positions are updated during each time step. So they could move from one box to another, resulting in changes in the task graph. (In the non-adaptive algorithm, the movement changes the weights. In the adaptive case, it can lead to task graph structure changes, if the space is re-partitioned, but we choose to maintain the same partition before doing any rescheduling, thus only the weights are changed.) Therefore, the schedule derived for the initial task graph, although very good for the initial iterations, may not be as good during later executions.

But still a static scheduling tool could be useful for such applications, because generally, the movement of particles is small during each time step. Even if the particles have high velocity, the time step δt has to be chosen small enough to ensure stability. On the other hand, PYRROS scheduler for coarse grain graphs does not need precise weight information, and we expect that small perturbations should not lead to significant deterioration in performance, as we saw in section 4 . Thus a schedule can be re-used for many steps, until its performance deteriorates to the point where rescheduling becomes beneficial. The interesting question is how frequent we need to reschedule, since this will determine the overall performance of PYRROS for the nbody problem. We investigate this

processors	slow-down(%)
p=8	4.0
p=16	4.9
p=32	6.1
p=64	7.5

TABLE 5. The increase in parallel time after 100 iterations.

next.

Using a uniform distribution of 10000 particles, we chose a $\delta t = 0.001$ which for this case is large enough to get sizable movement of particles in each step, but small enough to ensure stability. Our experiment runs iteratively on 8-64 processors. The same initial schedule is used for all iterations, and we calculated the deterioration of the performance: i.e. $\frac{PT_{100}-PT_1}{PT_1}$, where PT_1 is the parallel time for the first iteration and PT_{100} is the parallel time for the 100th iteration. The result are shown in table 5.

Note that the slow-down in parallel time is only 7.5% for 64 processors. (About 130 iterations are needed for the parallel time to increase 10%,in this particular case.) Of course, these data are only for this particular uniform distribution. But in general, if the change of the task graph happens slowly and gradually, we could say that a initial schedule can be reused for a substantial number of steps with reasonable performance. Combined with run-time reschedule to correct the imbalance when it gets too large, this approach could be useful in many real world applications, not just n-body. We plan to investigate this approach further.

7. CONCLUSIONS

We have presented the scheduling algorithms and schedule executing methods in PYRROS. We have provided an analysis on the run-time performance of static schedule. Our experiments with various applications show that the automatically scheduled task computation could attain a good performance provided that sufficient task parallelism is available.

There are several other systems related to PYRROS. PARAFRASE-2 [**27**] by Polychronopoulos et. al., is a parallelizing compiler system that performs dependence analysis, partitioning and dynamic scheduling on shared memory machines. SCHEDULER by Dongarra and Sorensen [**10**] uses centralized dynamic scheduling for a shared memory machine. PARTI by Saltz's group [**30**] focuses on run-time DOALL parallelism with irregular distribution of data and optimizes performance by precomputing data accessing patterns. HYPERTOOL by Wu and Gajski [**33**] and TASKGRAPHER by El-Rewini and Lewis [**12**] use the same task model as PYRROS, but they do not generate real message-passing parallel code for parallel machines.

References

1. F. L. Alvarado, The sparse matrix manipulation system users manual, University of Wisconsin, Oct. 1990.
2. S. Bokhari, On the mapping problem, *IEEE Trans. Comput.* C-30 3(1981), 207-214.
3. D. Callahan, and K. Kennedy, Compiling Programs for Distributed-memory Multiprocessors, *Journal of Supercomputing*, Vol. 2, 1988, pp. 151-169.
4. D. Y. Cheng and S. Ranka, An optimization approach for static scheduling of directed-acyclic graphs on distributed memory multiprocessors, Report, Syracuse Univ, 1992,
5. Frederic T. Chong, Shamik D. Sharma, Eric A. Brewer and Joel Saltz , Multiprocessor Runtime Support for Fine-Grained Irregular DAGs, Draft, 1994.
6. Ph. Chretienne, Task Scheduling over Distributed Memory Machines, *Proc. of Inter. Workshop on Parallel and Distributed Algorithms*, (North Holland, Ed.), 1989.
7. E.G. Coffman and R.L. Graham, Optimal scheduling for two-processors systems, *Acta Informatica*, 3 (1972), 200-213.
8. J. W. Cooley and J. W. Tukey, An algorithm for the machine calculation of complex Fourier series, *Mathematics of Computation*, Vol. 19, No. 90, 1965, pp. 297-301.
9. M. Cosnard, M. Marrakchi, Y. Robert, and D. Trystram, Parallel Gaussian Elimination on an MIMD Computer, *Parallel Computing,* vol. 6, pp. 275-296, 1988.
10. J.J. Dongarra, . and D. C. Sorensen, SCHEDULE: Tools for Developing and Analyzing Parallel Fortran Programs, in *The Characteristics of Parallel Algorithms,* D.B. Gannon, L.H. Jamieson and R.J. Douglass (Eds), MIT Press, 1987, pp363-394.
11. T.H. Dunigan, Performance of the INTEL iPSC/860 and nCUBE 6400 hypercube, ORNL/TM-11790, Oak Ridge National Lab., TN, 1991.
12. H. El-Rewini and T.G. Lewis, Scheduling parallel program tasks onto arbitrary target machines, *J. of Parallel and Distributed Computing*, 9(1990), 138-153.
13. A. George, , M.T. Heath, and J. Liu, Parallel Cholesky factorization on a shared memory processor, *Lin. Algebra Appl.,* 77(1986), pp. 165-187.
14. A. Gerasoulis, I. Nelken. Static scheduling for linear algebra DAGs. *Proc. of 4th Conf. on Hypercubes*, Monterey, Vol. 1, 1989, 671-674.
15. A. Gerasoulis and T. Yang, On the granularity and clustering of directed acyclic task graphs, *IEEE Trans. on Parallel and Distributed Systems.*, Vol 4, No 6, June 1993 pp. 686-701.
16. A. Gerasoulis and T. Yang, Static scheduling of parallel programs for message passing architectures.*Lecture Notes in Computer Science, No. 634, Parallel Processing: CONPAR 92 - VAPP V*, Springer-Varlag, 1992, pp. 601-612.
17. A. Gerasoulis and T. Yang, A comparison of clustering heuristics for scheduling DAGs on multiprocessors, *J. of Distributed and Parallel Computing,* special issue on scheduling and load balancing, Vol. 16, No. 4, pp. 276-291 (Dec. 1992).
18. M. Girkar and C. Polychronopoulos, Automatic extraction of functinal parallelism from ordinary programs, *IEEE Trans. on Parallel and Distributed Systems*, Vol. 3, No. 2, 1992, pp. 166-178.
19. Leslie Greengrad, The Rapid Evaluation of Potential Fields in Particle Systems Ph.D thesis, Yale University, 1987.

20. J.A. Hoogeveen, S.L. van de Velde, and B. Veltman, Complexity of scheduling multiprocessor tasks with prespecified processor allocations, CWI, Report BS-R9211 June 1992, Netherlands.
21. N. Karmarkar, A new parallel architecture for sparse matrix computation based on finite project geometries, *Proc. of Supercomputing '91*, IEEE, pp. 358-369.
22. S.J. Kim and J.C Browne, A general approach to mapping of parallel computation upon multiprocessor architectures, *Proc. of Int'l Conf. on Parallel Processing*, vol 3, pp. 1-8, 1988.
23. J. K. Lenstra and A.H.G. Rinnooy Kan, Complexity of Scheduling under Precedence Constraints, *Operation Research*, 26:1 (1978).
24. C.L. McCreary and D.H. Gill, Automatic determination of grain size for efficient parallel processing, *Communications of ACM,* vol. 32, pp. 1073-1078, Sept., 1989.
25. C. Papadimitriou and M. Yannakakis, Towards on an architecture-independent analysis of parallel algorithms, *SIAM J. Comput.,* vol. 19, pp. 322-328, 1990.
26. S. Pande, D. Agrawal and Jon Mauney A threshold Scheduling Strategy for Sisal on distributed Memory Machines *JPDC,* vol. 21, pp. 223-236, 1994.
27. C. D. Polychronopoulos, M. Girkar, M. Haghighat,C. Lee, B. Leung, and D. Schouten, The Structure of Parafrase-2: An advanced parallelizing compiler for C and Fortran, in *Languages and Compilers for Parallel Computing*, D. Gelernter, A. Nicolau and D. Padua (Eds.), 1990.
28. R. Pozo, Performance modeling of sparse matrix methods for distributed memory architectures, in *Lecture Notes in Computer Science, No. 634, Parallel Processing: CONPAR 92 - VAPP V*, Springer-Varlag, 1992, pp. 677-688.
29. Y. Saad, Gaussian elimination on hypercubes, in *Parallel Algorithms and Architectures*, Cosnard, M. et al. Eds., Elsevier Science Publishers, North-Holland, 1986.
30. Saltz, J., Crowley, K., Mirchandaney, R. and Berryman,H., Run-time scheduling and execution of loops on message passing machines, *J. of Parallel and Distributed Computing,* Vol. 8, 1990, pp. 303-312.
31. V. Sarkar, *Partitioning and Scheduling Parallel Programs for Execution on Multiprocessors*, The MIT Press, 1989.
32. V. Sarkar, Determining average program execution times and their variance, *Proc. of 1989 SIGPLAN*, ACM, pp. 298-312.
33. M. Y. Wu and D. Gajski, Hypertool: A programming aid for message-passing systems, *IEEE Trans. on Parallel and Distributed Systems,* vol. 1, no. 3, pp.330-343, 1990.
34. T. Yang and A. Gerasoulis, A fast static scheduling algorithm for DAGs on an unbounded number of processors, *Proc. of Supercomputing '91,* IEEE, Albuquerque, NM, 1991, pp. 633-642.
35. T. Yang and A. Gerasoulis, PYRROS: Static task scheduling and code generation for message-passing multiprocessors, *Proc. of 6th ACM Inter. Conf. on Supercomputing,* Washington D.C., July, 1992, pp. 428-437.
36. T. Yang and A. Gerasoulis, List scheduling with and without communication delay, *Parallel Computing*, Vol 19, 1993, pp. 1321-1344.
37. T. Yang and A. Gerasoulis, DSC: Scheduling parallel tasks on an unbounded number of processors, To appear *IEEE Trans. on Parallel and Distributed Systems.*, August 1994.

(1) DEPARTMENT OF COMPUTER SCIENCE, RUTGERS UNIVERSITY, NEW BRUNSWICK, NJ 08903.

(2) DEPARTMENT OF COMPUTER SCIENCE, UNIVERSITY OF CALIFORNIA, SANTA BARBARA, CA 93106.

DIMACS Series in Discrete Mathematics
and Theoretical Computer Science
Volume **21**, 1995

Routing in Optical Networks: The Problem of Contention

LESLIE ANN GOLDBERG

ABSTRACT. A Completely Connected Optical Communication Parallel Computer (OCPC) consists of n processors which are connected by a complete network. Communication on this network proceeds in a sequence of synchronous communication steps. During each communication step each processor can send one message to any other processor. If a processor is the destination of a single message during a communication step then it receives the message. However, if a processor is the destination of more than one message during a communication step then the messages are garbled and they are not delivered. We consider a class of message routing algorithms called *token routing* algorithms. An algorithm for routing messages on an optical network is said to be a *token routing* algorithm if it has the property that the only messages that are exchanged by the processors are the original messages to be routed, and these messages are treated as tokens. In this model each message is associated with a token which is passed on when the message is sent. Messages can be forwarded from one processor to another. However, a processor must have the token corresponding to a given message in order to send the message. We consider a contention resolution problem in which two processors (which are not known to each other) are given messages and the goal is to get both messages to processor 1. This problem can be solved in $O(\log \log n)$ steps by a deterministic token routing algorithm. We show that all deterministic token routing algorithms for solving this problem require $\Omega(\log \log n)$ steps. As a corollary, we find that every deterministic token-routing algorithm for routing h-relations on an n-processor OCPC requires $\Omega(h + \log \log n)$ communication steps.

1991 *Mathematics Subject Classification.* Primary 68Q22, 68Q25; Secondary 68R10.
This work was performed at Sandia National Laboratories and was supported by the U.S. Department of Energy under contract DE-AC04-76DP00789

1. Introduction

In massively parallel computers, processors communicate by exchanging messages over a network of communication links. In current machines the network is necessarily of low degree. Each processor can communicate directly with only a few other processors, and the remainder of the processors must be reached indirectly by routing messages along a sequence of links. Optical technology is likely to significantly speed up parallel computation because it makes it possible to build networks of high degree. In particular, the huge bandwidth of the optical medium can be divided so that each processor has its own channel for receiving messages and each processor can send on any channel[1]. Even though such an interconnection network is a complete graph, there remains the constraint that no processor can receive messages simultaneously from two other processors without corruption. That is, *contention* makes it difficult to route messages. Our goal is to understand the effect of contention on optical routing problems.

The mathematical model that we use for optical routing is the *Completely Connected Optical Communication Parallel Computer (OCPC)* model of Anderson and Miller [**1**]. In this model, n processors are connected by a complete network. Communication on this network consists of a sequence of synchronous communication steps. During each communication step each processor can send one message to any other processor. If a processor is the destination of a single message during a given communication step then it receives the message. However, if a processor is the destination of more than one message during a communication step then the messages are garbled and they are not delivered.

Previous papers related to the problem of routing on the OCPC include [**1, 5, 6, 7, 8, 9, 11** and **13**]. One problem that has been studied is that of routing an *h-relation* on an OCPC. An h-relation is a communication problem in which each processor has at most h messages to send and at most h messages to receive. Anderson and Miller [**1**] have observed that an h-relation can be routed in h communication steps if all of the processors are given total information about the h-relation to be realized. It is more usual to assume that initially each processor only knows about the messages that it wants to send and the processors learn about the h-relation only by receiving messages from other processors. In this case it is difficult to avoid contention and the problem of routing h-relations becomes (provably) more difficult. In particular, Goldberg, Jerrum and MacKenzie [**8**] have shown that (for $h > 1$) every randomized algorithm for routing arbitrary h-relations on an n-processor OCPC requires $\Omega(h + \sqrt{\log\log n})$ expected communication steps. The fastest known randomized algorithm for solving this problem (due to Goldberg, Jerrum, Leighton, and Rao [**7**]) takes $O(h + \log\log n)$ communication steps.

We believe that every randomized algorithm for routing h-relations on an n-

[1]The division of the bandwidth is typically done by *wavelength division multiplexing*. For more information about the implementation of such networks see [**10**], [**3**] and [**4**] and the references in [**6**] and [**12**].

processor OCPC requires $\Omega(h + \log\log n)$ communication steps. However, this conjecture has not been proved. More is known for specific classes of optical routing algorithms. In particular, several authors have studied a class of algorithms called *direct* algorithms for routing h-relations. An OCPC algorithm for realizing h-relations is said to be *direct* if it has the property that the only messages that are exchanged by the processors are the original messages of the h-relation and these messages are sent directly to their destinations with no forwarding. Geréb-Graus and Tsantilas [**6**] have shown that h-relations can be routed in $O(h + \log h \log n)$ expected communication steps by a randomized direct algorithm. MacKenzie, Plaxton and Rajaraman [**11**] have shown that every such algorithm requires $\Omega(h + (\log h/\log\log h)\log n)$ expected communication steps. MacKenzie, Plaxton and Rajaraman have also studied the problem of routing h-relations with deterministic direct algorithms. They showed that $\Omega(h \log n/\log h)$ steps are required and that $O(h \log n \log h)$ steps suffice.

The lower bounds of MacKenzie, Plaxton and Rajaraman demonstrate that direct algorithms are not very good at dealing with contention in optical networks. The goal of this work is to investigate the contention-resolution capability of a wider class of algorithms, which we call *token routing* algorithms. An algorithm for routing messages on an optical network is said to be a *token routing* algorithm if it has the property that the only messages that are exchanged by the processors are the original messages to be routed and these messages are treated as tokens. In this model, messages can be forwarded from one processor to another. However, a processor must have the token corresponding to a given message in order to send the message. (The token is passed on when the message is sent.) The motivation for considering token routing algorithms is that copying messages is expensive, so one would like to have message-routing algorithms which do not make many copies of the messages to be delivered. Token routing algorithms do not make copies of messages. The routing algorithm of Goldberg, Jerrum, Leighton, and Rao, by contrast, could make up to $\log n$ copies of any given message.

In order to investigate the contention-resolution capability of token-routing algorithms, we study a simple contention resolution problem, which we call the *2-contention resolution problem.* In this problem, two processors (which are not known to each other) are given messages and the goal is to get both messages to processor 1[2]. In this paper we confine our attention to deterministic algorithms. The 2-contention resolution problem can be solved in $O(\log\log n)$ steps by a deterministic token routing algorithm on an n-processor OCPC. We show that all such algorithms require $\Omega(\log\log n)$ steps[3]. As a corollary, we find that every

[2]The 2-contention resolution problem is similar to the 2-station "control tower problem" studied in [**11**]. The only difference is that the control tower problem requires algorithms to be direct. (A similar problem has been studied in a different context in [**2**].)

[3]In fact, we show that for any $x \geq 2$, the x-contention resolution problem, in which x processors are given messages to deliver to processor 1, requires $\Omega(x + \log\log n)$ communication steps. Surprisingly, this result is not implied by the fact that $\Omega(\log\log n)$ steps are required to

deterministic token-routing algorithm for routing h-relations on an n-processor OCPC requires $\Omega(h + \log\log n)$ communication steps.

2. Contention Resolution by Token Routing Algorithms

We begin this section by observing that the 2-contention resolution problem can be solved in $O(\log\log n)$ steps by a token routing algorithm on an n-processor OCPC. The algorithm, which was suggested to us by Satish Rao, is as follows. Suppose that the two messages are given to the ith processor and the jth processor. Let $i_1, \ldots, i_{\log n}$ be the bits in the binary representation of i and let $j_1, \ldots, j_{\log n}$ be the bits in the binary representation of j. Processors i and j use binary search to find an integer ℓ such that $i_\ell \neq j_\ell$. (The bits i_ℓ and j_ℓ will be used to break the contention.) To start the binary search, processor i sends its message to the processor whose binary representation is $i_1, \ldots, i_{(\log n)/2}, 0, \ldots, 0$ and processor j sends its message to the processor whose binary representation is $j_1, \ldots, j_{(\log n)/2}, 0, \ldots, 0$. If there is a collision then the two processors concentrate their search on the second half of their bits. Otherwise, the messages are returned, and the two processors concentrate their search on the first half of their bits.

In this section we will prove the following theorem.

THEOREM 1. *Every token routing algorithm requires* $\Omega(\log\log n)$ *steps to solve the* 2*-contention resolution problem on an* n*-processor* OCPC.

Since the 2-contention resolution problem is a particular example of a 2-relation, we obtain the following corollary.

COROLLARY 2. *Every token routing algorithm for routing* h*-relations on an* n*-processor* OCPC *requires* $\Omega(h + \log\log n)$ *communication steps.*

In order to get a result which is slightly stronger than Theorem 1 we will generalize the 2-contention resolution problem and we will consider the x-contention resolution problem (for $x \geq 2$), in which x processors (which are not known to each other) are given messages. The goal is to deliver all of the messages to processor 1. We will prove the following.

THEOREM 3. *Suppose that* $x \geq 2$. *Every token routing algorithm requires* $\Omega(\log\log n)$ *steps to solve the* x*-contention resolution problem on an* n*-processor* OCPC.

It is clear that Theorem 1 follows from Theorem 3. It is not obvious that the converse is true. In fact, Theorem 1 is somewhat easier to prove than Theorem 3 because in Theorem 1 one can make use of the fact that, since there are only two messages in the 2-contention resolution problem, either both messages collide on a given step, or neither collides. In the remainder of the section we will prove Theorem 3.

---

solve the 2-contention resolution problem.

We will say that an algorithm for the x-contention resolution problem is *weakly successful* on a given input in t steps if, when it is run with that input for t steps, at least one of the following is true.

1. There is some processor that sees more than one of the x messages, or
2. Every message visits a processor in $\{1, \ldots, \log n\}$.

We will prove the following lemma.

LEMMA 4. *Suppose that $x \geq 2$ and that $T \leq \log\log n/2$. Suppose that $\mathcal{A}$ is a token-routing algorithm that allegedly solves the x-contention resolution problem on an n-processor* OCPC. *Then there is an input on which $\mathcal{A}$ is not weakly successful in T steps.*

Clearly, Theorem 3 follows from Lemma 4. To prove Lemma 4 we use the following notation. A *t-step history* will be defined to be a pair (i, C) in which i is an integer in the range $1 \leq i \leq n$ and C is a subset of $\{1, \ldots, t\}$. (The t-step history (i, C) will be used in the proof to denote the fact that in a certain run of a certain algorithm a token starting at processor i had collisions on the steps in C (and not on the steps in $\{1, \ldots, t\} - C$).

We will use the following claim.

CLAIM 5. *Suppose that $t \geq 0$ and suppose that a token routing algorithm is run on an input on which it is not weakly successful in t steps. Then for any integer t' in the range $1 \leq t' \leq t+1$, and for any token in the input, the t-step history of the token contains enough information to deduce*

1. *which processor held the token at the start of step t',*
2. *whether (and where) the token was sent on step t', and*
3. *which processor held the token at the end of step t'.*

PROOF OF CLAIM 5. Claim 5 is proved by induction on t. The case $t = 0$ is trivial. Suppose that $t > 0$ and suppose that a token routing algorithm is run on an input on which it is not weakly successful in t steps. Consider any token in the input. By induction, the $(t-1)$-step history of the token contains enough information for us to deduce which processor (call it p) held the token at the start of step $t+1$. Since the algorithm is not weakly successful in t steps p did not see any other tokens during steps $1, \ldots, t$. Thus, all information available to p at the start of step $t+1$ is encoded in the t-step history of the token. □

It follows from Claim 5 that if a token routing algorithm is not weakly successful on a given input in t steps then the $t+1$st step of the algorithm is a function of the t-step histories of the tokens. Thus, for the purposes of proving Lemma 4, we can encode a T-step algorithm as a sequence of functions $f_1, \ldots, f_T$. Function f_t will map every $(t-1)$-step history to an integer in the range $\{0, ..., n\}$. If (i, C) is the $(t-1)$-step history of a given token and $f_t(i, C) = 0$ then the token will be held at step t. Otherwise, it will be sent to $f_t(i, C)$.

Suppose that $x \geq 2$ and that $T \leq \log\log n/2$. Fix a T-step token routing algorithm $\mathcal{A} = f_1, \ldots, f_T$. We wish to show that there is an input on which $\mathcal{A}$

is not weakly successful in T steps.

For each t in the range $\{0,\dots,T\}$ we will define a hypergraph H_t. Let the vertices of H_t be all possible t-step histories. For each size x subset $\{i_1,\dots,i_x\}$ of $\{1,\dots,n\}$ we will add up to one hyperedge to H_t as follows: Consider a run of t steps of $\mathcal{A}$ when the x tokens start at processors $i_1,\dots,i_x$. If during these t steps some processor sees more than one of the x messages then do not add a hyperedge corresponding to $\{i_1,\dots,i_x\}$. Otherwise, let (i_j, C_{i_j}) denote the t-step history of the token that started at i_j in this run of the algorithm and put the hyperedge containing the x vertices (i_j, C_{i_j}) in H_t.

Note that the set of hyperedges of H_t can be constructed from f_t and from the set of hyperedges of H_{t-1}. Suppose that H_t has a hyperedge corresponding to $\{i_1,\dots,i_x\}$. Then H_{t-1} will also have a hyperedge corresponding to $\{i_1,\dots,i_x\}$. Now, suppose that H_{t-1} contains a hyperedge consisting of the vertices $(i_1, C_{i_1}),\dots,(i_x, C_{i_x})$. To see whether H_t contains a hyperedge corresponding to $\{i_1,\dots,i_x\}$ and, if so, to determine which such hyperedge is in H_t, we proceed as follows. For each vertex (i_j, C_{i_j}) in the hyperedge in H_{t-1}: If there is another vertex (i_k, C_{i_k}) in the hyperedge in H_{t-1} such that $f_t(i_j, C_{i,j}) = f_t(i_k, C_{i_k})$ and this quantity is non-zero then let $C'_{i_j} = C_{i_j} \cup \{t\}$. Otherwise, let $C'_{i,j} = C_{i,j}$. Now look at the t-step histories $(i_1, C'_{i_1}),\dots,(i_x, C'_{i_x})$. By Claim 5 we can use these t-step histories to determine whether, during the t steps of a run of $\mathcal{A}$ with tokens starting at $\{i_1,\dots,i_x\}$, some processor sees more than one of the x messages. If not, the hyperedge consisting of vertices $(i_1, C'_{i_1}),\dots,(i_x, C'_{i_x})$ is put in H_t.

A *clique* in H_t is defined to be a subset S of the vertices such that every size x subset of S is a hyperedge of H_t. Consider the size of the largest clique in H_t.

CLAIM 6. *H_0 contains the n-clique consisting of the vertices* $(1,\emptyset),\dots,(n,\emptyset)$.

PROOF. Straightforward.

CLAIM 7. *Suppose that $\mathcal{A}$ is weakly successful on every input in T steps. Then the largest clique in H_T has size at most* $\log n$.

PROOF. To prove Claim 7 note that every hyperedge in H_T consists of x vertices $(i_1, C_{i_1}),\dots,(i_x, C_{i_x})$ such that when the algorithm is run for T steps with the tokens starting at $i_1,\dots,i_x$ every token visits a processor in $\{1,\dots,\log n\}$ but no processor sees more than one of the x tokens. □

CLAIM 8. *If H_t has a clique of size m for $m >$* $\log n$ *then H_{t+1} has a clique of size $m^{1/3}$.*

PROOF OF CLAIM 8. The proof of Claim 8 is based on an argument which is sometimes known as the "Dedekind box" argument. Suppose that H_t has a clique containing m vertices. Suppose that at least $m^{1/3}$ of these vertices (call them $(j_1, C_{j_1}),\dots,(j_{m^{1/3}}, C_{j_{m^{1/3}}}))$ are mapped to the same point by f_{t+1}. If this point is 0 then these same vertices form a clique in H_{t+1}. If it is non-zero then the vertices $(j_1, C_{j_1} \cup \{t\}),\dots,(j_{m^{1/3}}, C_{j_{m^{1/3}}} \cup \{t\})$ form a clique in H_{t+1}.

Otherwise, at least $m^{2/3}$ of the m vertices are mapped to distinct points by f_{t+1} (call the $m^{2/3}$ vertices $(j_1, C_{j_1}), \ldots, (j_{m^{2/3}}, C_{j_{m^{2/3}}})$). We will identify a large subset S of these vertices which forms a clique in H_{t+1}. To identify S, we construct a graph G on vertex set $(j_1, C_{j_1}), \ldots, (j_{m^{2/3}}, C_{j_{m^{2/3}}})$. For each vertex (j_i, C_{j_i}) add up to one edge to the graph as follows: If $f_{t+1}(j_i, C_{j_i}) > 0$ and there is another vertex (j_ℓ, C_{j_ℓ}) such that the t-step history (j_ℓ, C_{j_ℓ}) indicates that the token starting at j_ℓ visited $f_{t+1}(j_i, C_{j_i})$ in the first t steps then add the edge $((j_i, C_{j_i}), (j_\ell, C_{j_\ell}))$ to G. (There can be at most one such vertex (j_ℓ, C_{j_ℓ}) because the vertices form a clique in H_t.) It is clear that if we choose as S any set of vertices that is an independent set in G then the vertices in S will form a clique in H_{t+1}. Note that the maximum degree of a vertex in G is $t+2$. By Turán's theorem, G has an independent set of size $m^{2/3}/(t+3) \geq m^{1/3}$. □

Claim 6 and Claim 8 imply that H_T has a clique of size at least $n^{(1/3)^T}$ which is greater than $\log n$, as long as n is sufficiently large. Therefore, Lemma 4 follows from Claim 7.

3. Open Problems

This paper does not address the problem of solving the 2-contention resolution problem with *randomized* token routing algorithms. The obvious direct randomized algorithm for solving this problem runs in 3 expected steps, but the probability that more steps are required is large. A related problem is that of deriving a lower bound for the problem of routing h-relations with randomized token routing algorithms. It is possible that the technique of Goldberg, Jerrum and MacKenzie [**8**] could be used to convert Theorem 1 (or Theorem 3) into a randomized lower bound for h-relation routing. A much more interesting open problem is to extend our result to obtain an $\Omega(h + \log\log n)$ lower bound for routing h-relations with *general* OCPC algorithms. A first step towards proving this result might be to extend the proof of Theorem 1 to a similar theorem for a model which allows limited copying of messages.

Acknowledgments

I would like to thank Mark Jerrum, James Park and Jon Atkins for useful discussions concerning this problem.

References

1. R. J. Anderson and G. L. Miller, *Optical communication for pointer based algorithms*, Technical Report CRI 88-14, Computer Science Department, University of Southern California, Los Angeles, CA 90089-0782 USA (1988).
2. Robert Cypher and Smaragda Konstantinidou, *Bounds on the efficiency of message-passing protocols for parallel computers*, Proceedings of the 5th ACM Symposium On Parallel Algorithms and Architectures, ACM Press, 1993, pp. 173–181.
3. Patrick W. Dowd, *High performance interprocessor communication through optical wavelength division multiple access channels*, Proceedings of the 18th International Symposium on Computer Architecture, 1991, pp. 96–105.

4. Mary Mehrnoosh Eshaghian, *Parallel algorithms for image processing on OMC*, IEEE Transactions on Computers **40** (1991), no. 7, 827–833.
5. A. V. Gerbessiotis and L. G. Valiant, *Direct bulk-synchronous parallel algorithms*, Proceedings of the 3rd Scandinavian Workshop on Algorithm Theory, 1992.
6. M. Geréb-Graus and T. Tsantilas, *Efficient optical communication in parallel computers*, Proceedings of the 4th ACM Symposium On Parallel Algorithms and Architectures, ACM Press, 1992, pp. 41–48.
7. Leslie Ann Goldberg, Mark Jerrum, Tom Leighton and Satish Rao, *Doubly logarithmic communication algorithms for optical communication parallel computers*, pre-print (1994), (A preliminary version of this paper appeared in the Proceedings of the 5th ACM Symposium On Parallel Algorithms and Architectures in 1993.).
8. Leslie Ann Goldberg, Mark Jerrum and Philip D. MacKenzie, *An $\Omega(\sqrt{\log\log n})$ lower bound for routing in optical networks*, Proceedings of the 6th ACM Symposium on Parallel Algorithms and Architectures, ACM Press, 1994.
9. Leslie Ann Goldberg, Yossi Matias and Satish Rao, *An optical simulation of shared memory*, Proceedings of the 6th ACM Symposium on Parallel Algorithms and Architectures, ACM Press, 1994.
10. Alfred Hartmann and Steve Redfield, *Design sketches for optical crossbar switches intended for large-scale parallel processing applications*, Optical Engineering **28** (1989), no. 4, 315–327.
11. P.D. MacKenzie, C.G. Plaxton and R. Rajaraman, *On contention resolution protocols and associated probabilistic phenomena*, Proceedings of the 26th ACM Symposium on Theory of Computing, ACM Press, 1994.
12. W. F. McColl, *General purpose parallel computing*, Lectures on Parallel Computation, Proceedings of the 1991 ALCOM Spring School on Parallel Computation (A.M. Gibbons and P. Spirakis, ed.), Cambridge University Press, 1993, pp. 337–391.
13. L. G. Valiant, *General purpose parallel architectures*, Handbook of Theoretical Computer Science (J. van Leeuwen, ed.), Elsevier, 1990, (see especially p. 967).

Department of Algorithms and Discrete Mathematics, Sandia National Laboratories, MS 1110, PO Box 5800, Albuquerque, New Mexico 87185-1110

E-mail address: lagoldb@cs.sandia.gov

DIMACS Series in Discrete Mathematics
and Theoretical Computer Science
Volume **21**, 1995

Communications in Optically Interconnected Parallel Computer Systems

MOUNIR HAMDI

ABSTRACT. Optical technology has made significant contributions to the state-of-the-art for long distance communications, including high reliability, low interference, security benefits, and very high bandwidth. However, the technology has not been fully exploited in the domain of tightly-coupled parallel computer systems. This paper considers an array of processors with optical waveguides (optical buses), in which the processors in each row and in each column can access a waveguide. We introduce this model of parallel computation, and show efficient data communications on waveguides using the notion of pipelining achieved through space division multiplexing. We also show that this model is equivalent to other proposed models using wavelength division multiplexing or optical free-space communication. The pipelining properties of waveguides are then used to implement an $O(\log(N))$ sorting algorithm on these arrays of processors with waveguides based on the rotate-sort algorithm. Then, arrays of processors with waveguides are shown to emulate *deterministically* the CRCW PRAM model, with only $O(\log(N))$ degradation in time performance. Moreover, we demonstrate that if the size of these arrays of processors with waveguides are larger than that of a CRCW PRAM, then we could emulate it in constant time. There have been many experimental results carried out on waveguides which support our analysis, and suggest the practicality of designing hybrid optical-electronic parallel computers for next generation supercomputing systems. Thus, arrays of processors with waveguides appear to be particularly effective networks for PRAM emulation, and can be considered as *universal* interconnection networks.

1991 *Mathematics Subject Classification.* Primary 68Q22, 68Q25; Secondary 68Q10.
This research is supported in part by the Hong Kong Research Grant Council under the grant RGC/HKUST 100/92E.

1. Introduction

It has become very clear that significant improvements in computer performance in the future can only be achieved through exploitation of parallelism at all machine design levels. On the architectural side, communication among the elements of a high-performance computing system is recognized as the limiting and decisive factor in determining the performance and cost of the system. In recent years, there has been considerable efforts in the design of interconnection networks for parallel computers. Two of the most popular point-to-point interconnection networks for parallel computers today are the binary n-cube, also called hypercube, and the mesh interconnection networks. Several companies, including NCUBE, Connection Machine, FPS, Intel, and Ametek, are currently selling machines based on the hypercube topology. One major drawback of the hypercube network is its lack of scalability, which limits its use in building large size systems out of small size systems with little change in the configuration. On the other hand, mesh networks suffer from a major limitation which is their large diameters ($N^{1/2}$ for an N-node network). Moreover, a relatively small portion of algorithms for scientific and engineering problems efficiently fits the mesh topology.

On the technological side, optics, possesses the potential for a better solution to the communication problem in parallel and distributed computing. Consequently, recent advances in fiber-optic communication technology have made radically different approaches to parallel processor interconnection feasible. Optical interconnections for multiprocessor systems including both free-space and guided wave approaches have been studied extensively in the literature. These studies seek to exploit the benefits of optical communication including its inherent parallelism, high spectral and spatial bandwidth, and low signal crosstalk. In this paper, we propose a waveguide interconnection system with time-division communications. Time division communication is especially useful in systems with optical interconnections where high communication bandwidth can be exploited. Optical pulse transmissions on a single-mode waveguide have two distinct properties that are not shared by electronic signal transmissions, namely unidirectional propagation and predictable propagation delays per unit length. The second property states that the propagation delay is directly proportional to the length of a fiber. In a multiprocessor system connected with an optical waveguide (or bus), relationships between the spatial and temporal positions of transmitted pulses can be established. For example, if two processors transmit a pulse on the waveguide at the same time, the difference between arrival times of these two pulses at any checkpoint downstream, is equal to the propagation delay between the two processors. In other words, the spatial separation of the two processors determines the temporal separations between the pulses they transmit. By arranging spatial separations and controlling transmission (or receiving) times of processors, time-division multiplexing (demultiplexing) can be done without using multiplexors (or demultiplexors).

These powerful characteristics of optical communications coupled with the very high transmission bandwidth of optics lead us to consider the potential of designing an *ideal* parallel computer system using optical communications. However, if it is not possible to design this ideal parallel computer system, we want to know how far are we from this ideal machine. In order to answer these questions, we have to relate our design to the PRAM (Parallel random access machine) model [3]. A PRAM is an idealized parallel machine which consists of a set of processors all of which have unit-time access to a shared memory. In every step of the PRAM, each of its processors may execute a private RAM instruction. In particular, the processors may all simultaneously access (read from or write into) the common memory. Various types of PRAMs have been defined, differing in the conventions used to deal with read/write conflicts; i.e. attempts by several processors to access the same variable in the same step. In the most restrictive model, exclusive read-exclusive write or EREW PRAMs, no variable may be accessed by more than one processor in a given step. In contrast, CRCW (concurrent read-concurrent write) PRAMs allow simultaneous reading as well as simultaneous writing of each variable, with some rule defining how to handle simultaneous writing of distinct variables to the same location.

From a programmer's perspective, it would be ideal to develop parallel algorithms for the PRAM. PRAMs are very convenient for expressing parallel algorithms since one may concentrate on decomposing the problem at hand, without having to worry about the communication between the tasks. For this reason there are many parallel algorithms written for the PRAM [1], [3], [13], 15]. Moreover, given the fact that the biggest problem nowadays with parallel computers is their utilization, that is, using them effectively with enough ease seems to strongly support the efforts being conducted on the PRAMs as a parallel computer model. Unfortunately, the PRAM is not a very realistic model of parallel computation when the number of processors grows large. Present and foreseeable technology does not seem to make it possible to implement this model with more than a small number of processors.

How can we reconcile the convenience of CRCW PRAMs with the limitations of a real computer? The only alternative is to emulate a CRCW PRAM on a real network. Indeed, this school of thought has led many researchers to consider the emulation of the idealized parallel machine, the CRCW PRAM, on more realistic parallel machines using interconnection networks such as the hypercube, the mesh-of-trees, compound networks, the 2-D mesh, and the fully connected network [2], [10], [18], [21], [25], [30]. These emulations are either *deterministic* or *probabilistic*. While deterministic emulations typically take longer time than probabilistic emulations, they, on the other hand, have the characteristic that they give us an *exact* measure of how far are we from an ideal parallel computer system.

However, to the best of our knowledge, no practical interconnection network has been proposed to interconnect the processors of a parallel machine that can emulate *deterministically* any of the PRAM models of the same size in better than polylogarithmic degradation in time performance which render the emulated algorithms quite slow. The best deterministic emulation has been shown on the binary hypercube [7, 17]. In this paper, we show that by using optical communications, we can significantly improve upon the results obtained using electronic communications. We propose an array of processors with waveguides (APW), and we find that they are able to emulate any of the PRAM models with better than polylogarithmic degradation in time performance. This also means that the APW can emulate any interconnection network of the same size with better than polylogarithmic degradation in time performance since a PRAM can typically emulate any of these networks in constant time. Thus, the design of APW seem to be a step closer towards the realization of a *practical* PRAM. In other words, this research attempts to shorten the gap between theoretical and practical models of parallel computing. Moreover, it answers some of the questions regarding how close we can come in designing an *ideal* parallel machine. We should mention here the fact that there are some researchers who were able to show that certain networks are capable of emulating the PRAM in better than polylogarithmic degradation in time performance with high probability unlike our *deterministic* methods in this paper [14]. Further, there has been some research conducted that shows that the PRAM can be emulated on, for example, the reconfigurable mesh in constant time [21], [23]. However, in this case and many other related cases the size of the emulating machine is a lot bigger than that of the emulated machine (PRAM). This makes it quite impractical.

This paper is organized as follows. In Section 2 and in Section 3 we introduce the 1-D APW and the 2-D APW and we define some of their key communication characteristics. Then, we illustrate the implementation of some data movement operations on an APW. In Section 4, we introduce some equivalent networks to the APW which employ wavelength division multiplexing or free-space optical communication. In Section 5 we show the efficient implementation of the rotate-sort algorithm on the 2-D APW and subsequently demonstrate the efficient emulation of the PRAM models on the 2-D APW. We also show in Section 6 how we can emulate a PRAM model if we allow the number of processors in an APW to be larger than that of the PRAM. Finally, in Section 7, we present some concluding remarks.

2. Arrays of Processors With Waveguides

A unique property of optics is its ability to pipeline the transmission of signals through a channel. In electronic buses, signals propagate in both directions, while optical channels are inherently directional and have precise predictable path delays per unit distance [5], [9], [12], [20], [28]. Consider the system of Figure 1, where N processors, each having a constant number of registers, are connected through a single optical waveguide (bus). Each processor is coupled to the optical waveguide with two passive couplers, one for injecting (writing) signals on the waveguide and the other for receiving (reading) signals from the waveguide. As in the case of electronic buses, each processor j communicates with any other processor i by sending a message to i through the common bus. However, because optical signals propagate in one direction, a processor j may send signals to another processor i only if $i > j$.

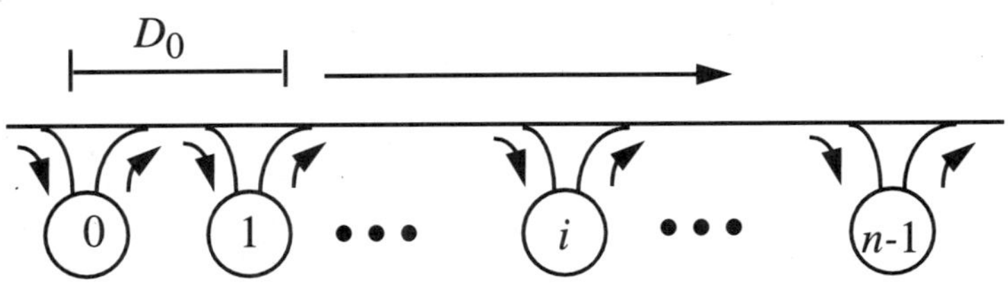

FIGURE 1. A system of n processors connected with a single optical waveguide.

Assume that a message on an optical bus consists of a sequence of pulses, each having a width w in seconds. The existence of an optical signal of width w represents a binary bit 1, and the absence of such a signal represents a 0. For analytical reasons, we let D_0 be the optical distance between each pair of adjacent nodes and τ be the time taken for an optical pulse to traverse the optical distance D_0. To transfer a message from a node j to node i, $i > j$, the sender j writes its message on the bus. After a time $(i - j)\tau$ the message will arrive at the receiver i, which then reads the message from the bus.

The properties of unidirectional propagation and predictable path delays of optical signals may be used advantageously. Specifically, unlike the electronic case, where the writing access to the bus by each node must be mutually exclusive, all nodes in the system of Figure 1 can write on the bus simultaneously, provided the following collision-free condition is satisfied [10], [12], [20],

$$D_0 > bwc_g \tag{1}$$

where b is the number of binary bits in each message, and c_g is the velocity of light in the waveguide. Clearly if this condition is satisfied and the system is synchronized such that every node starts writing a message on the bus at the same instant, then no two messages injected on the bus by any two distinct nodes will collide. Here by colliding we mean that two optical signals injected on the bus by any two distinct nodes arrive at some point on the bus simultaneously. With this condition satisfied, every node can, in parallel, send a message to some other node, and the messages will travel from left to right on the bus in pipelined fashion, as shown in Figure 2. Thus, we use the term *pipelined* bus. In the rest of the paper, we always assume that the collision-free condition (1) is satisfied. We term τ as *petit cycle*, and $n\tau$ as a *bus cycle*. Note that a bus cycle is the time taken for an optical signal to traverse the entire length of the optical bus.

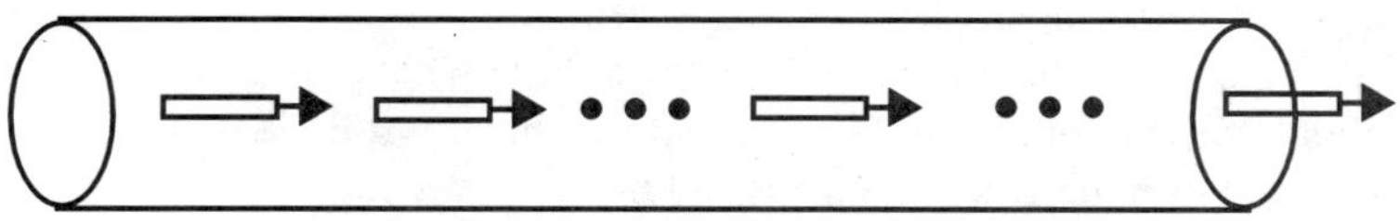

FIGURE 2. Message pipelining on the optical bus.

In the system of Figure 1, messages can be transmitted only from left to right. To allow message passing from right to left, another optical bus is used, as shown in Figure 3. In this figure, we have two optical buses; the upper one is used for sending messages from left to right, and the lower one is used for sending messages from right to left. Each node can write and read messages on either bus as desired. Obviously signals in different buses do not disturb one another; that is the two buses can support two separate pipelines. The system in Figure 3 is the architecture of a linear APW (1-D APW).

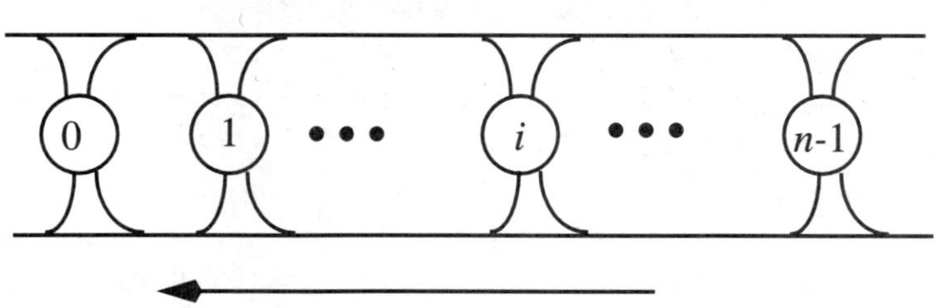

FIGURE 3. A 1-D APW with two optical waveguides.

To specify the time at which a node should receive a message, we introduce a function *twait*(i), which is defined as the time that node i should wait, relative to the beginning of a bus cycle, before reading the message sent to it from some other node j. Thus,

$$twait(i) = (i-j)\tau$$

If τ is considered as a time unit, then *twait* can be interpreted in terms of the number of such units and thus be written $twait(i) = i - j$. Clearly if $twait(i) > 0$, then the message is to be received from the left; if $twait(i) < 0$, then the message is to be received from the right. If $twait(i) = 0$, then no message should be received by node i. The value of $twait(i)$ can be stored in a *wait register*, and more than one such register may be used if a node is to receive more than one message in one bus cycle.

This *twait* control function, however, has the disadvantages that it depends crucially on timing accuracy and is sensitive to the optical distance D_0 between two adjacent nodes. An equivalent control function, *mwait*, that does not have these disadvantages may be defined if we require that each node inject a message, real or dummy, every bus cycle. In this case we define $mwait(i)$ as the number of messages that node i should skip before reading its message. For example, if $mwait(i) = \gamma$, then node i should receive $|\gamma|$th message that passes i on the bus. That is, it has to wait until $|\gamma| - 1$ messages have passed and then it reads its own message. The sign of γ determines on which bus the message should be received. Clearly *mwait* is equivalent to *twait* and either control function can be used. For convenience we simply write the control function as *wait*.

The control function *wait* can only be used when the communication pattern is known to the receiver in the sense that the receiver knows from which node the message is to be received. This is the case in most SIMD programming environments [11], [24], [29]. In cases where the communication pattern is unknown to the receiver, the coincident pulse techniques [5], [17] may be used such that an addressing pulse and a reference pulse coincide at the detector of the receiver, thereby addressing it. In this paper we use *wait* for addressing since the communication patterns are assumed to be known to the receiver.

2.1. Message Routing in a Linear APW

Various message routing patterns can be realized in a simple, straightforward way. Since a routing pattern is determined by the *wait* functions, we need only determine these *wait* functions for each routing pattern. The most common patterns are:

One-to-One. The system executes a *SEND*(j, i) instruction, which means that a message is to be transferred from node j to node i. Thus, $wait(i) = i - j$, where i is a single specific node.

Broadcast. The system executes *BROADCAST*(j) which means that node j broadcasts a message, and all other nodes i will receive that message. In this case, $wait(i) = i - j$ for all $i \neq j$.

Permutations. For each node j to send a message to node $i = PERM(j)$, where *PERM*( ) is an arbitrary permutation, we set $wait(i) = i - j$ for all i.

We see that the computation of $wait(i)$ is very simple and uniform. The only difference among the *wait* functions for different message routing patterns is that the nodes involved are different. It is clear that all these communication tasks can be performed using a single bus cycle. Note that, in a linear APW, message passing between two non-neighboring nodes is nearly as efficient as that between two neighbors. Specifically, a message takes τ more time to pass one more node on the optical bus. This is not the case in typical parallel computers with point-to-point connections, where to pass a node, en route to another node, a message has to go through a router. Moreover, τ can be extremely small for tightly coupled parallel computer systems because of the small distance between neighboring nodes and the high speed of optics.

Further, by looking at the *PERM*(j) routing patterns, it is clear that a linear APW can emulate a *module parallel computer* (MPC) [2], [18] in a single bus cycle [9], [10], [11].

An MPC computing model is a theoretical parallel computer model defined as having n processors (RAMs), each equipped with a memory module. Every processor may access every memory module via a complete network connecting the processors. However, the memory modules are *sequential* devices, i.e. able to satisfy only one request at a time. The MPC model has been used by many researchers as a more realistic model than the PRAM for the design and analysis of many parallel algorithms [2], [15].

3. Two-Dimensional APW

The above described linear APW has some practical limitations [9], [10], [20], [28]. First, the system size cannot be very large since it is limited by the minimum optical power that can be detected by the directional coupler of each PE which is inversely proportional to the number of optical path lengths an optical pulse would traverse. Second, when the number of processors, N, is very large, the end-to-end transmission delay (waveguide cycle) increases since the condition stated by Equation 1 has to be satisfied and the number of *petit* cycles is proportional to N. To overcome these shortcomings in a linear APW and be able to *practically* emulate PRAMs of large sizes, we consider a 2-D (two-dimensional) APW. In a 2-D APW, each node is coupled to four buses as shown in Figure 4, where the two horizontal buses are used for passing messages horizontally in the way as before, and the vertical buses are used for passing messages vertically in a similar way. Each node in a 2-D APW of size $N = m \times n$ will be given an identification (x, y), $0 \leq x < m$, $0 \leq y < n$, indicating the row-column position of the node.

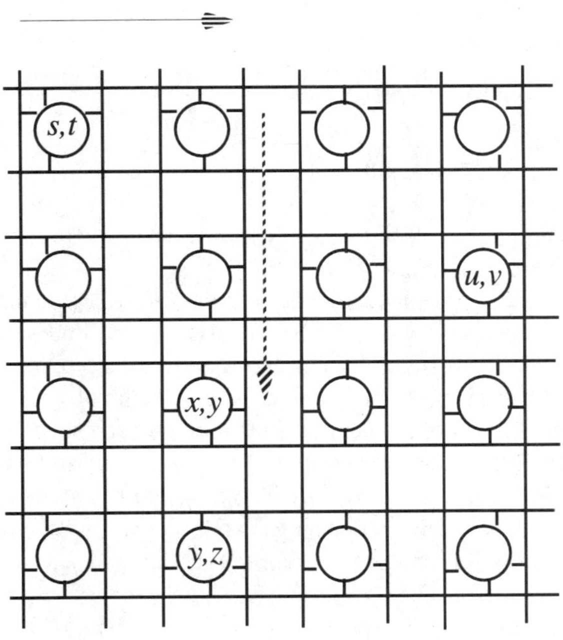

FIGURE 4. A 2-D APW where each processor is connected to four waveguides.

A unique issue that arises in the 2-D APW is the relay of messages. Specifically, suppose a message is to be transferred from node (x_1, y_1) to node (x_2, y_2), with $x_1 \neq x_2$ and $y_1 \neq y_2$. Then the message may first be sent from (x_1, y_1) to (x_1, y_2), which is the node at the intersection of row x_1 and column y_2, in the first bus cycle (a row bus cycle) and then from (x_1, y_2) to (x_2, y_2) in the second bus cycle (a column bus cycle). That is the message has to be buffered at node (x_1, y_2) at the end of the first bus cycle and then relayed to its destination in the second bus cycle. For the purpose of relaying messages, we define a control function *relay* for node (x_1, y_2) as

$$relay\,[\,(x_1, y_2)\,] \;=\; y_2 - y_1$$

which indicates that node (x_1, y_2) will read a message from a row bus at time $|y_2 - y_1|$ (relative to the start of the row bus cycle) and then write that message on the proper column bus at the beginning of the following column bus cycle.

Clearly, all the routing patterns described in the previous section for the 1-D APW (one-to-one, broadcast, permutations) can be similarly implemented on the 2-D APW using the *relay* function. For more details, the reader is referred to [10], [11], [12], [24]. However, they are omitted in this paper because they are not directly needed for the implementation of our algorithms. Rather, as will shown in the next section, the routing patterns presented for the 1-D APW are sufficient to implement our algorithms on the 2-D APW.

4. Alternative Optical Networks

In this section, we outline the principles involved in realizing an optical interconnection network using WDM-based fiber communication and free-space holographic optical communications [25]. Then, we show how these design choices can have the same communication power as an APW.

4.1. Wave Division Multiplexing

In wavelength division multiplexing the large capacity of the optical fiber is utilized by splitting the available bandwidth into independent, noninteracting channels, thereby, supporting simultaneous communication between many source-destination pairs on the communication medium. In WDM-based systems, wavelength encoded signals from the transmitting nodes are multiplexed onto the fiber using a device called the passive star coupler. The transmitters at the network nodes are connected to the input ports of the star coupler. The signal power at each input port is divided uniformly amongst the output ports. Thus, wavelengths from all the transmitters appear at each output port. Demultiplexing is performed at the receivers by recovering the desired input port signal from the common output port signal. The desired connectivity amongst the system nodes can be obtained by appropriate wavelength assignments to the transmitters and receivers. This is illustrated for a seven node binary tree in Figure 5. In the figure, all links are bidirectional and are implemented using fixed wavelength transmitters and receivers. Thus, the degree of a node determines the number of transmitters and receivers at that node..

By using tunable transmitters and receivers, versatile interconnections can be realized. As the logical connectivity between nodes can be achieved by tuning the transmitters and receivers to the desired wavelengths, reconfiguration of the interconnection network to support a different topology can be achieved by wavelength reassignment. WIth the availability of narrow line width semiconductor lasers and inexpensive passive star couplers, WDM-based optical interconnects that support reconfiguration seem feasible for multiprocessors. Rapid progress is being made in the development of tunable devices, both in the range over which they are tunable, and their tuning time. Limited tuning range restricts the kinds of topologies that can be supported, and the tuning times of devices constitute the

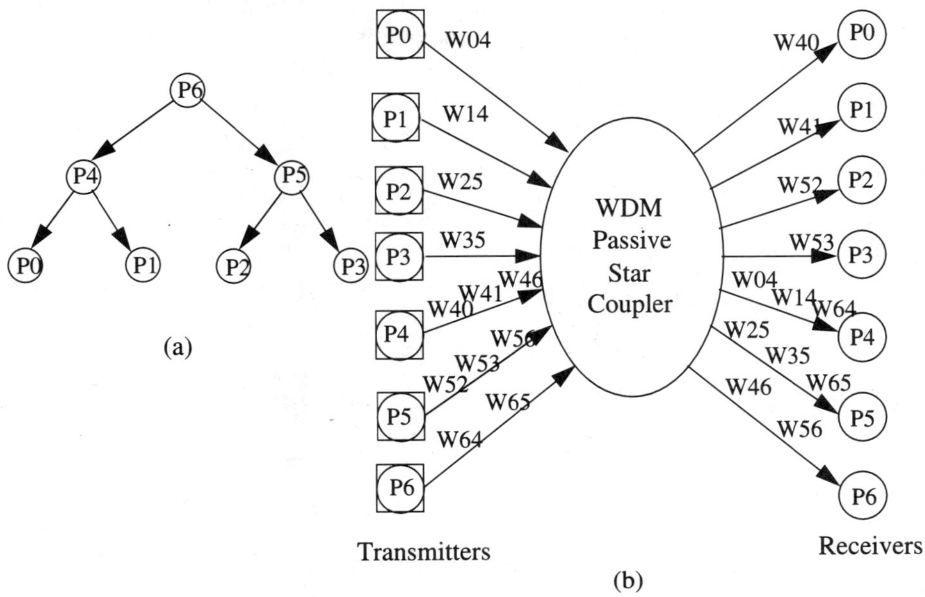

FIGURE 5. (a) A binary tree with seven nodes. (b) WDM Star embedding of the tree. W_{ij} is the wavelength assigned for communicating from node i to node j.

reconfiguration overheads. Both of these parameters are decided by the method of tuning used. Current tuning ranges are in the 4 - 10*nm* range and the tuning times vary from nanoseconds to milliseconds.

If we relate the communication of WDM-based optical networks to that on an APW, then we can establish a strong relation ship between the two schemes. In an APW, within each bus cycle, every node can communicate with any other attached node. Thus, if we assume that the tuning times of all wavelengths in an WDM-based interconnect is equivalent to a bus cycle in an APW, then each node would also be able to communicate with any other node attached to the same star coupler within a bus cycle by tuning to the appropriate wavelength of the receiver. In an APW system, if Equation (1) is satisfied, we are guaranteed that they will be no collision between message. However, in an WDM-based interconnect the collision avoidance of message is different. But, it can be achieved using various simple collision control means. Hence, we believe that the two communication systems are equivalent in terms of processing and communication power. Consequently, all the algorithms developed for an APW system, can be easily transformed to an WDM-based system.

4.2. Holographic Optical Interconnects

In the model for free-space holographic interconnection networks, light from transmitters is coupled into fibers. The ends of the fibers from all the transmitters are aligned into a vertical plane referred to as the transmitter plane. Similarly, there is a receiver plane parallel to the transmitter plane. Light incident on the fibers in the receiver plane is carried to the individual receivers at the nodes. The network connectivity can be achieved, if the light from a transmitter is coupled to the desired fiber in the receiver plane. This switching of the incident light direction is achieved by inserting a hologram in the path of light from the transmitter plane. As seen in Figure 6, the fiber from the transmitter at node i illuminates a portion of the hologram, referred here as subhologram i. Information on the positions of the receivers adjacent to node i in the desired topology is prerecorded in the subhologram. Light encoded data from the transmitter at node i is directed to all the desi8red receivers by the subhologram by space-variant imaging.

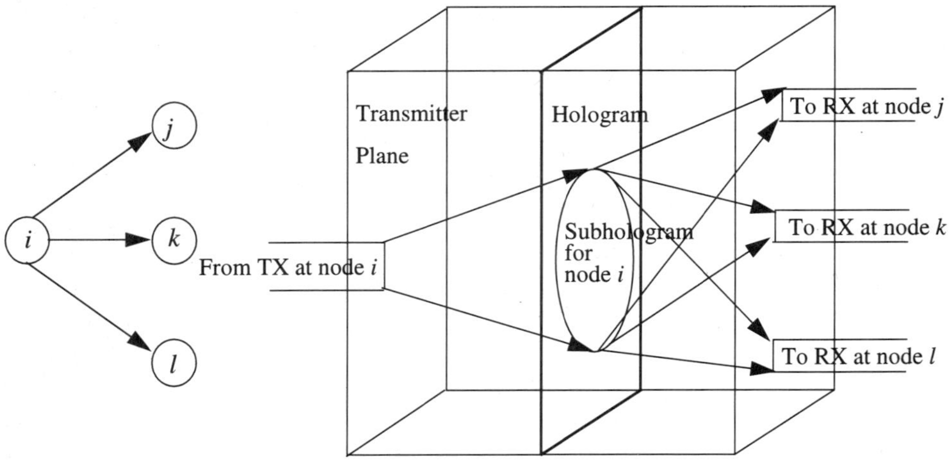

FIGURE 6. Holographic implementation of a free-space interconnects.

Note that the transmitters and receivers operate at a single wavelength, hence there is no wavelength limitation as in the WDM scheme. Also, if the degree of connectivity desired at each node is low, then these source-destination patterns for several topologies may be recorded on a single hologram. Thus, as all the desired connectivity could be established by the subhologram for the node, no reconfiguration overhead will be incurred. However, the transmitter and receiver power limitations may limit fan-in and fonout nodes.

Consequently, if we assume that recording all source-destination patterns take the equivalent of a bus cycle in an APW, the communication power of a free-space interconnection network would be equivalent to that of an APW. Hence, all the algorithm developed on an APW can be easily mapped to a free-space interconnect with the same complexity.

5. Emulation of The PRAM on APW

Most parallel algorithms in the literature are designed to run on a PRAM. PRAMs are very convenient for expressing parallel algorithms since one may concentrate on decomposing the problem at hand into simultaneously executable tasks, without having to worry about the communication between these tasks. For this reason, the implementation of certain data movement operations that will enable realistic parallel architectures to emulate a PRAM has been considered by many researchers in order to take advantage of all the parallel algorithms that have been written for the PRAM. These data movement operations are random access read (RAR) and random access write (RAW), also known as concurrent read and concurrent write, respectively [21], [22], [27]. These data operations are used to allow a given parallel architecture to emulate the concurrent read and the concurrent write capabilities of a CRCW PRAM. In case of multiple writes, only one processor succeeds according to some tie-breaking scheme such as minimum data value.

These two data movement operations are implemented using well-defined routines (sub-algorithms). We will analyze the time complexity of each of these routines on the 2-D APW to find the time complexities of RAR and RAW operations when performed on the 2-D APW. These routines are sorting, compression, ranking, distribution, and generalization [22], [27].

5.1. Sorting on APW

Now we will identify the time complexity of sorting on the 2-D APW which is based on the *rotate-sort* algorithm. Rotate-sort is a row-column sorting technique proposed originally for the two-dimensional mesh processor arrays [19]. As this paper is not particularly concerned with 2-D mesh arrays, the main interest is in rotate-sort as it applies to 2-D arrays of numbers; more precisely, the fact that rotate-sort partitions a set of N numbers into subsets (rows and columns), which can be sorted independently and efficiently on a 2-D APW while using just 1-D APW routing methods.

Given $N = RS$ elements arranged as a 2-D $R \times S$ array, the rotate-sort technique [19] sorts the N elements by alternately transforming the rows and columns of the array. The number of row and column phases is constant (14 or 16 phases). Each transformation phase consists of either performing a circular shift operation on the elements of each row or each column.

During rotate-sort, the $R \times S$ array of elements is partitioned as shown in Figure 7. A *vertical strip* (respectively, *horizontal strip*) is an $R \times S^{1/2}$ (respectively, $S^{1/2} \times S$) subarray of elements. Also, a *block* is a $S^{1/2} \times S$ subarray of elements. The algorithm as presented in [19] assumes that $R = 2^r$ and $S = 2^s$, where s is an even integer and $r \geq s/2$. However, other values of R and S can be used with little modification to the algorithm. The algorithm description can be facilitated by defining the following macros (operations):

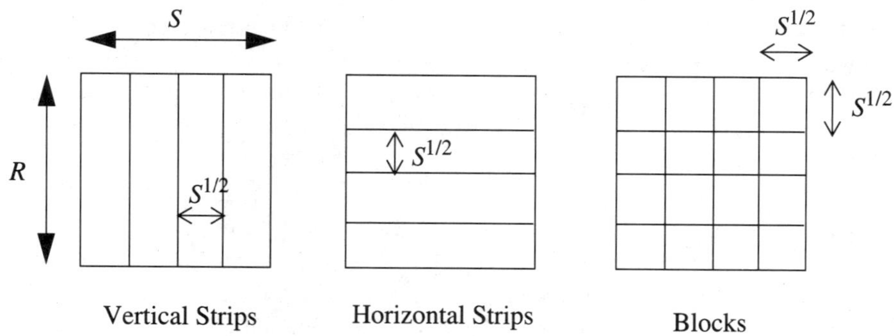

FIGURE 7. Partitions of the $R \times S$ array.

- Macro BALANCE: applies to a subarray of size $u \times v$ and consists of the following three steps:
  a. Sort all columns downward.
  b. Rotate each row i rightward by (i mod v) positions.
  c. Sort all columns downward.
- Macro UNBLOCK: distributes the elements of each block among all columns. It consists of the following two steps:
  a. Rotate each row i rightward by ($iS^{1/2}$ mod S) positions.
  b. Sort all columns downward.
- Macro SHEAR: equivalent to performing one iteration of the *shear-sort* algorithm [1].

It consists of the following two steps:

a. Sort all even-numbered rows rightward and all odd-numbered rows leftward.

b. Sort all columns downward.

The rotate-sort algorithm can be described in terms of these macros as follows:

Algorithm Rotate-Sort:

1) Perform BALANCE on each vertical slice.
2) Perform UNBLOCK on the entire array.
3) Perform BALANCE on each horizontal slice.
4) Perform UNBLOCK on the entire array.
5) Perform three iterations of SHEAR on the array.
6) Sort all rows rightward.

Following the applications of the above macros, the $R \times S$ array will be sorted in row-major order. It is clear that the above algorithm uses eight column phases and nine row phases, for a total of 17 phases. However, since Step 3c) (i.e. step (c) of the second BALANCE operation) and Step 4a) both involve row transformations, the two steps can be combined into one row transformations, thus reducing the total number of phases to 16. Furthermore, when $R \leq S^{1/2}$, then the number of phases can be reduced to 14.

Now, let us examine the time complexity of this algorithm when implemented on a 2-D APW. The BALANCE macro consists of two sorting steps along the columns of the 2-D APW. That is, we need to perform two sorting steps on a 1-D APW. It was shown previously that 1-D APW can emulate an MPC model in a single bus cycle through its *PERM* routing function. Further, it was proven by many researchers that an MPC model with N processor can sort N elements stored one element per processor in $O(\log(N))$ time [2], [6]. Thus, the two sorting steps of the BALANCE macro can be performed on a 2-D APW in $O(\log(N))$ bus cycles by directly emulating the MPC sorting algorithm. The second step needed in the BALANCE macro is the rotation of each row i by (i mod v) positions. This can be easily accomplished in one bus cycle on the rows of the 2-D APW by setting the control function *wait* as follows on all the processors:

$$wait(i) = i \bmod v.$$

Hence, the BALANCE macro can be executed on a 2-D APW in $O(\log(N))$ time. The UNBLOCK macro consists of one sorting step along the columns which takes $O(\log(N))$ bus cycles as shown above, and one rotation step of all the rows rightward which takes a single bus cycle as shown previously. Therefore, the UNBLOCK macro can be executed on a 2-D APW in $O(\log(N))$ time. Finally the SHEAR macro which consists of 2 sorting steps along the rows and along the columns can also be executed on a 2-D APW in $O(\log(N))$ time. Hence, the whole *Rotate-Sort Algorithm* can be executed on the 2-D APW in $O(\log(N))$. A step-by-step application of the *Rotate-Sort Algorithm* is shown in Figure 8.

This relatively fast execution time of sorting on the 2-D APW can be attributed to two factors. First, the capability of a 1-D APW to efficiently emulate an MPC computing model. Second, the suitability of the rotate-sort algorithm to the 2-D APW, that is, the independence of operations on the rows and columns which eliminated the usage of the relay function for the 2-D APW, and simply using the efficient and easy routing functions of the 1-D APW.

5.2. RAR and RAW Time Complexity

Here we develop the time complexities of the other sub-algorithms namely compression,

10	9	14	2
4	15	11	12
6	1	5	13
8	3	7	0

(a)

4	1	5	0
6	3	7	2
8	9	11	12
10	15	14	13

(b)

0	1	4	5
2	3	6	7
8	9	11	12
10	13	14	15

(c)

0	1	4	5
6	7	2	3
8	9	11	12
14	15	10	13

(d)

0	1	2	3
6	7	4	5
8	9	10	12
14	15	11	13

(e)

0	1	2	3
5	6	7	4
10	11	8	9
15	12	13	14

(f)

0	1	2	3
7	4	5	6
10	11	8	9
13	14	15	12

(g)

0	1	2	3
7	6	5	4
8	9	10	11
15	14	13	12

(h)

0	1	2	3
4	5	6	7
8	9	10	11
12	13	14	15

(i)

FIGURE 8. Application of the *Rotate-Sort Algorithm* on a 4 x 4 array of elements on a 2-D APW.

ranking, distribution, and generalization [22], [27] which, when added to the time complexity of *Rotate-Sort Algorithm*, would give us the time complexities of RAR and RAW on the 2-D APW. The above sub-algorithms are based on the class of *ascend* algorithms [22], [27]. An algorithm is said to be in the *ascend* class if it performs a sequence of operations on pairs of data that are successively 2^0, 2^1, ..., 2^{k-1} locations apart on a problem of size 2^k [31].

An algorithm of size $N = 2^k$ which is in the *ascend* class can be performed on an 2-D APW in the following manner (assuming that the 2-D APW is a square array of nodes):

1. Perform operations on pairs of data that are successively 2^0, 2^1, ..., $2^{k/2-1}$ locations apart (operations along the rows of the 2-D APW).
2. Perform operations on pairs of data that are successively $2^{k/2}$, $2^{k/2+1}$, ..., 2^{k-1} locations apart (operations along the columns of the 2-D APW).

Thus, it takes $h = \log(N)$ bus cycles to perform the above two steps on a 2-D APW. For example, at the *m*th bus cycle, a message is sent from each node to its neighbor which is at a distance 2^{m-1}. To accomplish this, the time that a node *i* has to wait during the *m*th bus cycle before receiving a message from its neighbor is

$$wait_h(i) = \pm 2^{h-1}.$$

Now, let us present a brief introduction of these sub-algorithms which are used in the implementation of RAR and RAW [22], [27]:

5.2.1. Ranking. When some PE's are selected in a network, the rank of a PE is equal to the number of selected PE's with a smaller index. The number of communication steps needed by this algorithm is equal to the number of communication steps needed by an *ascend* algorithm which is equal to log(N) for a 2-D APW. The number of computation steps is 2log(N) on the 2-D APW.

5.2.2. Compression. When the number of active PE's in a network is s, a proper subset of all the PE's in the network applying the compression algorithm on these active elements will move these active elements to the PE's indexed 0, 1, 2, ..., s-1. The number of communication steps needed by this algorithm is exactly equal to the number of communication steps needed by the *ascend* algorithm. It has no computation steps. thus, when implemented on a 2-D APW, it needs log(N) communication steps.

5.2.3. Distribution. Assume that some PE's m of a network each have a datum d_m and a PE destination h_m such that if $i < j$ then $h_i < h_j$. Executing the distribution algorithm on the network consists of routing, for each PE m the datum d_m to the PE h_m. The distribution algorithm is the inverse of the compression algorithm and requires the same number of communication steps.

5.2.4. Generalization. Given a set of PE's m of a network each has a datum d_m and a PE destination h_m such that if $i < j$ then $h_i < h_j$. The generalization algorithm consists of routing multiple copies of the datum d_m to the destination $h_{m-1} + 1$ through h_m. The number of communication steps is equal to that needed by the *ascend* algorithm which is log(N) steps for the 2-D APW. The number of computation steps needed is 3log(N).

Hence, compression, ranking, distribution, and generalization can each be executed on the 2-D APW in O(log(N)) time. A RAR is performed by executing the sorting sub-algorithm twice, the ranking sub-algorithm once, the compression sub-algorithm twice, the distribution sub-algorithm once, and the generalization sub-algorithm once [22], [27]. Thus, to perform a RAR operation, a 2-D APW requires O(log(N)) time. A RAW operation is performed by executing the sorting sub-algorithm once, the ranking sub-algorithm once, the compression sub-algorithm once, and the distribution sub-algorithm once [22], [27]. Thus, to perform a RAW operation, a 2-D APW requires O(log(N)) time. Hence, a 2-D APW of size N can emulate a CRCW PRAM of the same size with at most O(log(N)) degradation in time performance. This also means that O(log(N)) is an upper bound on the time needed for the APW to emulate arbitrary interconnection networks of the same size. Thus, in some sense the APW can be considered as a universal network [16].

6. Emulation of PRAM on a Larger APW

In this section we show how to emulate a CRCW PRAM model with N processors on an APW with N^2 in $O(1)$ time. The implementations of RAR and RAW data movement operations are slightly different for those implemented in the previous Section [21]. The RAW and RAW operations involve two sets of PEs, the *sources* and *destinations*. Source PEs send a fixed number of records, each consisting of a key and one or more data parts (a record may also be null). Destination PEs receive a fixed number of records sent by the source PEs. In order to maintain consistency during concurrent read and concurrent write operations, it will be assumed that there is at most one master record, stored in some processor, associated with each unique key.

In a RAR, every source PE creates no more than some fixed number of source records, where each source record consists of a key, the data value associated with the key, and some bookkeeping information. The purpose of the source records in the RAR is to make available the data value associated with every unique key to any processor to any processor that may want to read it. Each destination PE creates no more than some fixed number of destination (*request*) records, each of which specifies the key associated with a data value it wishes to receive. Several destination records can request a data value associated with the same key. A destination PE may create a destination record for which there is no source record, in which case it receives a null message. An implementation of a RAR in terms of fundamental data movement operations on an APW of size N^2 is given below.

Random Access Read

1. All PEs create the same fixed number, call it S, of source records corresponding to those keys for which the PE is responsible. Each record contains the key, associated data value, and ID of the PE creating the record. Some or all of the source records created by a PE may be dummy records, so as to balance the number of items per processor in subsequent sort steps. Notice that since there is only one master record maintained for each key, a key value may be represented by at most 1 source record somewhere in the mesh.
2. All PEs create the same fixed number, call it D, of destination (request) records with the desired key and the ID of the PE that created the record. Some or all destination records created may be dummy records, so as to balance the number of items per processor in subsequent sort steps.
3. Sort all $(D + S)N$ source and destination records together by key, where ties are broken in favor of source records, and ties between requests are broken arbitrarily.
4. By performing a broadcast operation within ordered intervals with respect to keys, every source record will inform all destination records that requested the value associated with its keys as to that value.
5. Sort the $(D + S)N$ source and destination records so that they are returned to the PE that originated the request. Notice that the source records are not conceptionally needed for the sort step, but are used so as to balance the number of items in each PE during this step. The source records are discarded after the source is complete.

The RAR is accomplished through a fixed number of sort and interval operations, and for fixed constants D and S. Obviously, the interval operations can be performed on an APW in $O(1)$ bus cycles using the *BROADCAST* and *PERM* operations presented previously. Later in this section we will show how to perform sorting of $O(N)$ elements on an APW of size N^2 in $O(1)$ time. Hence, the time complexity to perform RAR of N elements on an APW of size N^2 takes $O(1)$ time.

In RAW, every destination PE creates no more than some fixed number of destination records, consisting of a key and some bookkeeping information, for each of the master records that it maintains and is willing to receive an updated value for. At the end of the RAW, a destination PE will receive a record corresponding to each of the destination records it created, indicating the new value to be associated with that key. Each source PE creates no more than some fixed number of source records, each consisting of a key, a data value, and some bookkeeping information. If two or more source PEs send records in an attempt to update the data value associated with the same key, then the destination PE associated with that key, if one exists, will receive a record containing the minimum such source value. An implementation of a RAW in terms of fundamental data movement operations on an APW of size N^2 is given below.

Random Access Write

1. All PEs create the same fixed number, call it D, of destination records, corresponding to those master records for which the PE is willing to receive data values. Each record contains a key and the ID of the PE that created the record. Some or all of the destination records created by a PE may be dummy records, so as to balance the number of items per processor in subsequent steps.
2. All PEs create the same fixed number, call it S, of source records. Each record contains a key, associated data value, and the ID of the PE creating the record. SOme or all source records created by a PE may be dummy records, so as to balance the number of items per processor in subsequent sort steps.
3. Sort all SN source records by key, breaking ties of the same key arbitrarily.
4. Apply RAW tie-breaking mechanism within the ordered intervals. The tie breaker could be the minimum value which can be computed in $O(1)$ time in an APW using the inverse of the *BROADCAST* operation.
5. Sort all $(S + D)N$ source and destination records together by key, where ties are broken in favor of destination records, and ties between source records are broken in favor of the representative record.
6. All destination records obtain their updated value from their representative source record which is stored in the succeeding PE.
7. Sort all $(S + D)N$ records in such a way that the destination records take the required data back to the PE responsible for them. Notice that the source records are not conceptionally needed for the sort step, but are used so as to balance the number of items in each PE during this step. The source records are discarded after the sort is complete.

Like the RAR, the RAW is accomplished through a fixed number of sort and interval operations. The interval operation is accomplished in $O(1)$ time on an APW using the *BROADCAST* and the *PERM* operation. The sorting operation is accomplished in $O(1)$ time on an APW os size N^2, where the number of items to be sorted is $O(N)$. This sorting procedure is given next.

6.1 Sorting N Elements on an APW of Size N^2

In this section, an algorithm is presented for sorting N numbers on an $N \times N$ APW with optical delay lines. The algorithm takes 6 steps to complete. It is based on the idea of enumeration sort [1, 26]. The $N \times N$ 2-D APW used for sorting is similar to that of Figure 4 except each row is augmented with conditional delays. Let $P(x, y)$, $0 \leq x, y < N$, be a processor at row x and column y. The following algorithm *SORT* sorts N data item, d_x, $0 \leq x < N$, using a constant number of bus cycles. *SORT* uses another algorithm, denoted *BINARY-SUM*, to compute the rank of each data d_x in one cycle time. This algorithm is based on the work presented in [9].

Algorithm SORT

Initially, each processor $P(x, 0)$ holds a data item d_x, $0 \leq x < N$. At the completion of the algorithm, the sorted data are stored in column 0 such that if $P(x, 0)$ holds d_i and $P(x + 1, 0)$ holds d_j, then $d_i \leq d_j$.

1. $P(x, 0)$ broadcasts d_x in row x.
2. $P(x, x)$ broadcasts d_x in column x.
3. Each $P(x, y)$ sets $a_{x,y} = 1$, if $d_y < d_x$, or $d_y = d_x$ and $y < x$. It sets $a_{x,y} = 0$, otherwise.
4. Compute sum $\sigma_x = \sum_{y=0}^{n-1} a_{x,y}$ in each row x using the algorithm

BINARY-SUM given below. The result σ_x is stored at $P(x, \sigma_x)$.

5. $P(x, \sigma_x)$ sends d_x to $P(\sigma_x, \sigma_x)$.
6. $P(x, x)$ sends d_x to $P(x, 0)$.

The above sorting algorithm sort N elements on an $N \times N$ 2-D APW in 7 steps.

Next, we show how to implement the *BINARY-SUM* algorithm on an APW with conditional delays in just one bus cycle. The *BINARY-SUM* algorithm uses conditional delays on an APW. Here it is shown how they can be used for summing a binary sequence. The conditional delays can be implemented using 2×2 optical switches, as shown in Figure 9. This system is the same as the one presented earlier for an APW except that an optical switch $S(i)$ is inserted on the select bus at each processor $P(i)$, and each $S(i)$ is controlled by processor $P(i)$. Thus, the system is equivalent to the one presented previously if all the switches are set to straight. In this case, an optical signal incurs the same propagation delay on both the reference bus and the select bus between any two processors i and j in the two systems. However, in the system of Figure 9, an additional time delay equal to Δ can be introduced on the select bus by setting $S(i)$ to cross. In other words, when $S(i)$ is set to straight, it takes a time τ for an optical signal to propagate from $P(i)$ to $P(i+1)$ on the select bus, while when $S(i)$ is set to cross, such propagation will take a time $\tau + \Delta$.

In the following, the algorithm for computing the *BINARY-SUM* on the above system is presented. In this algorithm, it is assumed that the binary data are initially stored in the processors with processor $P(i)$ holding a_i, and all the optical switches are set to cross. Then, a reference pulse and a select pulse are inserted simultaneously on the reference bus and select bus, respectively, by processor $P(0)$.

If $\sigma = \sum_{i=0}^{n-1} a_i$, then σ delays (each delay equal to Δ) will be removed from the select bus such that the reference pulse and the select pulse coincide at processor P(σ).

Algorithm BINARY-SUM

Input: a binary sequence $a_i = 0$ or 1. Initially a_i is stored at $P(i)$.

Output: $\sigma = \sum_{i=0}^{n-1} a_i$.

1) $P(i)$ sets $S(i)$ to *straight* if $a_i = 1$, *cross* if $a_i = 0$.
2) At time 0, $P(0)$ injects a reference and a select pulse signals. If $P(j)$ is selected, then the sum is $\sigma = j$. That is, the index j of the processor that sees the coincidence of the reference pulse and the select pulse gives the sum σ.

Hence, It takes just one bus cycle to find the binary sum.

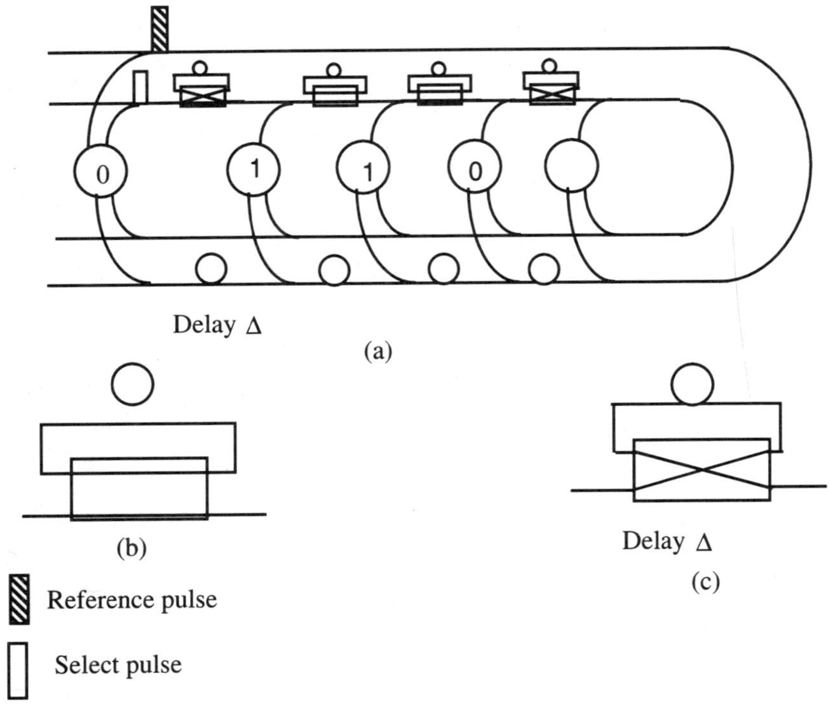

FIGURE 9. An optical delay line implemented with 2×2 optical switches. The propagation delay of an optical signal from $P(i)$ to $P(i+1)$ on the select bus is τ if the switch $S(i)$ is set to *straight* as shown in (b), and $\tau + \Delta$ if $S(i)$ is set to *cross* as shown in (c).

7. Conclusion

We have presented a new model of parallel computation using optical technology, APW, for the construction of large scale parallel distributed computer systems. The communication capability of this novel architecture has been shown using the notion of pipelining. Further, we have shown that this communication capability can be achieved using other techniques such as wavelength division multiplexing. An efficient sorting algorithm is shown for the APW which has *only* $O(\log(N))$ time complexity by taking advantage of the pipelining properties of the APW and using a constant number of column and row sorting phases well suited for such a layout. Further, an APW has been shown to emulate deterministically the CRCW PRAM model with only $O(\log(N))$ degradation in time performance using well defined sub-algorithms including the sorting sub-algorithm. Moreover, we have shown that an APW can emulate a CRCW PRAM in constant time if we allow the size of the APW to be larger than that of the emulated PRAM. APW appears to offer particularly good potential as an interconnection network for systems which emulate the PRAM models, and would make programming parallel computers more appealing. One future direction of this work is exploit probabilistic emulation techniques which typically take much less time that the deterministic techniques especially in a communication-efficient interconnection network such as APW.

ACKNOWLEDMENTS

I would like to thank Prof. Frank Hsu for his invitation to present some of this work at the DIMACS Workshop on Interconnection Networks and Mapping and Scheduling Parallel Computations. Also, I would like to thank Rami Melhem, Zicheng Guo, and Yi Pan for useful discussions of many aspects of this work.

REFERENCES

1. S. G. Akl, *The Design and Analysis of Parallel Algorithms*, Prentice Hall, Englewood Cliffs, NJ, 1989.
2. H. Alt, T. Hagerup, K. Mehlhorn, and F. Preparata, *Deterministic simulation of idealized parallel computers on more realistic ones*, SIAM J. Comput. **16** (1987), pp. 808-835.
3. A. Borodin, and J. Hopcroft, *Routing, merging, and sorting in parallel models of computation*, Proceedings ACM Symp. Theory Comput., pp. 338-344, 1982.
4. Y. C. Chen, W. T. Chen, and J. P. Sheu, *Designing efficient parallel algorithms on mesh-connected computers with multiple broadcasting*, IEEE Transactions on Parallel and Distributed Systems **1**, Vol. 1, pp. 241-245, 1990.
5. D. M. Chiarulli, S. P. Levitan, and R. G. Melhem, *Optical bus control for distributed multiprocessors*, Journal of Parallel and Distributed Computing **10** (1990), pp. 45-54.
6. R. Cole, *Parallel merge sort*, SIAM J. Comput. **17** (1988), pp. 770-785.
7. R. Cypher and C. G. Plaxton, *Deterministic sorting in nearly logarithmic time on the hypercube and related computers*, J. Computer and System Sciences **47** (1993), pp. 501-548.
8. S. Fortune and J. Wyllie, *Parallelism in random access machines*, Proceedings ACM Symp. Theory Comput., pp. 114-118, 1978.
9. Z. Guo, *Optically Interconnected Multiprocessor Architectures*, Ph.D. Dissertation, Department of Electrical Engineering, University of Pittsburgh, 1990.
10. M. Hamdi, *Communication-efficient interconnection networks for parallel computations*, Ph. D. Dissertation, Department of Electrical Engineering, University of Pittsburgh, 1991.
11. M. Hamdi and Y. Pan, *Efficient parallel algorithms for some numerical problems on arrays with pipelined optical buses*, 5th Australian Supercomputing Conference, 1992, pp. 205-214.
12. M. Hamdi, *Efficient parallel algorithms on optically interconnected computer systems*, IEE Proceedings - E, to appear (Feb. 1995).
13. C. P. Kruskal, L. Rudolph and M. Snir, *A Complexity theory of efficient parallel algorithms*, Theoretical Computer Science **71** (1990), pp. 95-132.
14. F. Luccio and G. Pucci, *An efficient probabilistic simulation of PRAMs in VLSI*, Information Processing Letters (1988), pp. 141-146.
15. C. E. Leiserson and B. M. Maggs, *Communication-efficient parallel algorithms for distributed random-access machines*, Algorithmica (1988), pp. 53-77.
16. C. E. Leiserson, *Fat-Trees: Universal networks for hardware efficient supercomputing*, IEEE Trans. Comput. **34** (1985), pp. 892-901, 1985.
17. F. T. Leighton. *Parallel Algorithms And Architectures: Arrays, Trees, and Hypercubes*. Morgan Kaufmann, 1992.

18. F. Luccio, A. Pietracaprina, and G. Pucci, *A new scheme for the deterministic simulation of PRAMs in VLSI*, Algorithmica **5** (1990), pp. 529-544, 1990.

19. J. M. Maberg and E. Gafni, *Sorting in constant number of row and column phases in a mesh*, Algorithmica (1988), pp. 561-572.

20. R. G. Melhem, D. M. Chiarulli, and S. P. Levitan, *Space multiplexing of waveguides in optically interconnected multiprocessor systems*, The Computer Journal **32** (1989), pp. 362-369, 1989.

21. R. Miller and Q. F. Stout, *Parallel Algorithms for Regular Architectures*, MIT Press, Cambridge, MA, 1990.

22. D. Nassimi and S. Sahni, *Data broadcasting in SIMD computers,* IEEE Trans. Comput. **30** (1981), pp. 101-107, 1981.

23. S. Olariu, J. L. Schwing, and J. Zhang, *On the power of two-dimensional processor arrays with reconfigurable bus systems*, Parallel Processing Letters **1** (1991), pp. 29-34, 1991.

24. Y. Pan and M. Hamdi, *Computation of the singular value decomposition on arrays with pipelined optical buses*, 1993 Symposium on Applied Computing, pp. 525-535, 1993.

25. T. M. Pinkston, *The GLORI strategy for multiprocessors: Integrating optics into the interconnect architecture*, CSL-TR-92-552, Stanford University, 1992.

26. Z. Paguo, *Sorting On Array Processors With Pipelined Buses*, 1992 International Conference on Parallel Processing, pp. 289-294, 1992.

27. S. Ranka and S. Sahni, *Hypercube Algorithms With Applications to Image Processing and Pattern Recognition*, Springer Verlag, New York, 1990.

28. C. Qiao and R. G. Melhem, *Time-division optical communications in multiprocessor arrays*, IEEE Trans. Computers **42** (1993), pp. 577-590.

29. Q. F. Stout and B. Wager, *Intensive hypercube communication: Prearranged communication in link-bound machines*, Journal of Parallel and Distributed Computing **10** (1990), pp. 167-181.

30. E. Upfal and A. Wigderson, *How to share memory in a distributed system*, J. ACM **34** (1987), pp. 116-127.

31. F. P. Preparata and J. Vuillemin, *The cube-connected cycles: A versatile network for parallel computations*, Commun. ACM **24** (1981), pp. 300-309, 1981.

DEPARTMENT OF COMPUTER SCIENCE, HONG KONG UNIVERSITY OF SCIENCE AND TECHNOLOGY, CLEAR WATER BAY, KOWLOON, HONG KONG

E-mail address: hamdi@cs.ust.hk

DIMACS Series in Discrete Mathematics
and Theoretical Computer Science
Volume **21**, 1995

FAULT-TOLERANT KAUTZ NETWORKS

RABAH HARBANE

ABSTRACT. Given a graph G_0 with n nodes, we say that a graph G_k with $n + k$ nodes is a k-fault-tolerant version of G_0, if G_k contains a subgraph isomorphic to G_0 after removal of any k nodes. Thus if G_0 models an interconnection network, then G_k represents an interconnection network capable of tolerating k-faults and still emulating G_0 without any performance degradation. In this paper, we will consider the case where G_0 is the Kautz graph $UK(d, D)$. For any integers d, D and k, we define a k-fault-tolerantgraph with maximum degree in $O(kd)$.

1. INTRODUCTION

Parallel machines are built with an underlying interconnection network of processors. The hypercube has been used for these networks (NCUBE, iPSC/860, ...). The hypercube has many good properties (symmetry of the topology, easy routing,...). However, its degree grows logarithmically with the number of nodes. Because technological considerations limit the degree of a node, there is a limit to the number of processors which can be connected in a hypercube topology. Several constant degree topologies have been proposed, which have many of the good properties of the hypercube. Most notable among these are de Bruijn and Kautz topologies [3]. In this paper, we focus on the Kautz networks which have a large number of nodes for given degree and diameter. As parallel computers with large number of processors are constructed, it becomes necessary to use these machines even when several components have failed. For this purpose a large amount of research has been devoted for proposing fault-tolerant parallel architectures. The techniques used in this research can be divided into two main classes. The first class (which is not considered here) consists of techniques which do not add redundancy to the desired architecture. Instead, these techniques attempt to mask the effects of faults by using the healthy part of the architecture to simulate the entire machine [2, 6, 10, 14]. These techniques do not cost anything for adding fault-tolerance but they can induce a significant degradation in performance (slowdown caused by the simulation of multiple nodes by a single one). The second class consists of techniques which do add redundancy to the desired architecture. These techniques attempt to isolate the faults, while maintaining the complete desired architecture. The goal is

1991 *Mathematics Subject Classification.* Primary 68, 94.
Key words and phrases. Fault-tolerant graph, subgraph, Kautz graph, bus implementation.
The author thanks M-C. Heydemann, D. Sotteau and P. Tvrdik for suggesting many improvements in the presentation of the paper.

to maintain the full performance of the desired architecture while minimizing the cost of the redundant components. One general technique for adding redundancy is based on a graph theoretic model of fault-tolerance [11]. In this model, the desired architecture is viewed as a graph (called *target* graph) in which the nodes represent the facilities and the edges represent the communication links between facilities. A *k-fault-tolerant graph* is created such that, after the removal of k faulty nodes, the target graph is still present as a subgraph. This technique yields fault-tolerant networks that can tolerate both node faults and edge faults (by viewing a node incident with the faulty edge as being faulty) and allows to run algorithms designed for the target network without any slowdown. Unfortunately, the degree of the fault-tolerant graph created with this model can be prohibitively large. For example, for any target graph on n nodes with maximum degree Δ, we can easily construct a k-fault-tolerant graph on $n+k$ nodes with maximum degree in $O(k^2\Delta)$. Adding exactly k nodes is minimal with respect to the number of nodes. Hence, minimizing the cost in this model amounts to constructing a fault-tolerant graph with a small maximum degree. A number of researchers have applied this graph model of fault-tolerance to various target graphs. Initially, Hayes [11] has used this model with cycles, linear arrays and trees as target graphs. In [8, 9] the authors studied the case of nonhomogeneous trees and the hypercube. Bruck and al. have used de Bruijn graphs, shuffle-exchange graphs [4], meshes and hypercubes [5] as target graphs. The main contribution of this paper is the creation of fault-tolerant Kautz graphs. We present a construction for fault-tolerant Kautz networks similar to the one given by Bruck and al. for fault-tolerant de Bruijn networks in [4]. Also, we extend the construction for generalized Kautz and de Bruijn graphs. However, these fault-tolerant graphs are relatively dense and thus not optimal. We show how buses can be used to reduce the degree of the fault-tolerant graphs constructed by almost a factor of 2. The paper is organised as follows. Definitions that are used throughout the paper are given in Section 2. In Section 3, we give lower and upper trivial bounds on the number of edges of the fault-tolerant graphs. Section 4 gives a simple scheme for creating k-fault-tolerant graphs for any arbitrary target graph. Fault-tolerant Kautz graphs and fault-tolerant generalized Kautz and de Bruijn graphs are defined in Section 5. In Section 6, we show how buses allow to create fault-tolerant architecture with small degree. Section 7 concludes the paper.

2. Definitions and notation

The following definitions will be used throughout this paper. Given an ordered set S of integers and x an element of S, the *rank* of x in S, denoted $rank(x, S)$, is the number of elements in S that are smaller than x. Let G be a graph and let $V(G)$ and $E(G)$ denote the vertex set and edge set of G, respectively. The degree of vertex $x \in V(G)$, denoted $d(x)$, is the number of edges incident to x. The minimum degree δ of G is $\delta = \min_{x\in V(G)} d(x)$ and the maximum degree Δ of G is $\Delta = \max_{x\in V(G)} d(x)$. If $F \subset V(G)$ then $G \setminus F$ is the graph obtained from G by deleting the vertices of F together with the edges incident to them. Let k be a nonnegative integer and let G, H be two graphs . We say that the graph G is *k-fault-tolerant graph* with respect to H, denoted *(k,H)-tolerant,* if the subgraph of G induced by any set of $|V(G)| - k$ nodes contains H as a subgraph. The graph H will be called the *target* graph and G the *supergraph* of H. Throughout this paper, we assume that the number of spare nodes is minimal, i.e., $|V(G)| = |V(H)| + k$. Let G_1, G_2 be two graphs. The *join* of G_1 and G_2, denoted $G_1 * G_2$, is the graph J such that $V(J) = V(G_1) \cup V(G_2)$, $E(J) = E(G_1) \cup E(G_2) \cup \{(u, v); u \in V(G_1), v \in V(G_2)\}$.

De Bruijn and Kautz graphs. De Bruijn and Kautz graphs can be defined in three different but equivalent ways: as graphs on an alphabet [7, 13], as iterated line digraphs of complete digraphs [1], and using arithmetic congruences [15, 12]. In this paper, we only use the arithmetic definition. Because it allows to define these graphs for any arbitrary number of vertices. The generalized de Bruijn graph, denoted $RPK_{d,n}$, is defined in [15] as follows. The vertex set of $RPK_{d,n}$ is the set of integers modulo n. The pair (x, y) is an edge if and only if there exists an integer $r \in \{0, 1, 2, \dots, d-1\}$ such that : $y = (dx + r) \mod n$ or $x = (dy + r) \mod n$. The graph $RPK_{d,n}$ has minimum degree $2d-2$ and maximum degree $2d$. Note that for $n = d^D$, we obtain the de Bruijn graph $UB(d, D)$. In [12], the authors defined the generalized Kautz graphs, denoted $II_{d,n}$ as follows. The vertex set of $II_{d,n}$ is the set of integers modulo n. The pair (x, y) is an edge if and only if there exists an integer $r \in \{1, 2, \dots, d\}$ such that : $y = (-dx-r) \mod n$ or $x = (-dy-r) \mod n$. The graph $II_{d,n}$ has minimum degree $2d-1$ and maximum degree $2d$. Note that if $n = d^D + d^{D-1}$, we obtain the Kautz graph $UK(d, D)$. Figure 1 shows the graph $UK(2, 3)$.

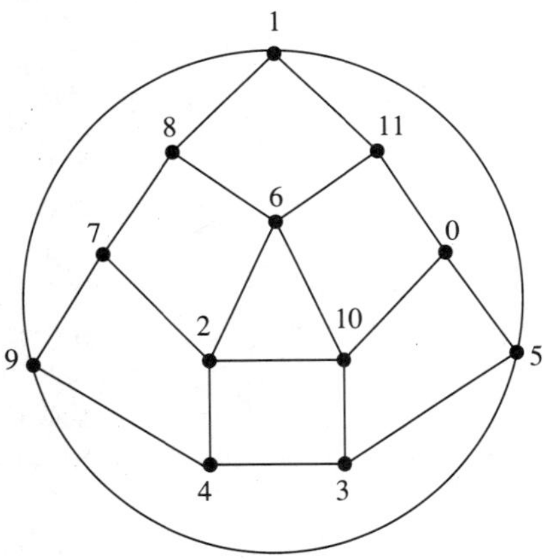

FIGURE 1. An example of Kautz graph $UK(2, 3)$

3. PRELIMINARY RESULTS

In this section, we give a lower bound and an upper bound on the number of edges of any fault-tolerant graph.

Proposition 3.1. *Let H be a graph with minimum degree δ. If G is (k,H)-tolerant, then* $|E(G)| \geq \lceil \frac{(\delta+k)(|V(H)|+k)}{2} \rceil$.

Proof. Since by definition $|V(G)| = |V(H)| + k$, it is sufficient to prove that the degree of every node in $V(G)$ is at least $\delta + k$. Suppose $d(x) \leq \delta + k - 1$ for some node x of $V(G)$. There exists a k-fault F in G which contains all the k neighbors of x but not x itself. The graph $G \setminus F$ must contain a subgraph T isomorphic to H.

Therefore x must be a node of T which is impossible since $d(x) = \delta - 1$ in $G \setminus F$. Hence $d(x) \geq \delta + k$ for every x in $V(G)$.

The aim is to built k-fault-tolerant graphs which have a number of edges not far from the lower bound. Unfortunately, there exists only a few target graphs for which the lower bound of Proposition 1 is reached. This is trivially the case if the target graph is the complete graph. It is also the case if the target graph is a path or a cycle [11]. Obviously, for any graph H, the complete graph on $|V(H)|+k$ nodes is always a (k, H)-tolerant graph. But there exists (k, H)-tolerant graphs with less edges than the complete graph $K_{|V(H)|+k}$. In particular the graph $H * K_k$, join of the target graph H and K_k, the complete graph on k nodes, gives a better upper bound on the number of edges of a (k, H)-tolerant graphs.

Proposition 3.2. *For any graph H of maximum degree Δ, the graph $G = H * K_k$ is (k,H)-tolerant and* $|E(G)| \leq \lceil \frac{|V(H)|(\Delta+2k)+k(k-1)}{2} \rceil$.

Proof. Note that G can be viewed as two components H and K_k connected by $k|V(H)|$ edges. Let F be any set of k faults in G. Let F_1 be the part of failed nodes in the component H and F_2 the part of failed nodes in the component K_k such that $|F| = |F_1| + |F_2| = k$. If $|F_1| = 0$, then all the faults occur in K_k, and the remaining graph is exactly H. If $|F_1| > 0$, then the set of $|F_1|$ failed nodes in H is replaced by the set of $k - |F_2|$ non-faulty nodes of K_k, so by construction of G, the remaining graph contains H as a subgraph. Thus the join graph $G = H * K_k$ is always (k, H)-tolerant. Also by definition of G, $|E(G)| = |E(H)| + |E(K_k)| + k|V(H)| \leq \lceil \frac{|V(H)|(\Delta+2k)+k(k-1)}{2} \rceil$.

4. Trivial scheme for creating fault-tolerant graphs

In this section, we give a general trivial scheme for constructing k-fault-tolerant graphs. Let G_0 be a graph on n nodes with maximum degree Δ . We will construct G_k, a (k, G_0)-tolerant graph on $n + k$ nodes. Without loss of generality we can label the nodes of G_0 with integers from 0 to $n - 1$, those of G_k with integers from 0 to $n + k - 1$. This induces an obvious order on the nodes of G_k. The idea is to construct G_k in such a way that, for any subset F with $|F| = k$ of $V(G_k)$, there exists a mapping $\varphi : V(G_0) \longrightarrow V(G_k \setminus F)$, such that the graph induced by $\varphi(V(G_0))$ is a subgraph of G_k isomorphic to G_0. The sample way to define the mapping is to construct a monotonically increasing one-to-one function from $V(G_0)$ into $V(G_k \setminus F)$ as follows. For any node x of G_0, $\varphi(x)$ is defined as the $(x+1)^{st}$ non-faulty node of G_k. For example, $\varphi(0)$ is the first node of G_k not in F, $\varphi(n-1)$ is the last node of G_k not in F. More generally, for any node x of G_0, if F contains no node of G_k with label less than or equal to x, then $\varphi(x) = x$, if F contains one node with label less than or equal to x, then $\varphi(x) = x + 1$, and so on. So that $x \leq \varphi(x) \leq x + k$, since $|F| = k$. Therefore, to insure that, for any F and any edge (x, y) of G_0, $(\varphi(x), \varphi(y))$ is an edge of G_k, the $(k+1)^2$ edges between the two sets $S(x) = \{x, x+1, \ldots, x+k\}$ and $S(y) = \{y, y+1, \ldots, y+k\}$ have to be present in G_k. Figure 2 shows an example of graph which is $(1, UK(2,3))$-tolerant constructed by the above trivial scheme and Figure 3 shows the scheme of $UK(2,3)$ after one fault.

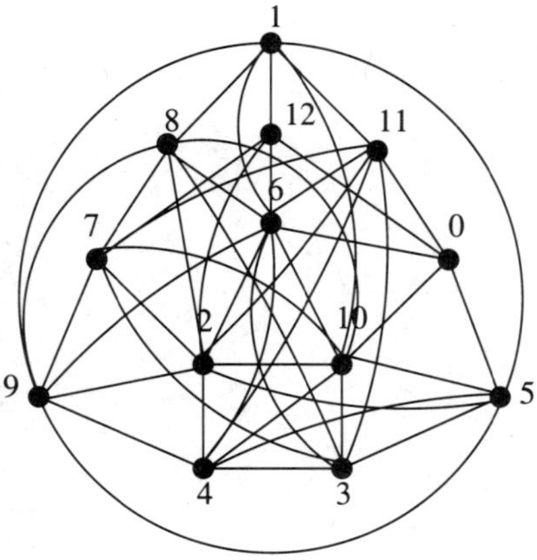

FIGURE 2. An example of graph which is $(1, UK(2,3))$-tolerant obtained by the trivial scheme.

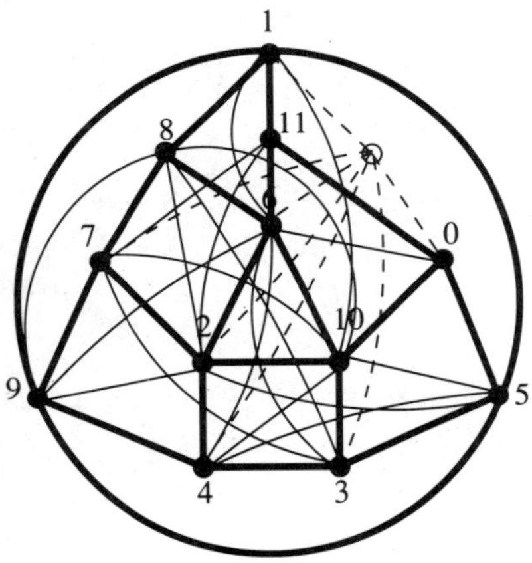

FIGURE 3. The scheme of $UK(2,3)$ after removal of one node in the graph of figure 2. The faulty node is shown as open circle and the edges of $UK(2,3)$ are shown with thick lines.

Theorem 4.1. *If the graph G_0 has maximum degree Δ, then the graph G_k constructed by the above trivial scheme is (k, G_0)-tolerant and has maximum degree in $O(k^2\Delta)$.*

Proof. We have to show that any graph obtained from G_k by deleting k arbitrary nodes contains a subgraph isomorphic to G_0. Let F be an arbitrary set of k nodes in G_k and let H be the subgraph of G_k induced by $G_k \setminus F$. Let us define the mapping $\varphi : V(G_0) \longrightarrow V(H)$ such that $rank(\varphi(x), H) = x$. Since $|F| = k$, we obviously have $\varphi(x) \in S(x)$. For any edge (x,y) in G_0, we have $\varphi(x) \in S(x)$ and $\varphi(y) \in S(y)$. Since, by construction of G_k there are all the edges between the sets $S(x)$ and $S(y)$. So, $(\varphi(x), \varphi(y))$ is an edge in H. On the other hand, each edge $(x,y) \in E(G_0)$ induces at most $(k+1)^2$ edges in G_k, so, if G_0 has maximum degree Δ, then the graph G_k will have maximum degree in $O(k^2\Delta)$.

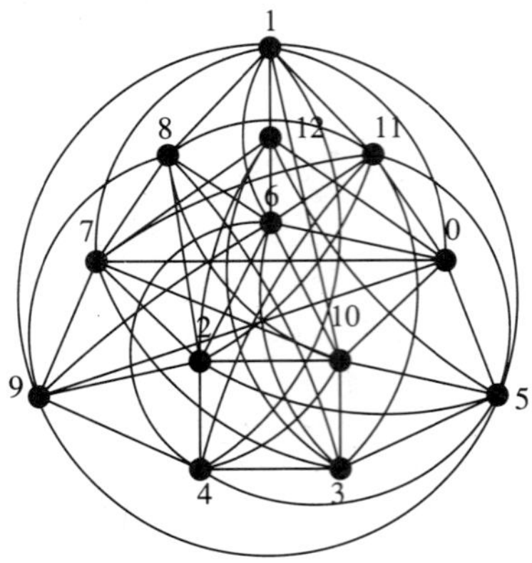

FIGURE 4. The graph $UK_1(2,3)$.

5. FAULT-TOLERANT KAUTZ AND DE BRUIJN GRAPHS

Obviously, we can get a $(k, UK(d,D))$-tolerant Kautz graph with maximum degree in $O(dk^2)$ by applying the above trivial scheme. But as we can see below, we can built a supergraph denoted $UK_k(d,D)$ which is $(k, UK(d,D))$-tolerant with maximum degree only in $O(dk)$.

Definition 5.1. For any integers $d \geq 2$, $n \geq 3$ and $k \geq 0$, let $II_{d,n,k}$ be the graph with nodes $\{0, 1, 2, \ldots, n+k-1\}$ and where the pair (x,y) is an edge if and only if there exists an integer $r \in \{d(-k)+1, d(-k)+2, \ldots, d(k+1)\}$ such that either $y = (-dx - r) \mod (n+k)$ or $x = (-dy - r) \mod (n+k)$.

By replacing n by $d^D + d^{D-1}$, we get the following definition of $UK_k(d,D)$.

Definition 5.2. For any integers $d \geq 2$, $n \geq 3$ and $k \geq 0$, let $UK_k(d,D)$ be the graph with nodes $\{0, 1, 2, \ldots, d^D + d^{D-1} + k - 1\}$ and where the pair (x,y) is an edge if and only if there exists an integer $r \in \{d(-k)+1, d(-k)+2, \ldots, d(k+1)\}$ such that either $y = (-dx - r) \mod (d^D + d^{D-1} + k)$ or $x = (-dy - r) \mod (d^D + d^{D-1} + k)$.

Let us remark that from the definition given above, $UK_k(d,D)$ has a structure which is very similar to the structure of the target graph $UK(d,D)$. The only differences are that $UK_k(d,D)$ has $(d^D + d^{D-1} + k)$ nodes and each node is connected to $d(2k+1)$ consecutive nodes rather than to d consecutive nodes. In particular note that $UK_k(d,D)$ has degree at most equal to $d(4k) + 2d$ and $UK_0(d,D)$ is isomorphic to $UK(d,D)$. Figure 4 shows as an example the graph $UK_1(2,3)$ and Figure 5 shows $UK(2,3)$ as a subgraph of $UK_1(2,3)$ after removal of one node. We will now prove that the graph $II_{d,n,k}$ is $(k, II_{d,n})$-tolerant.

Theorem 5.3. *For any integer $d \geq 2$, $D \geq 3$, and $k \geq 0$, the graph $II_{d,n,k}$ is $(k, II_{d,n})$-tolerant.*

Proof. Let H be the subgraph of $II_{d,n,k}$ induced by an arbitrary set of n nodes. We will show that $II_{d,n}$ is a subgraph of H. Let (x,y) be an arbitrary edge in $II_{d,n}$ and assume without loss of generality that there exists an integer $r \in \{1, 2, \ldots, d\}$ such that $y = (-dx - r) \mod n$. There exists $t \in \{1, 2, \ldots, d\}$ such that $y = (-dx - r) + tn$. We define the embedding function $\varphi : V(II_{d,n}) \longrightarrow V(H)$ as

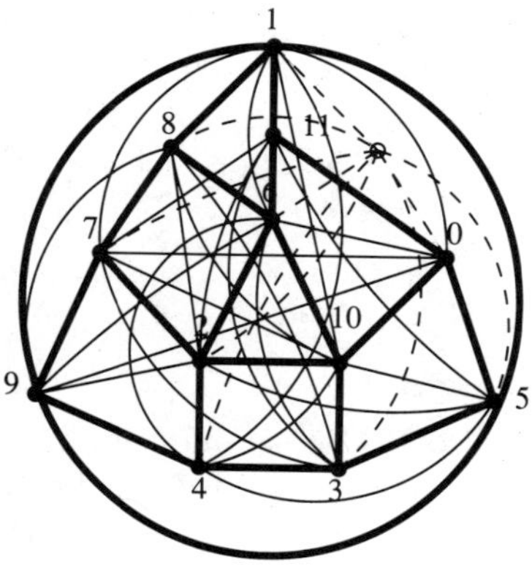

FIGURE 5. New labels of $UK_1(2,3)$ after one fault.

follows. For each $x \in V(II_{d,n})$, $\varphi(x) = u$ where $rank(u,H) = x$. Note that φ is a monotonically increasing 1-to-1 function. Let $\delta_x = \varphi(x) - x$, $\delta_y = \varphi(y) - y$, $\Omega = \{d(-k)+1, d(-k)+2, \dots, d(k+1)\}$, $s = r + tk - \delta_y - d\delta_x$. The fact that $x \le \varphi(x) \le x+k$ and $y \le \varphi(y) \le y+k$ implies that $0 \le \delta_x \le k$ and $0 \le \delta_y \le k$. Therefore $\varphi(y) = y+\delta_y = (-dx-r)+tn+\delta_y = -d(\varphi(x)-\delta_x)-r+t(n+k-k)+\delta_y = -d\varphi(x)+d\delta_x - r + \delta_y - tk + t(n+k) = (-d\varphi(x) - s) + t(n+k)$.

Recall that $1 \le r, t \le d$ and $0 \le \delta_x, \delta_y \le k$, thus $s = r + tk - \delta_y - d\delta_x \le r + tk \le d + dk = d(k+1)$, $s = r + tk - \delta_y - d\delta_x \ge 1 + k - k - dk = d(-k)+1$, so $s \in \Omega$ and $(\varphi(x), \varphi(y))$ is an edge in H. As a result, $II_{d,n}$ is a subgraph of H.

By replacing n by $d^D + d^{D-1}$ in the proof of Theorem 5.3, we get the following result for the Kautz graphs.

Corollary 5.4. *For any integer $d \ge 2$, $D \ge 3$, and $k \ge 0$, the graph $UK_k(d,D)$ is (k, UK(d,D))-tolerant.*

Also, we can extend the result given for de Bruijn graphs $UB(d,D)$ in [4] to the generalized de Bruijn graphs as follows.

Definition 5.5. For any integers $d \ge 2$, $n \ge 3$ and $k \ge 0$, let $RPK_{d,n,k}$ be the graph with nodes $\{0,1,2,\dots,n+k-1\}$ and where the pair (x,y) is an edge if and only if there exists an integer $r \in \{(d-1)(-k), (d-1)(-k)+1, \dots, (d-1)(k+1)\}$ such that either $y = (dx+r) \mod (n+k)$ or $x = (dy+r) \mod (n+k)$.

Note that the graph $RPK_{d,n,k}$ has degree at most equal to $(d-1)(4k)+2d$. By replacing n by d^D, we get the definition of $UB_k(d,D)$.
Before proving that the graph $RPK_{d,n,k}$ is $(k, RPK_{d,n})$-tolerant, we prove the following lemma.

Lemma 5.6. *Let (x,y) be an arbitrary edge in the graph $RPK_{d,n}$. Assume without loss of generality that there exists an integer $r \in \{0,1,2,\dots,d-1\}$ such that $y = (dx+r) \mod n$, and let t be an integer such that $y = dx+r-tn$. Then either $x<y$ and $t \in \{0,1,\dots,d-2\}$ or $x>y$ and $t \in \{1,2,\dots,d-1\}$.*

Proof. Note that $dx+r \ge 0$ and $y<n$ imply that $t>-1$. Also, $y \ge 0$, implies that $tn \le dx+r$. But $x \le n-1$ and $r \le d-1$, so $dx+r \le dn-1$, which implies that

$t < d$. Thus $t \in \{0, 1, \ldots, d-1\}$. Let us consider the case where $x < y$. Assume that $t = d - 1$. Then $x \leq n - 2$ and $x < y = dx + r - (d-1)n$, so $0 < dx + r - x - (d-1)n = (d-1)x + r - (d-1)n \leq (d-1)(n-2) + r - (d-1)n = -2(d-1) + r$, so $r > 2d - 2 \geq d$ which is a contradiction. Now consider the case where $x > y$. Assume that $t = 0$, because $d \geq 2$, $r \geq 0$, we have $y = dx + r > x > y$ which is a contradiction.

Theorem 5.7. *For any integer $d \geq 2$, $n \geq 3$, and $k \geq 0$, the graph $RPK_{d,n,k}$ is (k,$RPK_{d,n}$)-tolerant.*

Proof. Let H be a subgraph of $RPK_{d,n,k}$ induced by an arbitrary set of n nodes. We will show that $RPK_{d,n}$ is a subgraph of H. Let (x, y) be an arbitrary edge in $RPK_{d,n}$. Assume without loss of generality that there exists an integer $r \in \{0, 1, \ldots, d-1\}$ such that $y = (dx + r) \mod n$, and let t be an integer such that $y = dx + r - tn$. The embedding function is defined as in the proof of Theorem 5.3 $\varphi : V(RPK_{d,n}) \longrightarrow V(H)$ such that for each $x \in V(RPK_{d,n})$, $\varphi(x) = u$ where $rank(u, H) = x$. Let $\delta_x = \varphi(x) - x$, $\delta_y = \varphi(y) - y$, $\Omega = \{(d-1)(-k), (d-1)(-k) + 1, \ldots, (d-1)(k+1)\}$, $s = r + tk + \delta_y - d\delta_x$. $\varphi(y) = y + \delta_y = (dx + r) - tn + \delta_y = d(\varphi(x) - \delta_x) + r - t(n + k - k) + \delta_y = d\varphi(x) - d\delta_x + r + \delta_y + tk - t(n+k) = (d\varphi(x) + s) - t(n+k)$. Note that $0 \leq \varphi(y) < n + k$, so $\varphi(y) = (d\varphi(x) + s) \mod (n+k)$. We distinguish two cases.

Case 1 : $x < y$. In this case, we have $\delta_x \leq \delta_y$, and thus $s \geq \delta_y - d\delta_x \geq (1-d)\delta_x \geq (d-1)(-k)$. Also, in this case it follows from Lemma 10 that $t \in \{0, 1, \ldots, d-2\}$. So $s = r + tk + \delta_y - d\delta_x \leq r + tk + \delta_y \leq (d-1) + (d-2)k + k = (d-1)(k+1)$, and therefore $s \in \Omega$ and $(\varphi(x), \varphi(y))$ is an edge in H.

Case 2 : $x > y$. In this case, it follows from Lemma 10 that $t \in \{1, 2, \ldots, d-1\}$. Also, $\varphi(x) > \varphi(y)$ implies that $\delta_x \geq \delta_y$. Thus $s \leq k(d-1) + (d-1) + \delta_x - d\delta_x \leq k(d-1) + (d-1) = (d-1)(k+1)$, and $s \geq k - dk = (d-1)(-k)$, hence $s \in \Omega$ and $(\varphi(x), \varphi(y))$ is an edge in H. As a result, $RPK_{d,n}$ is a subgraph of H.

Corollary 5.8. [4] *For any integer $d \geq 2$, $n \geq 3$, and $k \geq 0$, the graph $UB_k(d, D)$ is (k,$UB(d, D)$)-tolerant.*

Remark. Note that from the inequality $k^2 \leq k + 1$ follows that the $(1, UK(d, D))$-tolerant graphs constructed by trivial scheme have smaller degree than the fault-tolerant graphs defined by Definition 5.2 (see Figures 2 and 4 for examples). In fact, there is a little difference between the degrees of graphs constructed by trivial scheme and those given by Definition 5.2 for small values of k. The same remark can be made for the generalized graphs $II_{d,n,k}$ and $RPK_{d,n,k}$.

The challenge is to built k-fault-tolerant graphs with even smaller maximum degree than the graphs defined in Sections 4 and 5 have. To fill this gap, we suggest an implementation with buses.

6. Implementation with buses

In this section, we show how fault-tolerant architecture with smaller degree can be obtained by using buses. The technique is similar to that used in [4]. Consider a Kautz graph $UK(d, D)$. We associate to $UK(d, D)$ a bus network structure $BK(d, D)$ in the following way. The vertices of $V(UK(d, D))$ represent the processing elements PE of the bus network. For each vertex $i \in V(UK(d, D))$

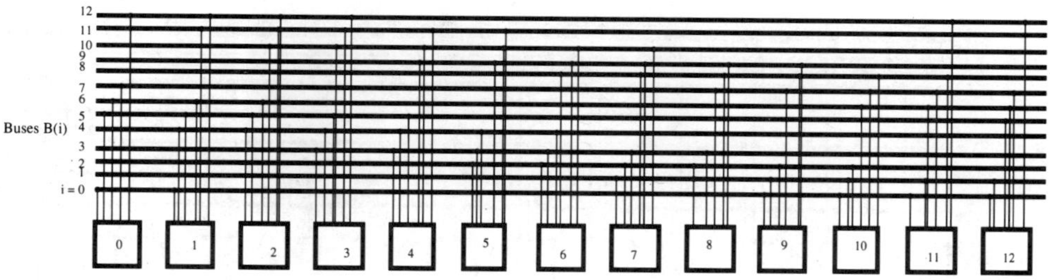

FIGURE 6. The scheme of bus network $BK_1(2,3)$.

there exists a bus $B(i)$ in the bus network representing the set of successors of i. Hence, a processing element $PE(i)$ is connected to a bus $B(j)$ in the bus network if j is successor of i in $UK(d,D)$. Therefore, the number of buses is equal to the number of nodes of $UK(d,D)$. In this bus structure, the degree of $PE(i)$ is the number of buses connected to $PE(i)$. Recall that in the Kautz graph $UK(d,D)$, node i is connected to a set of d nodes $(-di-1) \bmod (d^D+d^{D-1})$, $(-di-2) \bmod (d^D+d^{D-1})$, ... , $(-di-d) \bmod (d^D+d^{D-1})$. Therefore all the connections of $UK(d,D)$ will be maintained if we replace all the d successors of i by a single bus $B(i)$. The same technique can be used to reduce the degree of the fault-tolerant graph $UK_k(d,D)$ (and similarly the degree of $II_{d,n,k}$ and $RPK_{d,n,k}$). Recall that in the fault-tolerant graph $UK_k(d,D)$, each node i is connected to a set of $2dk+d$ nodes namely nodes labelled $(-di+dk-1) \bmod (d^D+d^{D-1}+k)$, $(-di+dk-2) \bmod (d^D+d^{D-1}+k)$, ... , $(-di-dk-d) \bmod (d^D+d^{D-1}+k)$. Therefore, in $BK_k(d,D)$ a single bus could be used to connect each processing element $PE(i)$ to all $d(2k)+d$ processing elements. This use of buses results in a fault-tolerant architecture with degree $d(2k)+d$ which is equal to the half of the degree required in the graph $UK_k(d,D)$ given in Definition 5.2. Figure 6 shows $BK_1(2,3)$ and Figure 7 shows $BK(2,3)$, resulting from $BK_1(2,3)$ after one fault. Finally note that with this bus architecture, even bus faults can be tolerated. Indeed if the bus connecting $PE(i)$ to the set of $d(2k)+d$ consecutive PE is faulty, then the use of this bus can be avoided by viewing $PE(i)$ as being faulty.

7. CONCLUSION

In this paper, constructions are given for adding fault-tolerance to Kautz networks and also to the generalized Kautz and de Bruijn networks. We have defined supergraphs denoted by $UK_k(d,D)$, $II_{d,n,k}$ and $RPK_{d,n,k}$ with the property that given any set of k faults, the remaining graphs contain $UK(d,D)$, $II_{d,n}$ and $RPK_{d,n}$, respectively, as subgraphs. The graphs constructed have the minimum number of nodes required for the given level of fault-tolerance. Furthermore, the fault-tolerant graphs have a degree which is linear in the number of tolerated faults. But, it has not been proven that the given constructions have the smallest possible degree. It would be interesting to prove that, or to improve the results, by giving other constructions. Finally, we believe that other techniques, such as creating fault-tolerant graphs with more than $N+k$ nodes, can be used to reduce the degree.

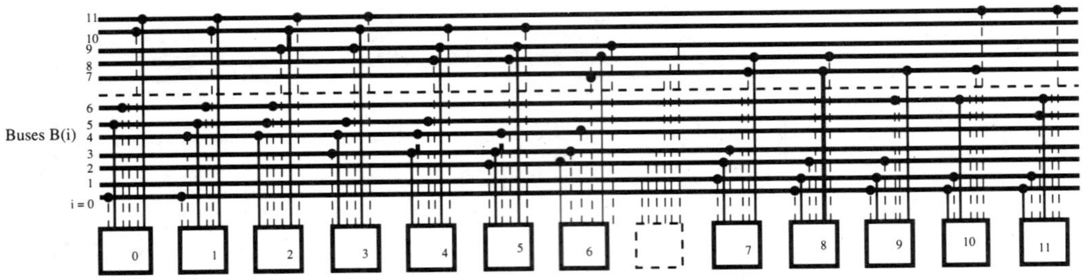

FIGURE 7. With solid lines the scheme of bus network $BK(2,3)$ after one fault in $BK_1(2,3)$.

REFERENCES

[1] I. Alegre, M. A. Fiol and J.L. Yebra, *Line digraph iterations and the (d,k) digraph problem*, IEEE Transcations on Computers **C-33** (1984), 400–403.

[2] F. Annexstein, *Fault-Tolerance in Hypercube-Derivative networks*, Proceedings of the 1st Annual ACM Symposium on Parallel Algorithms and Architectures **4** (1989), 179–188.

[3] J- C. Bermond and C. Peyrat, *De Bruijn and Kautz networks : a competitor for the hypercube?*, Hypercube and distributed computers (1989), 279–293.

[4] J. Bruck, R. Cypher and C. T. Ho, *Fault-tolerant de Bruijn and shuffle-exchange networks*, Tech. Report IBM Almaden Research Center (1992).

[5] J. Bruck, R. Cypher and C. T. Ho, *Fault-tolerant Mesh and Hypercube architectures*, Techn. Report IBM Almaden Research Center.

[6] J. Bruck, R. Cypher and D. Soroker, *Running Algorithms Efficiently on Faulty Hypercubes*, Proceedings of the 2nd Annual ACM Symposium on Parallel Algorithms and Architectures (1990), 37–44.

[7] N.G. deBruijn, *A combinatorial problem*, Koninklijke Nederlandsche Akad.Van Wetenschappen Proc. **49** (1946), 758–764.

[8] S. Dutt and J.P. Hayes, *Designing fault-tolerant systems using automorphisms*, Journal of Parallel and Distributed Computing **12** (1991), 249–268.

[9] S. Dutt and J. P. Hayes, *On designing and reconfiguring k-fault-tolerant tree architectures*, IEEE Transactions on Computers **39** (1990), 490–503.

[10] J. Hastad, F. T. Leighton and M. Newman, *Fast Computations using Faulty Hypercubes*, Proceedings of 21st Annual ACM Symposium on Theory of Computing (1989), 251–284.

[11] J.P. Hayes, *A graph model for fault-tolerant computing systems*, IEEE Transactions on Computers **C-25** (1976), 875–883.

[12] M. Imase and M. Itoh, *A design for directed graphs with minimum diameter*, IEEE Transactions on Computers **C-32** (1983), 782–784.

[13] W. H. Kautz, *Bounds on directed (d,k) graphs. Theory of cellular logic networks and machines*, AFCRL-68-0668 Final report (1968), 20–28.

[14] T. Leighton, B. Maggs and R. Sitaraman, *On the Fault-Tolerance of Some Popular Bounded-Degree Networks*, Proceedings of 33rd Annual IEEE Symposium on Foundations of Computer Science (1992), 542–552.

[15] S. M. Reddy, D. K. Pradhan and J. G. Kuhl, *Directed graphs with minimum diameter and maximum connectivity*, School Engineering, Oakland University, Rochester, MI, July (1980).

LABORATOIRE DE RECHERCHE EN INFORMATIQUE BÂT. 490, UNIVERSITÉ DE PARIS-SUD, 91405 ORSAY CEDEX, FRANCE

Current address: Laboratoire de Recherche en Informatique Bât. 490, université de Paris-Sud, 91405 Orsay cedex, France

E-mail address: harbane@lri.fr

DIMACS Series in Discrete Mathematics
and Theoretical Computer Science
Volume **21**, 1995

Asynchronous Packet Routers

CHRIS JESSHOPE, IVAILO NEDELCHEV

ABSTRACT. The paper describes the asynchronous implementation of four basic building blocks for asynchronous packet routers for direct networks. Combining them, one can design various routing architectures. In this paper we demonstrate how two different types or routing devices for mesh topologies can be constructed using these four basic blocks.

Another major objective of this paper is also to demonstrate asynchronous design techniques for VLSI based upon a novel set of asynchronous library cells.

1. Introduction

The paper presents the implementation of four asynchronous blocks with which we have constructed packet switching nodes for direct networks, i.e. we consider networks where there is a routing node (RN) attached to each processing element (PE) and this RN is also connected to several other routing nodes. The network comprising the latter RNs provides transport mechanisms to the PEs such as point to point communications, broadcast over a region, etc. In this paper we consider only mesh topologies, so each RN is connected to 4 neighbours thus conforming a grid.

Each of these basic building blocks

- asynchronous multiplexor - MUX
- asynchronous demultiplexor - DEMUX
- address decrementor - DEC
- asynchronous vertical pipeline - FIFO

has been specified in a CSP [**8**] like notation called Delay Insensitive Algebra (DIA). A compilation scheme has been developed that translates expressions in the algebra into a netlist which comprises only basic library cells for asynchronous

1991 *Mathematics Subject Classification.* Primary 68M07, 68M10, 68M20.
We gratefully acknowledge the EPSRC for funding to support this research under grant # H78986.

design [**3, 4**]. It has been applied to the specification of the different building blocks and thus the final circuit has been produced.

In the following section we will introduce the DIA notation used for specification of these four blocks. In the third section we will briefly introduce the set of library elements as a set of subcomponents for asynchronous design. We will also discuss the implementation of the four different blocks and focus on two routing node architectures based on these blocks.

2. Delay Insensitive Algebra

As a specification language for the behaviour description of the four different blocks (MUX, DEMUX, DEC and FIFO) we use a CSP based notation [**8**]. References [**10, 11, 9**] propose such an algebraic notation, called Delay Insensitive Algebra, in which delay-insensitive circuit specifications can be expressed concisely, including obligations to be met by the environment. We will briefly focus on the operational semantics of some basic DIA operators and constructs which we use further in this paper.

By a *process* in DIA we mean a mathematical model at a certain level of abstraction of the way in which a delay-insensitive circuit interacts with its environment. Typical processes are P and Q. A circuit receives signals from its environment on its input wires and sends signals to its environment on its output wires; with each process are associated an alphabet of input wires I and an alphabet of output wires O, with typical wire names a, b, c, d.

The syntax of the algebra is very similar to the CSP syntax

(i) $\perp$, *chaos*, denotes the possibility of error.

(ii) $a?x; P$, *input prefix*, denotes a process that must wait for a data to arrive with a request on a before it can behave like P.

(iii) $c!x; P$, *output prefix*, denotes a process that outputs x on c and then behaves like P.

(iv) $[g_n \& a_0?x_0 \rightarrow P_0 \;\Box\; \ldots \;\Box\; g_n \& a_n?x_n \rightarrow P_n]$, *guarded choice*
An alternative of the form $a_i?x_i \rightarrow P_i$ is selected only if a request has been received on a and its corresponding boolean guard g_i is true. (If g_i is missing for some i then it is assumed that g_i is true.)

The environment must ensure that a process never gets into a $\perp$ state. A common use is in the alternative $a? \rightarrow \perp$ of a guarded choice, in which case the environment is obliged not to signal on a until it first receives an output.

Mutually exclusive guarded choice. The environment of the process P

$$P = [a? \rightarrow [b? \rightarrow \perp \Box S_0] \;\Box\; b? \rightarrow [a? \rightarrow \perp \Box S_1]$$

must not send a signal on a followed by another on b, or vice versa, without receiving an acknowledgement from P of the first signal generated by S_0 or S_1. Such inputs are called *mutually-exclusive* and we will use the following syntax as an abbreviation for the above expression

$[a? \to [S_0] \mid b? \to [S_1]]$

3. Asynchronous Library Cells

Before we present the different blocks we will also briefly describe the set of library cells used throughout the asynchronous design of these building blocks. These library cells are major subcomponents of the netlist. The latter is a result of the syntax driven compilation technique[**3, 4**] applied to the DIA specification.

Merge: The MERGE element is usually implemented as a logical XOR gate. Each incoming event on a input is copied out onto its output.

Sparse Decision Wait: The Sparse Decision Wait (SDW) element is based on the definition of Decision Wait element described in [**10**]. Some implementations of SDW at a transistor level can be found in [**4**].

Its DIA specification is as follows:

$$SDW_{n,m} = [\|_{i=1}^{n} r_i? \to [\|_{i=1}^{m} s_{i,j} \& c_j? \to o_{i,j}!; SDW_{n,m}]]$$

where $s_{i,j}$ is a vector of boolean values and if some $s_{i,j}$ is $FALSE$ the cell $_{i,j}$ is missing in the graphical symbol for the DW.

The circuit symbol for the Decision-Wait is shown in figure 1 where $s_{i,i}$ is $FALSE$ for $1 \leq i \leq 3$.

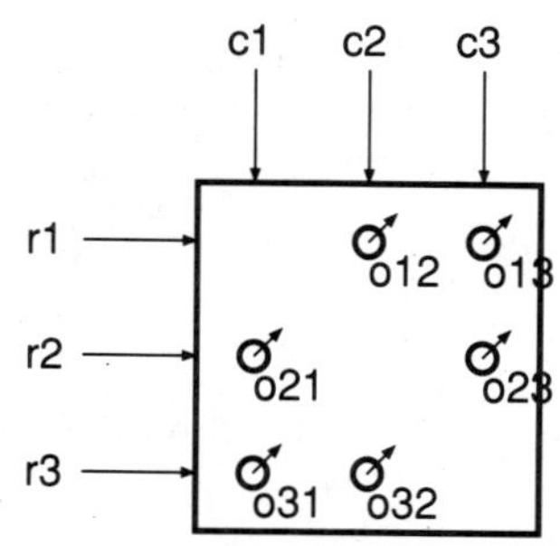

FIGURE 1. The Decision-Wait

Toggle: is an asynchronous library cell with one input and two outputs[**13**]. It copies each odd input onto the bullet signed output and the other even events - onto the second output.

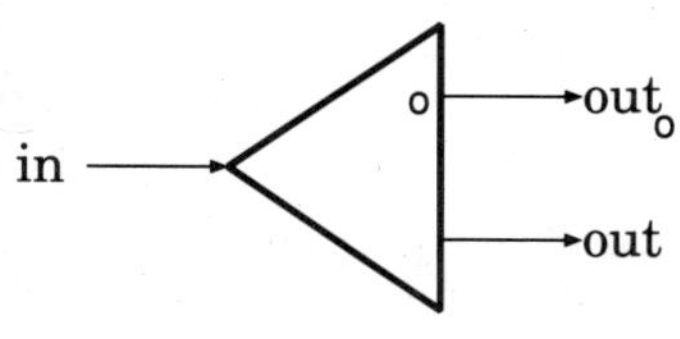

FIGURE 2. Toggle

Two-way Arbiter: A two-way arbiter is a device that gives two processes mutually-exclusive access to a single shared resource. Its circuit symbol is given below

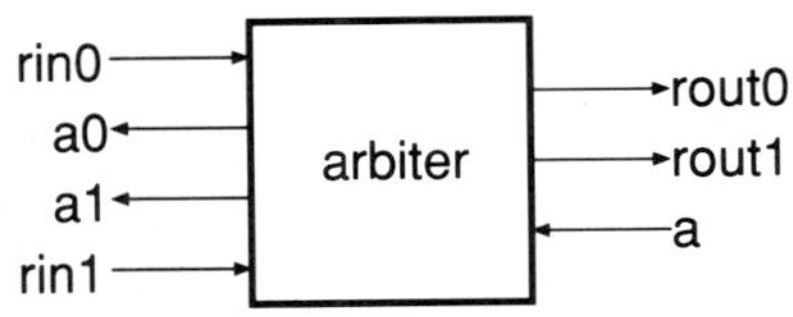

FIGURE 3. The Arbiter element

A pair of request and acknowledge wires rin_i, a_i connects the two processes to the arbiter. The $rout_i$ and a wires are used for resource request and acknowledge. The behaviour of the arbiter is defined by the following algebraic expression:

$$A = [\Box_{i=0}^{1} rin_i? \rightarrow rout_i!; a?; a_i!; A_i]$$
$$A_i = rin_i?; rout_i!; a?; a_i!; A$$

The request signals rin_0 and rin_1 can be concurrent. In this case one of them is blocked until the resource has been released by the other process.

Select: is a device with one input in and n outputs out_i [**13**]. s is a binary value which specifies which output is currently selected. The DIA specification is as follows:

$$Select = in?; out_s!; Select$$

An incoming event on in is copied on out_s.

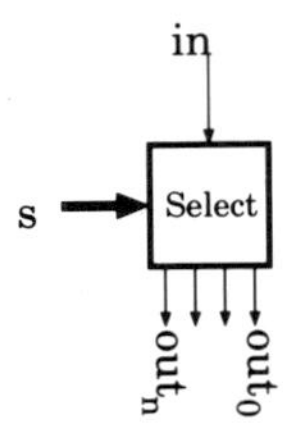

FIGURE 4. Select element

Register: is a device with n *store* and m *pass* inputs and corresponding acknowledges for both sets [**13**]. When there is an input *store* on some of the set, the register holds the values on *data.in* valid on *data.out* until there is another consequent event on *pass* input where the register simply copies the values from *data.in* on *data.out*.

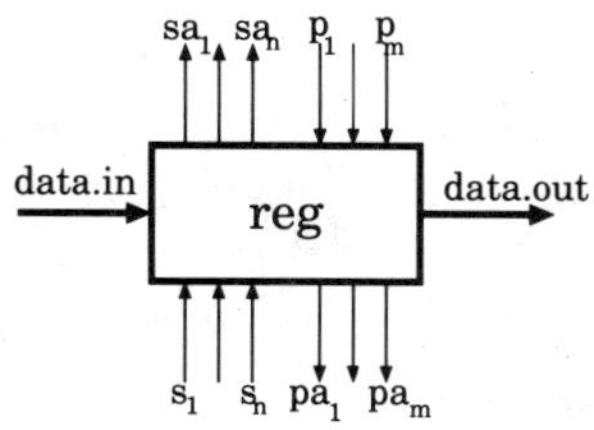

FIGURE 5. Asynchronous register

4. Communication channels and packets format

All building blocks in this paper communicate on channels of one and the same type. Some of the channel's wires carry the logical values of the data, the rest are control wires which provide a correct clocking scheme for node-to-node communication. In this section we describe the meaning of each wire of the communication channel.

We use a bundled data convention for data transmission and 2 phase signaling [**13**] for the control wires. By 2 phase signaling we mean that an asynchronous event is a voltage transition ($0 \rightarrow 5V$, $5V \rightarrow 0$).

Each channel is physically represented by several control/data wires:

- $data_i :_{1 \leq i \leq n}$: are the logical values of the current data word.
- *req* is an event which latches the valid data on $data_i$.
- e is an event which indicates *end of packet* and it also latches the initial control word on $data_i$. In the latter case e might be regarded as a start of the packet event.
- *ack* is an event which acknowledges the data from the recipient. The existence of two acknowledge type of wires simplifies the design therefore we will use ack_e and ack_r for the corresponding acknowledges for data with event e and *req*. In fact we will design our blocks using these two wires and only for the external interface we will use one wire *ack* instead of these two. Such an acknowledge *ack* is a result of simply merging ack_e and ack_r.

The routing nodes receive and send out packets on the described above communication channels. The format of these packet can be formally expressed by the following expression: **e.req***.**e**. The first event e carries the control word which consists of a control field - n bits and relative address of the destination - m bits. Each address field contains the number of the hops to be traveled on the current dimension. On each hop this relative address is decremented by 1. When some relative address becomes zero, the control field in the control word indicates where the rest of the message should be copied to, i.e. reinjected into another dimension or consumed by the processing element.

After the initial control word the packet continues with data words latched by *req*. In fact these data words can also be control words but they will be interpreted as such by some routing nodes further on in the message's path. The

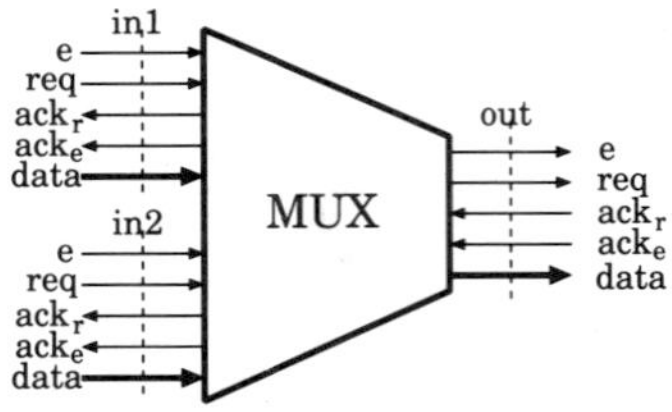

FIGURE 6. MUX block

second incoming event e indicates the end of the packet. Each event e or req is acknowledged on the corresponding channel ack_e/ack_r or if it is an off-chip interface on channel ack.

The blocks, we present in the following three sections, accept and send out data on packet channels of the format described above.

5. Asynchronous Packet Multiplexor - MUX

The first building block for asynchronous packet routers is the asynchronous multiplexor MUX (figure 6). Block MUX serialises the concurrent request from its two input channels in_1 and in_2 on the output channel out. The specification of MUX process is as follows:

$$MUX \quad = \quad [\Box_{inpCh \in (in_1, in_2)} e_{inpCh}?x \rightarrow e!x; ack_e?; ack_{inpCh,e}!; Copy_{inpCh}]$$

$$Copy_{inpCh} = \quad [req_{inpCh}?x \rightarrow req!x; ack_r?; ack_{inpCh,r}!; Copy_{inpCh}$$
$$|e_{inpCh}?x \rightarrow e!x; ack_e?; ack_{inpCh,e}!; MUX]$$

Actually only the initial $start\ of\ packet$ events e are concurrent therefore there is an arbitration for resolving possible conflicts for only these signal events. All subsequent requests on req lines are mutual exclusive and they can be simply merged. After the arbitration process has been resolved, the direction which input data is copied on the output channel *out* is determined. Therefore the result from the arbitration process controls the multiplexion for the data inputs. There is no data processing in MUX element. Its task is to transmit one incoming packet from an input channel to its output *out* in a non-interleaving fashion with respect to the other input.

One possible implementation of the MUX process is shown in figure 7.

If a MUX block with more inputs is required it can be built out as a composition of several MUX blocks with two inputs as shown in figure 8.

6. Address Decrementor - DEC

In this section we present the address decrementor(figure 9). As we mentioned in the previous sections, we use relative addressing. On each hop the relative address of the target is decremented by 1. As the relative address and the control

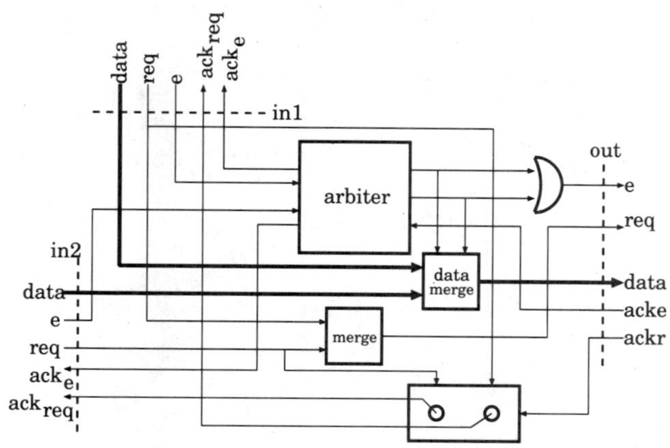

FIGURE 7. The implementation of MUX

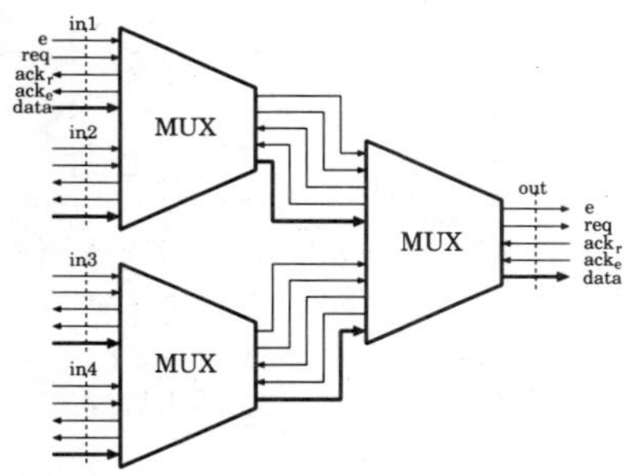

FIGURE 8. Composition of MUX blocks

field are in the initial control word, DEC has to decrement only the address field from the current incoming word latched with event e. The specification of DEC follows:

$$DEC = [e_{in}?[y:z]; ack_e!; e_{out}![y-1:z]; Copy$$

$$\begin{aligned} Copy = & [e_{in}?x \rightarrow ack_e!; e_{out}!x; DEC \\ & |req_{in}?x \rightarrow ack_r!; req_{out}!x; Copy \end{aligned}$$

Here we specify by $[y : z]$ a variable which contains fields: y - the address field and z - the control field. With the first event e, DEC accepts the control field and the relative address. It sends out the control field unchanged but the relative address is decremented by 1. With each consequent request on req and e DEC transmits the data unchanged on the output channel. DEC acknowledges

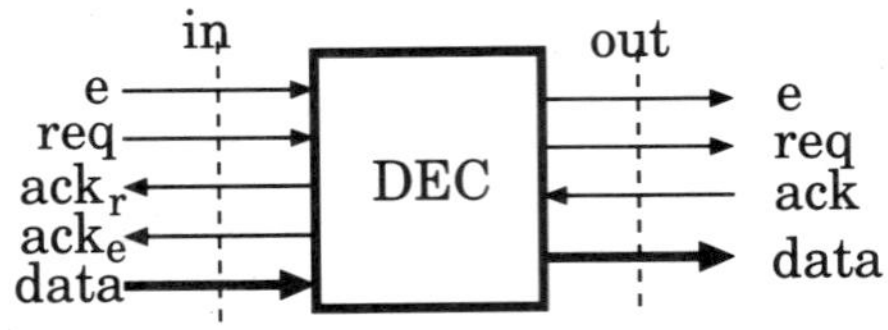

FIGURE 9. *DEC* block

incoming events on *req* and *e* on the corresponding ack_e and ack_r. The second incoming *e* event initialises the process *DEC*.

One possible implementation is shown in figure 10.

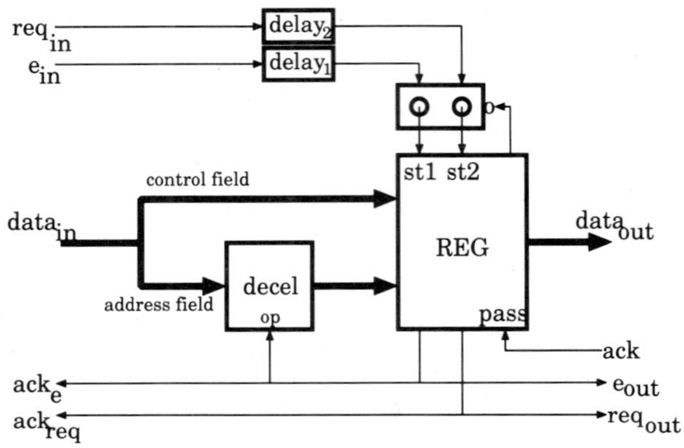

FIGURE 10. Implementation of DEC process

The element *decel* is a functional block which decrements the input data by 1 if the input *op* is logical false. Otherwise it leaves the data unchanged. Because of these two modes of operation the delay lines ($delay_1$ and $delay_2$) for *e* and *req* are different. These two delay lines are necessary to preserve the correct timing on the bundle of wires: (*e*, *req*, *data*).

7. Asynchronous Demultiplexor - DEMUX

The block *DEMUX* (figure 11) is the most complex of all building blocks. It has one input channel and as many output channels as necessary for the specific implementation. The DIA specification of *DEMUX* routing block follows:

$$
\begin{aligned}
DEMUX \quad &= \quad e?x; \mathbf{store!?x}; ack_e!;\\
&\quad\; \mathbf{if}\ (\neg zh)\ \mathbf{then}\ e_{ch_1}!x; ack_{ch_1,e}?; Copy_{ch_1}\\
&\quad\; \mathbf{else}\ (zh)\ \mathbf{then}\ req?x; ack_r!; e_{ch_{cf}}!x; ack_{ch_{cf},e}?; Copy_{ch_{cf}}
\end{aligned}
$$

$$
\begin{aligned}
Copy_{chn} \quad &= \quad [req?x \rightarrow ack_r!; req_{chn}!x; ack_{chn,req}?; Copy_{chn}\\
&\quad\; |e?x \rightarrow ack_e!; e_{chn}!x; ack_{chn}?; DEMUX]
\end{aligned}
$$

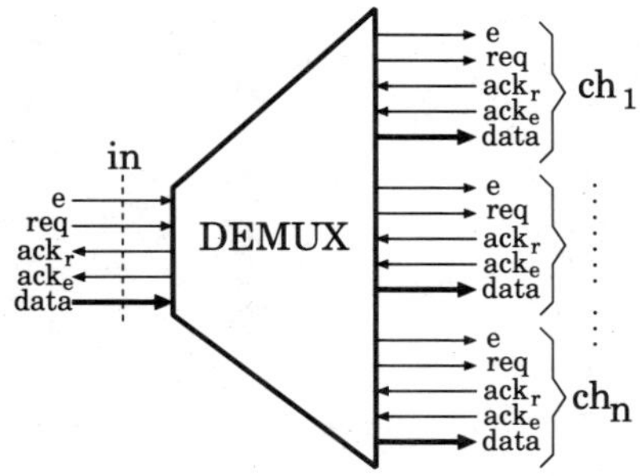

FIGURE 11. *DEMUX* block

It accepts the control word of the packet with the first incoming event e and if the relative address is not zero ($zh = false$) then $DEMUX$ copies the control word and the rest of the message on the direct channel ch_1. (We stipulate that channel ch_1 is the direct one and its binary representation in the control field cf from the first control word is zero. The direct channel is the one that forwards the data on the same dimension.) If the relative address is zero ($zh = true$) then the routing block $DEMUX$ opens channel ch_{cf} with event e where cf is the control field from the control word as a binary value. This control word is saved by operation **store**!?**x** in an additional control register as we can see later from the implementation.

One possible implementation is shown in figure 12.

The boolean value zh (zero header) is a logical OR of all bits in the binary representation of the relative address. It is saved in the single bit latch register reg. The boolean vector sc is defined as $(zh \wedge cf_1, ..., zh \wedge cf_n)$, where cf_i is the ith bit from the control field. As one can notice this boolean vector is valid through the whole transmission of the packet because the control field is saved in register *ctrl reg* and zh is stored in the bit-latch reg.

8. Packet switching device using oblivious routing

Oblivious routing for grid topology is a strategy for deadlock avoidance [**7**]. Basically the routing is initially performed on dimension X and then on dimension Y ($first\ X\ then\ Y$). Each routing node is connected to its neighbours with 4 pairs (input and output) of channels. There are another two channels for injection and consumption of the message. For each input channel there is a routing process with an assigned buffer and because the routing is performed in $first\ X\ then\ Y$ way there is no possibility for channel dependency cycles [**6**]. The latter is a sufficient condition for deadlock freedom. One possible implementation based on the basic blocks is shown in figure 13.

On injection the address field of the first control word is zero. The control field contains the 'address' of the direction to be injected - (+X, -X, +Y, -Y) which is the address of an output channel for the injection *RT* node. Each consequent control word of the message carries the relative address of the destination for the

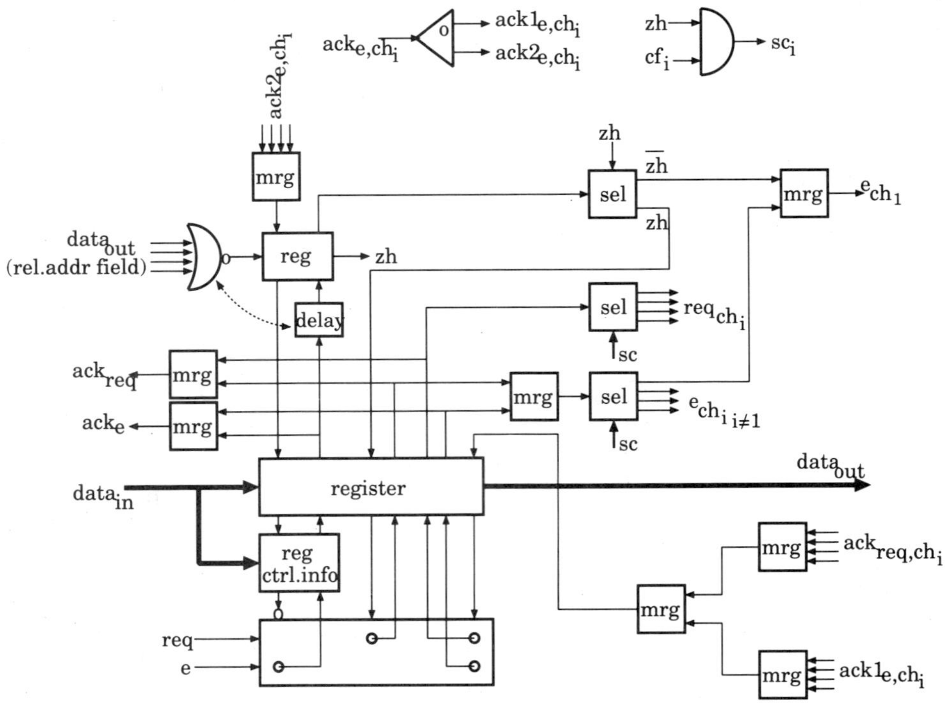

FIGURE 12. Implementation of *DEMUX* process

current dimension. The control field shows on which channel the message should be delivered after it satisfies the current dimension - whether it is reinjected in (+Y, -Y) or it is consumed by the recipient. Thus the first control word indicates on which dimension the message is to be injected. The next control word (the second word from the packet) shows how many hops it has to travel on this dimension and when the address is satisfied - to where the message should be reinjected. When the packet arrives at the destination RN, the last control word must address channel *consume*.

9. Restricted routing packet switching device

Another strategy for deadlock avoidance is using virtual networks [**5**]. There are 4 virtual planes for the grid topology where the routing is performed in (+X,+Y),(+X,-Y),(-X,+Y),(-X,-Y) as shown in figure 14. Obviously there can not be channel cycles in any of these 4 planes so any type of routing is deadlock free. Depending on the address of the destination the message is initially injected in one of these 4 planes and it is routed only in this plane.

One possible implementation for a single routing node of a single plane is shown in figure 15.

On injection the address field of the first control word is zero. The control field contains the 'address' of the direction to be injected - (X or Y). Each consequent control word of the message carries the relative address of the destination (in its

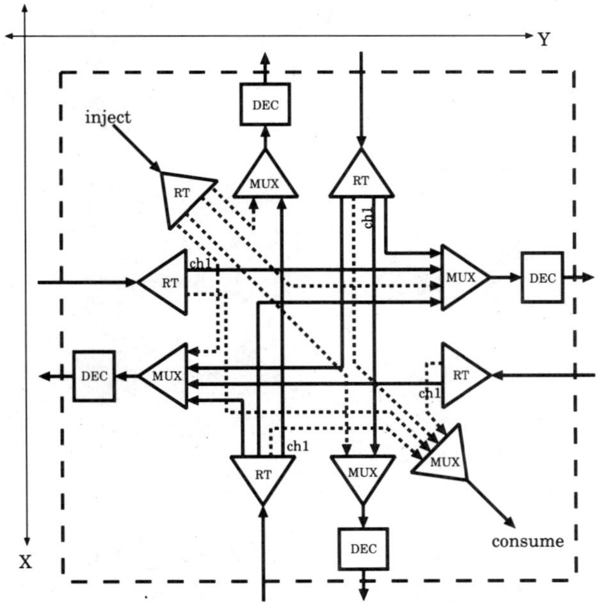

FIGURE 13. Oblivious routing packet switching device

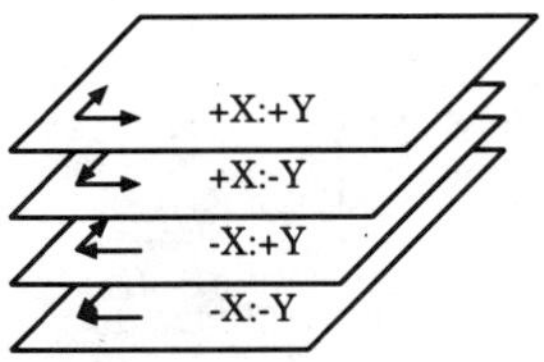

FIGURE 14. Deadlock free routing using virtual nets

address field) for each dimension. The corresponding control field shows on which channel the message should be delivered after it satisfies the current dimension - whether it is reinjected in X or Y or it is consumed by the recipient on channel *consume*.

10. Asynchronous buffering

As one can envisage from the designs presented in the previous sections, all blocks communicate in a strict loop-manner $req \rightleftharpoons ack$, i.e. there are two phases in each single communication cycle.

During the *req* phase the sender is active and signals to the receiver the valid data. During the second phase the sender is inactive and waits for an acknowledge in order to continue further. While such a scheme is feasible for on-chip communications, it is obvious that applying it for off-chip is unacceptable. The delays in off-chip wires are significant and this has led us to investigate a novel scheme for the off-chip interfacing which we describe in this section[1]. To avoid

[1]The work on asynchronous buffering is a result of a joint project with Oxford University

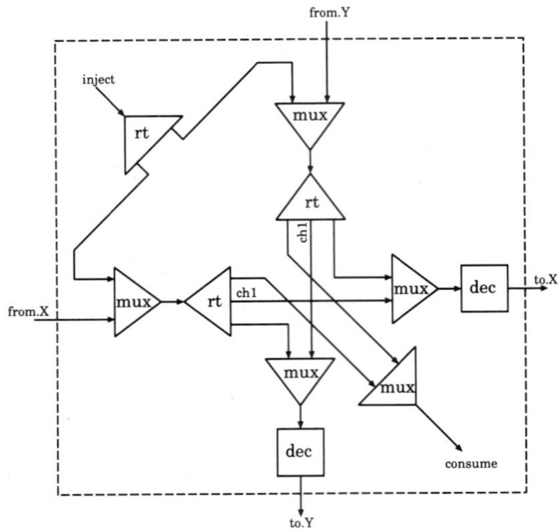

FIGURE 15. Restricted routing packet switching device

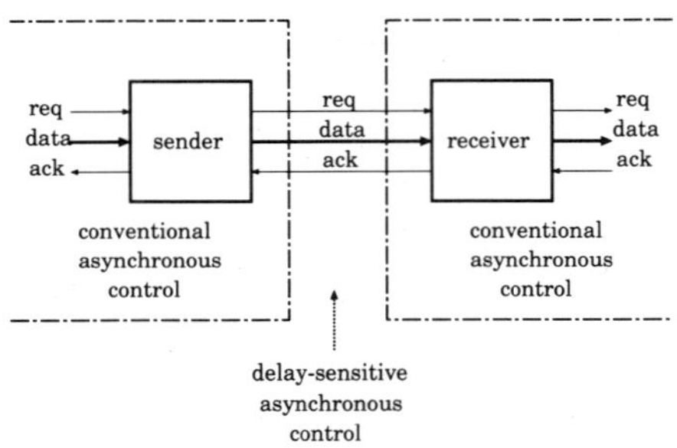

FIGURE 16. $req \leftrightarrow ack$ loop

the problem described above, we would like such an implementation to minimise the number of inactive *ack* phases for off-chip communications.

Instead of waiting for the corresponding *ack*, the sender can continue signaling data with a certain delay between each subsequent words, providing that the receiver can accommodate each incoming word within the duration of this delay. There is a buffer of size *sz* words on the receiver side, thus the sender can continuously send *sz* words (figure 16). When the receiver empties the buffer, it acknowledges the whole window with only one *ack*.

The specification of the sender is as follows:

$$\begin{aligned} SFIFO &= S_0/pa_{out}? \\ S_i &= \textbf{if } (i=0) \textbf{ then } ack_{out}? \textbf{ fi}; pa_{out}?; req_{in}?x; req_{out}!x; ack_{in}!; S_{i+1\%sz} \end{aligned}$$

The specification of the receiver is as follows:

$$RFIFO(ap) = \quad R^{ap}_{0,0}/ack_{out}?$$

$$R^{ap}_{r,w} = \quad [req_{in}?mem[w] \rightarrow pa_{in}!; R^{ap}_{r,w+1\%sz} \Box$$
$$ack_{out}? \rightarrow \textbf{if } (r = ap) \textbf{ then } ack_{in}! \textbf{ fi};$$
$$\textbf{if}(r = w)\textbf{then } req_{in}?mem[w]; pa_{in}!; req_{out}!mem[r]; R^{ap}_{r+1\%sz,w+1\%sz}$$
$$\textbf{else } req_{out}!mem[r]; R^{ap}_{r+1\%sz,w}]$$

One implementation of the sender is shown in figure 17.

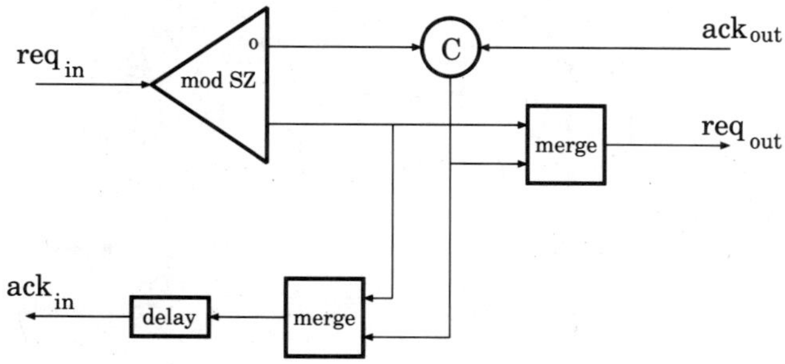

FIGURE 17. Implementation of the *SFIFO* process

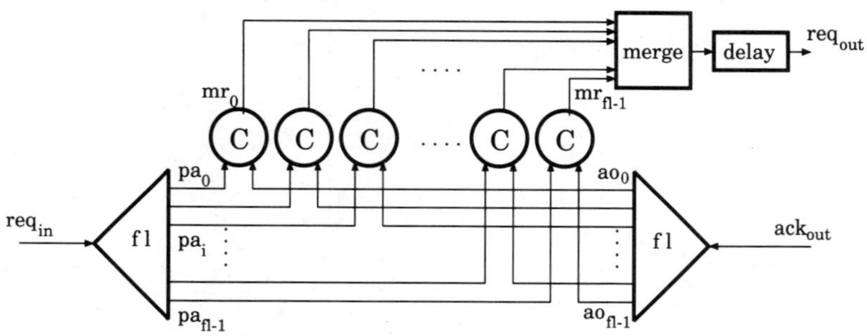

FIGURE 18. Implementation of the $R^0_{0,0}$ - control part

One possible implementation of the process $R^0_{0,0}$ is shown in figure 18 and figure 19, where signals are defined as $e_i = mr_i \oplus mr_{i+1}$ and the acknowledge ack_{in} is identical to signal ao_0. The counters acknowledge each incoming event on a different output (see figure 20). Such an efficient implementation for *nCounter* is possible for only odd numbers n $(= 2 \times k + 1)$. The task of the sender is to signal the valid data with a certain delay between two consequent words. After sz words it awaits an acknowledge in order to further continue. The control part of the receiver ensures that the reader side will not overtake the writer one. The two counters synchronise with each other using a set of muller C elements providing the above constraint.

The data path of the receiver stores each incoming word in the current latch. The merge and the delay line provide a pulse for the latch to accept the current data value.

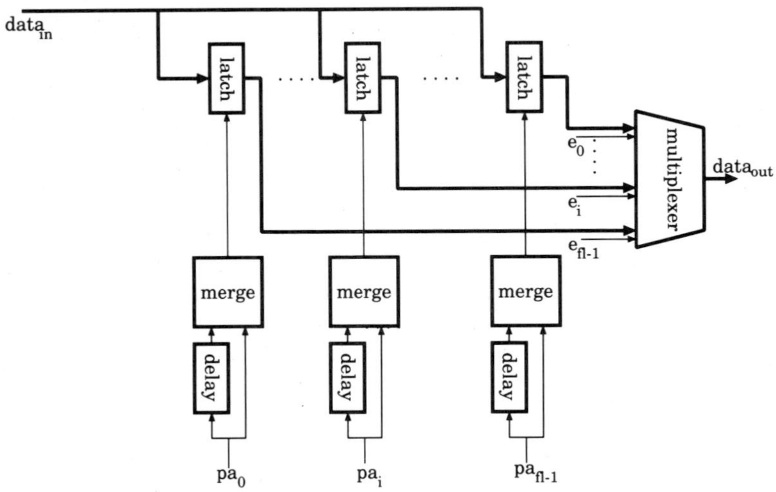

FIGURE 19. Implementation of the $R^0_{0,0}$ / data path

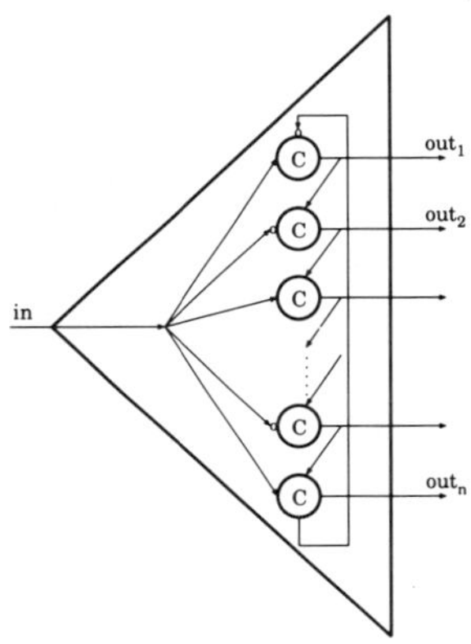

FIGURE 20. Implementation of the *nCounter*

Introducing two additional blocks in the routing node implementation - sender and receiver, improves the latency of the off-chip communications by approximately $T_{improve}$ where

$$T_{improve} = sz \times (DELAY_{ack,off-chip} - DELAY_{ack,on-chip})$$

Another advantage of using such an interfacing scheme is hidden in the fact that coupled together $RFIFO(0) \| SFIFO$ constitute a new process $FIFO(sz)$ where $FIFO(sz)$ (see figure 21) is the elastic micropipeline described in the well-known Sutherland's article [**13**]. This additional buffering improves the bandwidth when the network is heavily loaded.

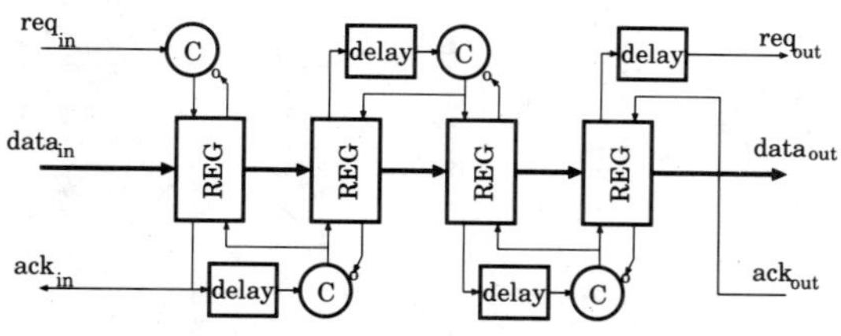

FIGURE 21. Sutherland's FIFO

$$FIFO(sz) \quad = \quad MPL_{[]}/ack_{out}?$$

$$MPL_x \quad = \quad [(\sharp x \neq 0)\&\ ack_{out}? \rightarrow req_{out}!head(x); MPLtail(x)\Box$$
$$(\sharp x \neq sz)\&req_{in}?y \rightarrow ack_{in}!; MPL_{cat(x,y)}]$$

where by $\sharp x$ we denote the length of the list of values x; $head, tail, cat$ are the well-known list functions returning correspondingly the head, the tail and the concatenation of the list parameters. One can argue that the implementation of such an elastic micropipeline suffices for the asynchronous buffering and also the design in Sutherland's article [**13**] is simpler. Since we send the data in burst mode and the receiver should ensure the accommodation of all *sz* incoming words, the current receiving buffer needs to be initially empty. If such a buffer is implemented as an elastic micropipeline, the data has to ripple through the whole pipeline until it reaches the output. The latter is unacceptable for us; it is imperative that we reduce the latency between the input of the sender and the output of the receiver.

One can notice from the specification that the sender is blocked once it fills the buffer on the receiver side, awaiting for an acknowledge. In order to avoid this, we would like to introduce another additional buffer on the receiver side (a receiver with 2 buffers), where the sender fills these buffers in a wagging manner. It could happen in such a case that while the receiver empties the first buffer, the sender fills the second one. Such an implementation will provide a feature of possible contiguous writing if the data is available on the sender side and if the processes further along the pipeline stages are ready to proceed.

The behaviour of the sender is unchanged - it implements $SFIFO/ack_{out}?$

We use two *RFIFO* blocks in order to implement the receiver side (figure 22). The sender initially fills the first two halfs of both *RFIFO* and continues writing in the second two halfs of them since the later are empty. Simultaneously with that, the receiver empties the current set of two halfs *RFIFOs* and acknowledges on completion.

The trick here is that each R block should send an acknowledge ack_{in} only after the previous one has been consumed, i.e. after there is an acknowledge for the acknowledge ack_{in}. That is how ack_{in} is a result of merging the two signals mr_0 and mr_k respectively from block R_0 and R_k. The signal mr_0 appears on ack_{in} only if the receiver has consumed the second set of half *RFIFOs* and the

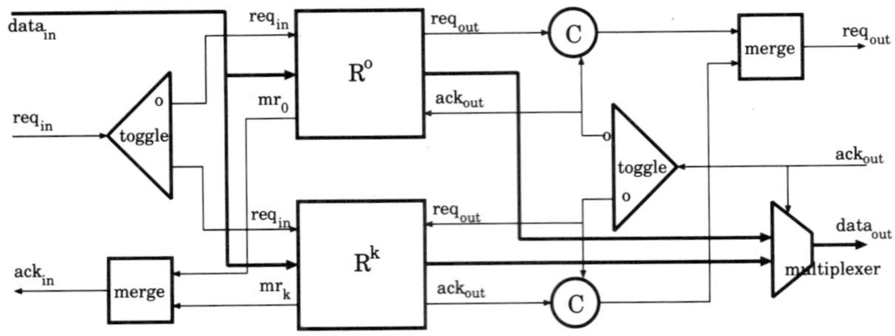

FIGURE 22. Receiver using two RFIFO blocks

sender has received the previous signal mr_k from R_k. The same is valid for the signal mr_k. Therefore mr_0 and mr_k form acknowledges respectively for an empty first/second set half *RFIFOs*.

11. Conclusions

This paper presents an asynchronous implementation of some basic blocks for packet routing switches. It also demonstrates an asynchronous design for creating a correct clocking scheme for the different functional blocks.

The asynchronous control provides a self-clocking scheme inside the chip. We use Delay Insensitive Algebra as a top-specification language, but as the reader can envisage from the implementation that it is not strictly speaking delay insensitive. Our aim was to preserve the delay insensitivity only for the external behaviour of the building blocks regarding their control wires - req, e, ack_e, ack_r. Signals req and e are bundled with the data wires and they have to obey the timing constraints as described in [**13**], i.e. the data must become valid before e or req appear. Certainly it requires the use of *delay* lines for parts of the implementation so that the timing is preserved.

Although we confine ourselves only on mesh topologies in this paper, these are just examples of one use of the universal building blocks we has proposed. Packet routers for other architectures can be constructed using the same design techniques described in the paper. In fact the basic blocks can easily be enhanced so that some other facilities like a broadcast option can be also implemented.

REFERENCES

1. J.Yantchev, I.Nedelchev, *"Implementation of a routing device as a delay-insensitive circuit"*, ACiD-WG/EXACT Workshop on Asynchronous Data Processing, Veldhoven, Netherlands, Dec 10-12, 1992
2. J.Yantchev, I.Nedelchev, *"Implementation of a packet switching device as a delay-insensitive circuit"*, Research on Integrated Systems, the MIT press 1993
3. I.M.Nedelchev, C.R.Jesshope, C.G.Huang, *"Compilation Techniques for Asynchronous Controllers"*, ACiD-WG/EXACT Workshop on Asynchronous Controllers and Interfacing, Leuven, Belgium, Sep 14-16, 1992
4. C.R.Jesshope, I.M.Nedelchev, C.G.Huang, *"Compilation of Process Algebra Expressions into Delay Insensitive Circuits"*, IEE - Computers and Digital Techniques, vol 140, No 5,

Sep 1993

5. Yantchev JT, Jesshope CR, *"Adaptive, Low-Latency, Deadlock-Free Packet-Routing for Networks of Processors"*, IEE Proc.E, Vol.136, No.3, May 1989, pp 178-186.
6. W.J.Dally, C.L.Seitz, *"Deadlock-Free Routing in Multiprocessor Interconnection Networks"*, IEEE Trans on Computers, Vol.C, No.5 May 1987
7. C.J.Seitz, *"The Cosmic Cube"*, CASM Vol. 28, Jan 1985
8. Tony Hoare, "Communicating Sequential Processes", Prentice-Hall 1985
9. Josephs, M., *"Receptive Process theory"*, Eindhoven Uni. Tech. Rep. 90/8, 1990.
10. Josephs, M.B., Udding, J.T., *"Delay-Insensitive Circ uits: An Algebraic Approach to their Design"*, Lect. Notes in Comp. Sci., Vol. 458, pp. 342-366, Springer-Verlag, 1990.
11. Josephs, M.B., Udding, J.T., *"The Design of a Delay-Insensitive Stack"*, In: Jones, G., Sheeran, M. (eds.) Designing Correct Circuits, pp. 132-152, Springer-Verlag, 1990.
12. Josephs, M., Udding, J.T., *"An algebra for delay-insensitive circuits"*, Proceedings DIMACS/IFIP Workshop on Computer-Aided Verification, Rutgers University, New Jersey 1990, DIMACS Series in Discrete Mathematics and Theoretical Computer Science, Vol. 3, pp 147-175, AMS-ACM 1991.
13. Ivan Sutherland, *"Micropipelines"* CASM 6 1989

Department of Electronic Engineering, Univerity of Surrey, Guildford GU2 5XH, England

E-mail address: {C.Jesshope,I.Nedelchev}@EE.Surrey.AC.UK

DIMACS Series in Discrete Mathematics
and Theoretical Computer Science
Volume **21**, 1995

Cayley Digraphs of Finite Cyclic Groups With Minimal Average Distance

XINGDE JIA

Revised on October 30, 1994

ABSTRACT. For any integer $k \geq 1$ and any real number $r > 0$, let $m^*(r,k)$ denote the largest positive integer m such that the Cayley digraph $\mathrm{Cay}(m, A)$ of the cyclic group $\mathbb{Z}_m$ of integers modulo m has average distance at most r for some k-element set A of integers. It is proved in this paper that, for any fixed $k \geq 1$ as $r \to \infty$,

$$\beta_k \left(\frac{10\sqrt[4]{5}}{7}\right)^k \left(\frac{r}{k}\right)^k + O(r^{k-1}) \leq m^*(r,k) \leq \left(\frac{k^2-1}{k^2-k+1}\right)^k \frac{r^k}{(k-1)(k-1)!} + o(r^k),$$

where β_k is a constant close to 1 depending only on k. This improves earlier results by Wong and Coppersmith [**33**] and Jia [**23**]. Applications to distributed loop network and a problem in number theory will also be discussed.

1. Introduction and remarks

Let Γ be any finite group with a subset A. The *Cayley digraph* of Γ generated by A, denoted $\mathrm{Cay}(\Gamma, A)$, is a digraph whose vertex set is Γ, and (x, y) is a (directed) edge if and only if $yx^{-1} \in A$. Let m be a positive integer. Let $\mathbb{Z}_m$ denote the finite cyclic group of residue classes modulo m. We use $\mathrm{Cay}(m, A)$ to denote the Cayley digraph $\mathrm{Cay}(\mathbb{Z}_m, A)$. Cayley digraphs have been studied extensively. In this paper, we focus only on Cayley digraphs of finite cycle groups $\mathbb{Z}_m$.

1991 *Mathematics Subject Classification.* Primary 05C25, 05C35; Secondary 11B13.

This research was supported in part by SWTSU Faculty Research Enhancement Grant 3-2082.

Let m be any positive integer, $A \subseteq \mathbb{Z}_m$ be any subset. Since Cayley digraph $\mathrm{Cay}(m, A)$ is vertex-symmetric, we may define the *average distance* $r(m, A)$ of $\mathrm{Cay}(m, A)$ as follows:

$$r(m, A) = \frac{1}{m} \sum_{n \in \mathbb{Z}_m} d_{\mathrm{Cay}(m,A)}(0, n),$$

where $d_{\mathrm{Cay}(m,A)}(x, y)$ is the (directed) distance from x to y in $\mathrm{Cay}(m, A)$. Given any positive real number r, let $m^*(r, A)$ denote the greatest positive integer m such that $r(m, A) \leq r$. For any positive real number r and any integer $k \geq 1$, define

$$m^*(r, k) = \max_{|A|=k} m^*(r, A),$$

where $|A|$ denotes the cardinality of the set A. In other words, $m^*(r, k)$ is the largest positive integer m such that $r(m, A) \leq r$ for some k-element set A.

Wong and Coppersmith [**33**] proved that

$$\left(\frac{2r}{k} + 1\right)^k \leq m^*(r, k) \leq \left(1 + \frac{1}{k}\right)^k \frac{r^k}{k!}. \tag{1.1}$$

In particular, they proved that

$$\begin{aligned} 8\left(\frac{r}{3}\right)^3 + O(r^2) \leq \quad & m^*(r, 3) \quad \leq 10.6667\left(\frac{r}{3}\right)^3, \\ 16\left(\frac{r}{4}\right)^4 + O(r^3) \leq \quad & m^*(r, 4) \quad \leq 26.0417\left(\frac{r}{4}\right)^4. \end{aligned}$$

Recently, Jia [**23**] proved that

$$m^*(r, 2) = \frac{27}{25} r^2 + O(r) \quad \text{as} \quad r \to \infty.$$

A similar result can be found also in [**9**]. Jia also proved that, for any fixed $k \geq 3$ as r tends to infinity,

$$m^*(r, k) \geq \alpha_k \left(\frac{4\sqrt[3]{4}}{3}\right)^k \left(\frac{r}{k}\right)^k + O(r^{k-1}),$$

where α_k is a constant defined as follows:

$$\alpha_k = \begin{cases} 1 & \text{if } k \equiv 0 \pmod 3 \\ \dfrac{3}{2\sqrt[3]{4}} \approx 0.945 & \text{if } k \equiv 1 \pmod 3 \\ \dfrac{243}{200\sqrt[3]{2}} \approx 0.964 & \text{if } k \equiv 2 \pmod 3 \end{cases},$$

which improves the lower bound (1.1) of Wong and Coppersmith. In this paper, we shall prove the following lower and upper bounds of $m^*(r,k)$ for any fixed $k \geq 1$ as r tends to infinity:

$$\beta_k \left(\frac{10\sqrt[4]{5}}{7}\right)^k \left(\frac{r}{k}\right)^k + O(r^{k-1}) \leq m^*(r,k)$$
$$\leq \left(\frac{k^2-1}{k^2-k+1}\right)^k \frac{r^k}{(k-1)(k-1)!} + o(r^k),$$

where β_k is close to 1, which is defined in Theorem 2 on page 9. This improves the lower bound of Jia and the upper of Wong and Coppersmith. In particular, we prove that

$$9.4824\left(\frac{r}{3}\right)^3 + O(r^2) \leq \quad m^*(r,3) \quad \leq 10.0758\left(\frac{r}{3}\right)^3 + o(r^3),$$
$$20.8246\left(\frac{r}{4}\right)^4 + O(r^3) \leq \quad m^*(r,4) \quad \leq 25.2092\left(\frac{r}{4}\right)^4 + o(r^4).$$

This paper is arranged as follows. In section 2, we prove a lower bound for $m^*(r,4)$, and use a recursive theorem to establish the lower bound for $m^*(r,k)$ for any fixed $k \geq 2$ as $r \to \infty$. In section 3, we shall prove the upper bound of $m^*(r,k)$, where we shall use a method of Wong and Coppersmith [**33**] to visit the vertices of the Cayley digraph $\mathrm{Cay}(m,A)$. We shall discuss the average distance of Cayley digraph in section 4. And in section 5, we shall discuss an application of the result to the average order of asymptotic bases in additive number theory.

We close this section with few remarks. The "non-average" version of $m^*(r,k)$ is defined as follows. Given any positive integers d, k,

$$m(d,k) = \max\{m \mid d(m,A) \leq d \text{ for some } A \text{ with } |A| = k\},$$

where $d(m,A)$ is the diameter of Cayley digraph $\mathrm{Cay}(m,A)$. This extremal function and other related problems have been studied by many authors, see, for instance, Wong and Coppersmith [**33**], Fiol, Yebra, Alegre and Valero [**9**], Erdös and Hsu [**8**], Hsu and Jia [**15**], Hsu and Shapiro [**17, 18**], Jia [**22**], and a recent survey by Bermond, Comellas and Hsu [**3**]. In particular, Wong and Coppersmith [**33**] proved that

$$m(d,k) \geq \left(\frac{d}{k}+1\right)^k.$$

Hsu and Jia [**15**] proved among other results that

$$m(d,2) = \left\lfloor \frac{d(d+4)}{3} \right\rfloor + 1$$

for all $d \geq 2$, and as $d \to \infty$,

$$\frac{1}{16}d^3 + \frac{3}{8}d^2 + O(d) \leq m(d,3) \leq \frac{1}{14-3\sqrt{3}}(d+3)^3.$$

Jia [**22**] proved that, for fixed $k \geq 4$ as $d \to \infty$,

$$m(d,k) \geq \gamma_k \left(\frac{256}{125}\right)^{\lfloor k/4 \rfloor} \left(\frac{d}{k}\right)^k + O(d^{k-1}),$$

where

$$\gamma_k = \begin{cases} 1 & \text{if } k \equiv 0 \text{ or } 1 \pmod 4 \\ \frac{4}{3} & \text{if } k \equiv 2 \pmod 4 \\ \frac{27}{16} & \text{if } k \equiv 3 \pmod 4 \end{cases}$$

Chen and Gu improved this lower bound in [**6**]. The best known lower bound is due to Su [**31**], who proved that

$$m(d,k) \quad \geq \quad \delta_k \left(\frac{5^5 \cdot 7^4}{17^5}\right)^{\lfloor k/5 \rfloor} \left(\frac{d}{k}\right)^k + O(d^{k-1}),$$

where

$$\delta_k = \begin{cases} 1 & \text{if } k \equiv 0, 1 \pmod 5 \\ 4/3 & \text{if } k \equiv 2 \pmod 5 \\ \frac{4752}{2197} \approx 2.163 & \text{if } k \equiv 3 \pmod 5 \\ \frac{165888}{50625} = 3.2768 & \text{if } k \equiv 4 \pmod 5 \end{cases}$$

This lower bound was constructed by using computer. Recently, Hsu and Jia [**16**] proved that

$$m(d,k) \leq \frac{1}{(k-1)!} \left(\frac{k-1}{k^{1-1/k} - 1}\right)^{k-1} (d+5)^k.$$

There is still a big gap between the lower and the upper bounds of $m(d,k)$. In particular, the best known estimate for $m(d,3)$ is

$$0.08010929r^3 + O(r^2) \leq m(d,3) \leq 0.113587r^3 + O(r^2).$$

It is very desirable to have an approximate formula for $m(d,3)$.

If we use the interval $[1,m]$ instead of $\mathbb{Z}_m$ in the definition of $m(d,k)$, we obtain the extremal function $n(d,k)$. Namely, $n(d,k)$ is the largest positive integer m such that there exists a k-element subset A of positive integers with the property that every integer in $[1,m]$ is sum of d not necessarily distinct elements of A. The problem of computing $n(d,k)$ is sometimes referred as *the postage stamp problem,* due to a rather obvious combinatorial interpretation. This is an old problem in additive number theory, and has received extensive study (see, for example, Selmer's research monograph [**29**]). Obviously, $m(d,k) \geq n(d,k)$ for all $d \geq 1$ and $k \geq 1$. Stöhr [**30**] and several later writers proved that

$$n(d,2) = \left\lfloor \frac{d^2 + 6d + 1}{4} \right\rfloor,$$

and Hofmeister [**13, 14**] showed that

$$n(d,3) = \frac{4}{81}d^3 + O(h^2).$$

The postage stamp problem was studied by Stöhr [**30**], Mrose [**24**], Windecker [**32**], Mossige [**25**], and many others (see, for example, Guy's problem book [**12**]). We may study the "average" version of $n(d,k)$, which is defined as follows. For any real number $r > 0$, and any positive integer k, let $n^*(r,k)$ be the largest positive integer m such that there exists a k-element subset A of positive integers with

$$\frac{1}{m}\sum_{i=1}^{m} f(A,i) \leq r,$$

where $f(A,i)$ is the least number of elements (repetitions allowed) of A with sum i. Clearly, $n^*(d,k) \geq n(d,k)$. Estimates of $n(d,k)$ and $m(d,k)$ have been established for any fixed d as k tends to infinity. However, as far as I know, there are no analogue estimates for $n^*(r,k)$ and $m^*(r,k)$ for fixed r as $k \to \infty$. Obviously, for any fixed $r \geq 1$,

$$n(\lfloor r \rfloor, k) \leq n^*(r,k) \leq m^*(r,k) \leq \frac{k^{\lceil r \rceil}}{\lceil r \rceil !} + O(k^{\lceil r \rceil - 1}).$$

It would be interesting to improve these bounds.

2. A lower bound

Let $\mathrm{Cay}(m,A)$ be any Cayley digraph. For any integer n, let $f(A,n)$ be the least number of elements of A with sum congruent to n modulo m. In other words, $f(A,n)$ is the distance from 0 to n in the digraph $\mathrm{Cay}(m,A)$. Thus

$$r(m,A) = \frac{1}{m}\sum_{n \in \mathbb{Z}_m} f(A,n).$$

THEOREM 1. *For $r \to \infty$,*

$$m^*(r,4) \geq \frac{50000}{2401}\left(\frac{r}{4}\right)^4 + O(r^3).$$

Proof. Let $r \geq \frac{19}{5}$ be any real number. Define

$$\begin{aligned} t &= \left\lfloor \frac{5(r-1)}{14} \right\rfloor \\ a &= 5t \\ b &= at + t = 5t^2 + t \\ c &= bt + 2t = 5t^3 + t^2 + 2t \\ m &= ct + 3t = 5t^4 + t^3 + 2t^2 + 3t. \end{aligned}$$

Then

$$m = \frac{50000}{2401} \left(\frac{r}{4}\right)^4 + O(r^3).$$

Let $A = \{1, a, b, c\}$. We only need to show that the average distance of the Cayley digraph $\mathrm{Cay}(m, A)$ is at most r, i.e., $r(m, A) \leq r$.

First consider

$$n = p + a(x-1) + by + cz,$$

where

$$0 \leq p < a = 5t, \quad 1 \leq x, y, z \leq t.$$

Regarding p, we consider the following four cases.

I. $0 \leq p < 2t$. Since

$$n = p \cdot 1 + a(x-1) + by + cz,$$

we see that

$$f(A, n) \leq p + (x-1) + y + z.$$

Thus

$$\begin{aligned} \sum_{n=a(x-1)+by+cz}^{2t-1+a(x-1)+by+cz} f(A,n) &\leq \sum_{p=0}^{2t-1} (p + (x-1) + y + z) \\ &= 2t^2 - 3t + 2t(x+y+z). \end{aligned}$$

II. $2t \leq p < 3t$. Since $p - 2t \geq 0$ and

$$n \equiv (p-2t) \cdot 1 + ax + by + c(z+t) \pmod{m},$$

we have that

$$f(A, n) \leq (p-2t) + x + y + (z+t) = p - t + x + y + z.$$

Thus

$$\begin{aligned} \sum_{n=2t+a(x-1)+by+cz}^{3t-1+a(x-1)+by+cz} f(A,n) &\leq \sum_{p=2t}^{3t-1} (p - t + x + y + z) \\ &= \frac{3}{2}t^2 - \frac{1}{2}t + t(x+y+z). \end{aligned}$$

III. $3t \leq p < 4t$. Noting that $p - 3t \geq 0$ and

$$n = (p - 3t) \cdot 1 + ax + b(y + t) + c(z - 1),$$

we see that

$$f(A < n) \leq (p - 3t) + x + (y + t) + (z - 1) = p - 2t - 1 + x + y + z.$$

Thus

$$\begin{aligned} \sum_{n=3t+a(x-1)+by+cz}^{4t-1+a(x-1)+by+cz} f(A,n) &\leq \sum_{p=3t}^{4t-1} (p - 2t - 1 + x + y + z) \\ &= \frac{3}{2}t^2 - \frac{3}{2}t + t(x + y + z). \end{aligned}$$

IV. $4t \leq p < 5t = a$. Noting that $p - 4t \geq 0$ and

$$n = (p - 4t) \cdot 1 + a(x + t) + b(y - 1) + cz,$$

we see that

$$f(A, n) \leq (p - 4t) + (a + t) + (y - 1) + z = p - 3t - 1 + x + y + z.$$

Thus

$$\begin{aligned} \sum_{n=4t+a(x-1)+by+cz}^{ax+by+cz-1} f(A,n) &\leq \sum_{p=4t}^{5t-1} (p - 3t - 1 + x + y + z) \\ &= \frac{3}{2}t^2 - \frac{3}{2}t + t(x + y + z). \end{aligned}$$

Therefore,

$$\begin{aligned} \sum_{n=a(x-1)+by+cz}^{ax+by+cz-1} f(A,n) &= \left\{2t^2 - 3t + 2t(x + y + z)\right\} \\ &\quad + \left\{\frac{3}{2}t^2 - \frac{1}{2}t + t(x + y + z)\right\} \\ &\quad + \left\{\frac{3}{2}t^2 - \frac{3}{2}t + t(x + y + z)\right\} \\ &\quad + \left\{\frac{3}{2}t^2 - \frac{3}{2}t + t(x + y + z)\right\} \\ &= \frac{13}{2}t^2 - \frac{13}{2}t + 5t(x + y + z). \end{aligned}$$

Since x is arbitrary in $[1, t]$, we see that

$$\begin{aligned} \sum_{n=by+cz}^{at+by+cz-1} f(A,n) &\leq \sum_{x=1}^{t} \left\{\frac{13}{2}t^2 - \frac{13}{2}t + 5t(x + y + z)\right\} \\ &= 9t^3 - 4t^2 + 5t^2(y + z). \end{aligned} \tag{2.1}$$

When $at+by+cz \le n < b(y+1)+cz$, we may write n as

$$n = (n-(at+by+cz))\cdot 1 + at+by+cz,$$

which implies that

$$F(A,n) \le (n-(at+by+cz)) + t+y+z.$$

Hence,

$$\begin{aligned} \sum_{n=at+by+cz}^{b(y+1)+cz-1} f(A,n) &\le \sum_{p=0}^{t-1}(p+t+y+z) \\ &= \frac{3}{2}t^2 + \frac{1}{2}t + t(y+z). \end{aligned} \tag{2.2}$$

Therefore,

$$\begin{aligned} \sum_{n=by+cz}^{b(y+1)+cz-1} f(A,n) &= \sum_{n=by+cz}^{at+by+cz-1} f(A,n) + \sum_{n=at+by+cz}^{b(y+1)+cz-1} f(A,n) \\ &\le \{9t^3 - 4t^2 + 5t^2(y+z)\} + \left\{\frac{3}{2}t^2 + \frac{1}{2}t + t(y+z)\right\} \\ &= 9t^3 - \frac{5}{2}t^2 + \frac{1}{2}t + (5t^2+t)(y+z) \end{aligned}$$

Since y is arbitrary in $[1,t]$, we see

$$\begin{aligned} \sum_{n=b+cz}^{b(t+1)+cz-1} f(A,n) &\le \sum_{y=1}^{t}\left\{9t^3 - \frac{5}{2}t^2 + \frac{1}{2}t + (5t^2+t)(y+z)\right\} \\ &= \frac{23}{2}t^4 + \frac{1}{2}t^3 + t^2 + (5t^3+t^2)z \end{aligned} \tag{2.3}$$

When $b(t+1)+cz \le n < b+c(z+1)$, n can be written as

$$n = (n-b(t+1)-cz)\cdot 1 + b(t+1)+cz,$$

thus

$$f(A,n) \le (n-b(t+1)-cz) + t+1+z,$$

which implies that

$$\begin{aligned} \sum_{n=b(t+1)+cz}^{b+c(z+1)-1} f(A,n) &\le \sum_{p=0}^{3t-1}(p+t+1+z) \\ &= \frac{15}{2}t^2 + \frac{3}{2}t + 3tz. \end{aligned} \tag{2.4}$$

It follows from (2.3) and (2.4) that

$$\sum_{n=b+cz}^{b+c(z+1)-1} f(A,n) \le \frac{23}{2}t^4 + \frac{1}{2}t^3 + \frac{17}{2}t^2 + \frac{3}{2}t + (5t^3+t^2+3t)z.$$

Again, noting the arbitrariness of $z \in [1, t]$, we see that

$$\begin{aligned}\sum_{n=b+c}^{b+c(t+1)-1} f(A,n) &\le \sum_{z=1}^{t}\left\{\frac{23}{2}t^4+\frac{1}{2}t^3+\frac{17}{2}t^2+\frac{3}{2}t+(5t^3+t^2+3t)z\right\}\\ &= 14t^5+\frac{7}{2}t^4+\frac{21}{2}t^3+3t^2. \qquad (2.5)\end{aligned}$$

When $b + c(t+1) \le n < m + b + c = b + c(t+1) + 3t$, we write n as

$$n = (n - b - c(t+1)) \cdot 1 + b + c(t+1),$$

which implies that

$$\sum_{n=b+c(t+1)}^{m+b+c-1} f(A,n) \le \sum_{p=0}^{3t-1}(p+1+(t+1)) = \frac{15}{2}t^2+\frac{9}{2}t. \qquad (2.6)$$

Hence it follows from (2.5) and (2.6) that

$$\sum_{n=b+c}^{m+b+c-1} f(A,n) \le 14t^5+\frac{7}{2}t^4+\frac{21}{2}t^3+\frac{21}{2}t^2+\frac{9}{2}t.$$

Therefore, when t is large,

$$\begin{aligned} r(m,A) &= \frac{1}{m}\sum_{n\in\mathbb{Z}_m} f(A,n)\\ &\le \frac{14t^5+\frac{7}{2}t^4+\frac{21}{2}t^3+\frac{21}{2}t^2+\frac{9}{2}t}{5t^4+t^3+2t^2+3t}\\ &\le \frac{14t^5+5t^4}{5t^4} = \frac{14}{5}t+1 \le r.\end{aligned}$$

The proof of Theorem 1 is complete.

THEOREM 2. *For any fixed $k \ge 1$ as $r \to \infty$,*

$$m^*(r,k) \le \beta_k\left(\frac{10\sqrt[4]{5}}{7}\right)^k\left(\frac{r}{k}\right)^k + O(r^{k-1}), \qquad (2.7)$$

where

$$\beta_k = \begin{cases} 1 & \textit{if } k \equiv 0 \pmod 4\\ \dfrac{7}{5\sqrt[4]{5}} \approx 0.9362 & \textit{if } k \equiv 1 \pmod 4\\ \dfrac{1323}{625\sqrt{5}} \approx 0.9467 & \textit{if } k \equiv 2 \pmod 4\\ \dfrac{10976}{3375\sqrt[4]{125}} \approx 0.9726 & \textit{if } k \equiv 3 \pmod 4\end{cases}.$$

To prove this theorem, we need the following two lemmas.

LEMMA 1. *For r large,*
(i) $m^*(r,1) = \lfloor 2r \rfloor$,
(ii) $m^*(r,2) = \frac{27}{25}r^2 + O(r)$,
(iii) $m^*(r,3) \geq \frac{256}{729}r^3 + O(r^2)$.

Proof. (i) is obvious, and (ii) and (iii) were proved in [**23**].

LEMMA 2 (JIA, [**23**]). *Let $r_1 \geq 1$ and $r_2 \geq 1$ be any real numbers, k_1 and k_2 any positive integers. Then*

$$m^*(r_1 + r_2, k_1 + k_2) \geq m^*(r_1, k_1) \cdot m^*(r_2, k_2).$$

Proof of Theorem 2. Let $k = k' + u$ where $k' = 4q$, $q \in \mathbb{Z}$. Let $r' = r/k$, then

$$\begin{aligned}
m^*(r,k) &= m^*(4qr' + ur', 4q + u) \\
&\geq m^*(4qr', 4q) m^*(ur', u) \\
&\geq m^*(4r', 4)^q m^*(ur', u) \\
&\geq \left\{ \frac{50000}{2401} \left(\frac{4r'}{4} \right)^4 + O(r'^3) \right\}^q m^*(ur', u) \\
&= \left\{ \left(\frac{10\sqrt[4]{5}}{7} \right)^{4q} \left(\frac{r}{k} \right)^{4q} + O(r^{4q-1}) \right\} m^*(ur/k, u) \\
&\geq c_u u^u \left(\frac{7}{10\sqrt[4]{5}} \right)^u \left(\frac{10\sqrt[4]{5}}{7} \right)^k \left(\frac{r}{k} \right)^k + O(r^{k-1}),
\end{aligned}$$

where

$$c_0 = 1, \quad c_1 = 2, \quad c_2 = \frac{27}{25}, \quad \text{and} \quad c_3 = \frac{256}{729}.$$

(2.7) follows immediately from a simple calculation.

3. An upper bound

LEMMA 3. *Let $\{D_n \mid n = 1, 2, \ldots,\}$ be a sequence of measurable solids in the first quadrant of $\mathbb{R}^k$. Suppose that either $f(\xi) = 1, (\xi \in \mathbb{R}^k)$ or $f(\xi) = x_1 + \cdots + x_k, (\xi = (x_1, \ldots, x_k) \in \mathbb{R}^k)$. Let S_n denote the the boundary of D_n and T_n denote the set of integral lattice points in D_n. If*

$$\lim_{n \to \infty} \frac{\oint_{S_n} f(\xi)\, d\xi}{\int_{D_n} f(\xi)\, d\xi} = 0, \tag{3.1}$$

then

$$\lim_{n \to \infty} \frac{\sum_{\xi \in T_n} f(\xi)}{\int_{D_n} f(\xi)\, d\xi} = 1.$$

Proof. Let $C_\xi \subseteq \mathbb{R}^k$ be the unit cube in $\mathbb{R}^k$ centered at ξ. Clearly,

$$f(\xi) = \int_{C_\xi} f(\zeta)\, d\zeta.$$

Let L_k denote the set of all integral lattice points in $\mathbb{R}^k$. Then

$$\begin{aligned}\left|\sum_{\xi\in T_n} f(\xi) - \int_{D_n} f(\xi)d\xi\right| &= \left|\sum_{\xi\in T_n}\int_{C_\xi} f(\zeta)\, d\zeta - \int_{D_n} f(\xi)\, d\xi\right| \\ &= \left|\sum_{\substack{\xi\in T_n\\ C_\xi\cap S_n\neq\emptyset}}\int_{C_\xi} f(\zeta)\, d\zeta - \sum_{\substack{\xi\in L_k\\ C_\xi\cap S_n\neq\emptyset}}\int_{C_\xi\cap D_n} f(\zeta)\, d\zeta\right| \\ &\leq \sum_{\substack{\xi\in L_k\\ C_\xi\cap S_n\neq\emptyset}}\int_{C_\xi} f(\zeta)\, d\zeta.\end{aligned}$$

Let $n(S_n)$ be the union of all balls with center on S_n and diameter $\frac{1}{2}$. Noting that S_n is the surface of a solid, we see that

$$\int_{n(S_n)} f(\xi)\, d\xi \leq \oint_{S_n} f(\xi)\, d\xi. \tag{3.2}$$

Let C_ξ be such that $\xi \in L_k$ and $C_\xi \cap S_n \neq \emptyset$. We divide C_ξ into $\lceil 2\sqrt{k}\rceil^k$ equal subcubes. Let C'_ξ be a subcube in C_ξ that intersects S_n. Since $f(x_1, \ldots, x_k) = 1$ or $f(x_1, \ldots, x_k) = x_1 + \cdots + x_k$, there exists an absolute constant $c = c(k)$ depending only on k such that

$$\int_{C_\xi} f(\zeta)\, d\zeta \leq c\int_{C'_\xi} f(\zeta)\, d\zeta. \tag{3.3}$$

Noting that the diameter of C'_ξ is $\leq \frac{1}{2}$, we see that $C'_\xi \subseteq n(S_n)$. Therefore, it follows from (3.2) and (3.3) that

$$\begin{aligned}\left|\sum_{\xi\in T_n} f(\xi) - \int_{D_n} f(\xi)\, d\xi\right| &\leq \sum_{\substack{\xi\in L_k\\ C_\xi\cap S_n\neq\emptyset}}\int_{C_\xi} f(\zeta)\, d\zeta \\ &\leq c\sum_{\substack{\xi\in L_k\\ C_\xi\cap S_n\neq\emptyset}}\int_{C'_\xi} f(\zeta)\, d\zeta \\ &\leq c\int_{n(S_n)} f(\xi)\, d\xi \\ &\leq c\oint_{S_n} f(\xi)\, d\xi.\end{aligned}$$

This together with (3.1) will imply the required property.

LEMMA 4. *Let $D = D(\alpha)$ be the solid in the k-dimensional Euclidean space bounded by*

$$x_1 \geq 0, x_2 \geq 0, \dots, x_k \geq 0,$$
$$\sum_{j\neq i} x_j \leq \alpha \quad \textit{for} \quad i = 1, 2, \dots, k$$

Then the volume of D is

$$V(D) = \frac{\alpha^k}{(k-1)(k-1)!}.$$

Proof. See [**16**].

LEMMA 5. *Let E be the solid bounded by*

$$\sum_{i=1}^{k} x_i \leq \alpha, \quad x_i \geq 0.$$

Then

$$\int_E (x_1 + \cdots + x_k)\, dx_1 \cdots dx_k = \frac{k\alpha^{k+1}}{(k+1)!}. \tag{3.4}$$

Proof. For $\mu \geq 1$, let $E_\mu(t)$ be the solid in $\mathbb{R}^\mu$ bounded by

$$\sum_{i=1}^{\mu} x_i \leq t, \quad x_i \geq 0.$$

Assume that (3.4) is true for $\mu - 1$. Then it follows from Lemma 4 that

$$\begin{aligned}
&\int_{E_\mu(t)} (x_1 + \cdots + x_\mu)\, dx_1 \cdots dx_\mu \\
&= \int_0^t \int_{E_{\mu-1}(t-x_\mu)} (x_1 + \cdots + x_{\mu-1})\, dx_1 \cdots dx_\mu \\
&= \int_0^t \left\{ \int_{E_{\mu-1}(t-x_\mu)} (x_1 + \cdots + x_{\mu-1})\, dx_1 \cdots dx_{\mu-1} \right\} dx_\mu \\
&\quad + \int_0^t \left\{ \int_{E_{\mu-1}(t-x_\mu)} x_\mu\, dx_1 \cdots dx_{\mu-1} \right\} dx_\mu \\
&= \int_0^t \frac{(\mu-1)(t-x_\mu)^\mu}{\mu!}\, dx_\mu + \int_0^t \frac{(t-x_\mu)^{\mu-1}}{(\mu-1)!} \cdot x_\mu\, dx_\mu \\
&= \frac{(\mu-1)t^{\mu+1}}{(\mu+1)!} + \frac{t^{\mu+1}}{(\mu+1)!} \\
&= \frac{\mu t^{\mu+1}}{(\mu+1)!},
\end{aligned}$$

i.e., (3.4) is true for μ. Obviously, it is true for $\mu = 1$. Thus it is true for every μ. In particular, we have

$$\int_E (x_1 + \cdots + x_k)\, dx_1 \cdots dx_k = \frac{k\alpha^{k+1}}{(k+1)!}.$$

LEMMA 6. *Let $D = D(\alpha)$ be as defined in Lemma* 4. *Then*

$$\int_D (x_1 + \cdots + x_k)\, dx_1 \cdots dx_k = \frac{(k^3 - k^2 + k)\alpha^{k+1}}{(k-1)^2(k+1)!}. \tag{3.5}$$

Proof. The proof will follow the idea in the proof of Lemma 4. It is easy to see that the solid D has the following vertices:

$$(0, \dots, 0), \left(\frac{\alpha}{k-1}, \dots, \frac{\alpha}{k-1}\right), \quad (\alpha, 0, \dots, 0), \dots, (0, \dots, 0, \alpha).$$

It is clear that the distance between $(0, \dots, 0)$ and $\left(\frac{\alpha}{k-1}, \dots, \frac{\alpha}{k-1}\right)$ is $\frac{\sqrt{k}\alpha}{k-1}$.

Let $0 \le t \le \frac{\sqrt{k}\alpha}{k-1}$. Let $P(t)$ denote the $(k-1)$-dimensional hyperplane:

$$P(t): \quad x_1 + \cdots + x_k = \sqrt{k}t.$$

Clearly, $P(t)$ is perpendicular to the line

$$L: \quad x_1 = x_2 = \cdots = x_k,$$

and the intersection point is $\left(\frac{t}{\sqrt{k}}, \dots, \frac{t}{\sqrt{k}}\right)$. Note that the distance from $(0, \dots, 0)$ to $P(t)$ is t. The cross-section of the solid D by $P(t)$ is a regular k-gon, whose vertices are

$$(\sqrt{k}t, 0, \dots, 0), \dots, (0, \dots, 0, \sqrt{k}t)$$

if $0 \le t \le \frac{\alpha}{\sqrt{k}}$. When

$$\frac{\alpha}{\sqrt{k}} < t < \frac{\sqrt{k}\alpha}{k-1},$$

the vertices of the k-gon are determined by the following k systems of linear equations:

$$\begin{cases} \sum\limits_{j \ne i} x_j = \alpha \quad i = 1, \dots, k, \; i \ne r; \\ \sum\limits_{i=1}^{k} x_i = \sqrt{k}t \end{cases} \tag{3.6}$$

for $r = 1, 2, \dots, k$. In the case $r = k$, the above system can be written as follows:

$$\begin{bmatrix} 0 & 1 & \cdots & 1 & 1 \\ 1 & 0 & \cdots & 1 & 1 \\ \vdots & & \ddots & & \vdots \\ 1 & 1 & \cdots & 0 & 1 \\ 1 & 1 & \cdots & 1 & 1 \end{bmatrix} \begin{bmatrix} x_1 \\ x_2 \\ \vdots \\ x_{k-1} \\ x_k \end{bmatrix} = \begin{bmatrix} \alpha \\ \alpha \\ \vdots \\ \alpha \\ \sqrt{k}t \end{bmatrix}.$$

A simple calculation shows that the solution to this system is

$$x_1 = x_2 = \cdots = x_{k-1} = \sqrt{k}t - \alpha, \quad x_k = (k-1)\alpha - (k-2)\sqrt{k}t.$$

Let v_k denote this vertex. It is clear that $v'_k = (0, \dots, 0, k\alpha - (k-1)\sqrt{k}t)$ is the projection of v_k onto the boundary of D along the line L. v'_k is a vertex of the cross-section of D by

$$P(\sqrt{k}\alpha - (k-1)t) : \sum_{i=1}^{k} x_i = k\alpha - (k-1)\sqrt{k}t.$$

Because of the symmetry of (3.6), the projection points of the vertices of the cross-section by $P(t)$ along L are the vertices of the cross-section of D by $P(\sqrt{k}\alpha - (k-1)t)$. Let $S(t)$ be the area of the cross-section of the solid D by $P(t)$. Noting that $P(t)$ and $P(\sqrt{k}\alpha - (k-1)t)$ are parallel, we have

$$S(t) = S\left(\sqrt{k}\alpha - (k-1)t\right) \quad \text{for} \quad 0 \le t \le \frac{\sqrt{k}\alpha}{k-1}. \tag{3.7}$$

Let $\|\xi\| = x_1 + \cdots + x_k$. Then

$$\begin{aligned}
\int_D \|\xi\|\, d\xi &= \int_0^{\frac{\sqrt{k}\alpha}{k-1}} S(t)\sqrt{k}t\, dt \\
&= \int_0^{\frac{\alpha}{\sqrt{k}}} S(t)\sqrt{k}t\, dt + \int_{\frac{\alpha}{\sqrt{k}}}^{\frac{\sqrt{k}\alpha}{k-1}} S(t)\sqrt{k}t\, dt \\
&= \int_E \|\xi\|\, d\xi + \int_{\frac{\alpha}{\sqrt{k}}}^{\frac{\sqrt{k}\alpha}{k-1}} S\left(\sqrt{k}\alpha - (k-1)t\right)\sqrt{k}t\, dt.
\end{aligned}$$

In the second integral, let

$$s = \sqrt{k}\alpha - (k-1)t,$$

then it follows from Lemmas 4 and 5 that

$$\begin{aligned}
\text{Second Integral} &= \frac{\sqrt{k}}{(k-1)^2}\int_0^{\frac{\alpha}{\sqrt{k}}} S(t')(\sqrt{k}\alpha - s)\, ds \\
&= \frac{k\alpha}{(k-1)^2}\int_0^{\frac{\alpha}{\sqrt{k}}} S(t)\, dt - \frac{1}{(k-1)^2}\int_0^{\frac{\alpha}{\sqrt{k}}} S(t)\sqrt{k}\alpha\, dt \\
&= \frac{k\alpha}{(k-1)^2}\cdot\frac{\alpha^k}{k!} - \frac{1}{(k-1)^2}\cdot\frac{k\alpha^{k+1}}{(k+1)!} \\
&= \frac{k^2\alpha^{k+1}}{(k-1)^2(k+1)!}.
\end{aligned}$$

Therefore,

$$\int_D \|\xi\|\, d\xi = \frac{k\alpha^{k+1}}{(k+1)!} + \frac{k^2\alpha^{k+1}}{(k-1)^2(k+1)!} = \frac{(k^3 - k^2 + k)\alpha^{k+1}}{(k-1)^2(k+1)!}.$$

The proof is complete.

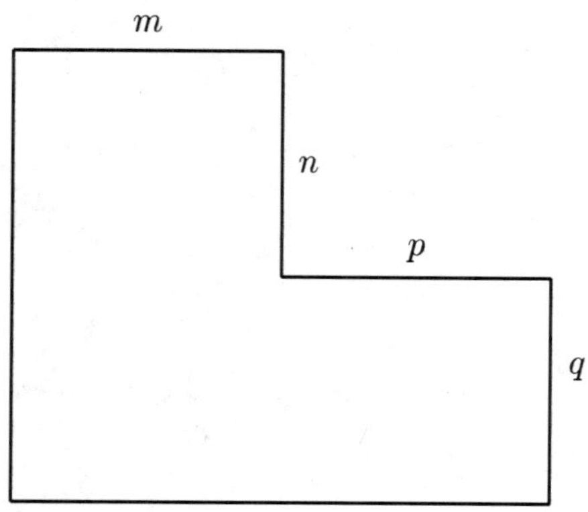

FIGURE 1. The general pattern

For a given $k \geq 2$, we are interested in computing $m^*(r,k)$ in terms of r. First of all, we study the case when $k=2$. Wong and Coppersmith [**33**] provided the construction of a pattern for the Cayley digraph Cay(m, A).

In the computation of the average distance of $G(m, A)$, where $A = \{1, s\}$, one proceeds to fill the lattice point (x, y) ($x \geq 0, y \geq 0$ are integers) of the Euclidean plane with an integer n ($0 \leq n < m$) if

$$x \cdot 1 + y \cdot s \equiv n \pmod{m}.$$

We start from the origin $(0,0)$, then the points on the line $(1,0)$, $(0,1)$, and then the points on the line $(2,0)$, $(1,1)$, $(0,2)$, and so on. At each point (x, y) if the value n has not appeared so far, we write it down, otherwise we just leave a blank. The process ends when all values of n in $\mathbb{Z}_m$ have been used. Wong and Coppersmith [**33**] proved the following lemma.

LEMMA 7 (WONG AND COPPERSMITH [**33**]). *The filled pattern is always of the form shown below in Figure 1, where $m \geq 0$, $n \geq 0$, $p > 0$ and $q > 0$. Clearly, the diameter $d(G(N, A))$ of $G(N, A)$, where $A = \{1, s\}$, is equal to $m + q + \max\{n, p\} - 2$.*

In the general k-dimensional case, Wong and Coppersmith had the following visitation method. For convenience, we only consider the first quadrant in the k-dimensional Euclidean space. Let u be a nonnegative integer. Define

$$S_u^{(k)} = \left\{ (x_1, \ldots, x_k) \;\middle|\; x_i \in \mathbb{Z},\ \sum_{i=1}^{k} x_i = u \right\}.$$

For any given m, we shall visit $S_u^{(k)}$ in the order of $u = 0, 1, 2, \ldots$. Within each $S_u^{(k)}$, the visitation order $O_u^{(k)}$ can be defined recursively as follows:

(i) Let U_i be the set of points in $S_u^{(k)}$ which have the kth components equal to i. They will be visited in the order of $i = 0, 1, 2, \ldots$.

(ii) In U_i, regard the first $k-1$ components of each point as a point in $S^{(k-1)}_{u-i}$. Visit U_i in the order $O^{(k-1)}_{u-i}$.

(iii) $O^{(2)}_u$ is defined as follows: first visit $(u,0)$, then $(u-1,1)$, $(u-2,2)$, $\ldots$, $(0,u)$.

On each visit to an integral lattice point $(x_1,\ldots,x_k)$ in $S^{(k)}_u$, we write

$$n \equiv x_1a_1 + \cdots + x_ka_k \pmod{m},$$

where $A=\{a_1,\ldots,a_k\}$ is the generating set of the Cayley digraph. We leave a blank if the number in $\mathbb{Z}_m$ has been appeared earlier according to visitation order defined above. We stop when all the numbers in $\mathbb{Z}_m$ are appeared. The resulted pattern, i.e., the portion of lattice points filled with numbers is contained in the first quadrant of the k-dimensional space. We are going to show that the resulted pattern has certain shape.

Let $\xi=(x_1,\ldots,x_k), \zeta=(y_1,\ldots,y_k)\in\mathbb{R}^k$. We say $\xi\le\zeta$ if $x_i\le y_i$ for $i=1,\ldots,k$. Let $A=\{a_i,\ldots,a_k\}$. Let D_n denote the resulted pattern from the Cayley digraph $\mathrm{Cay}(m,A)$. Let $\xi=(x_1\ldots,x_k)\notin D_n$. Then there exists $\zeta=(y_1,\ldots,y_k)\in D_n$ such that $\zeta<\xi$ and

$$\sum_{i=1}^k x_ia_i \equiv \sum_{i=1}^k y_ia_i \pmod{m}.$$

Noting that

$$(x_1+1)a_1+x_2a_2+\cdots+x_ka_k = (y_1+1)a_1+y_2a_2+\cdots+y_ka_k,$$

we see that $\xi'=(x_1+1,x_2,\ldots,x_k)\notin D_n$. In general, we have

LEMMA 8. *If $\xi\notin D_n$, then $\zeta\notin D_n$ for every $\zeta\ge\xi$.*

A corner is called *complete* if its vertex is not on any coordinate (hyper) plane. By a similar argument as used in the proof of Lemma 2 in Hsu and Jia [**15**], we can prove the following Lemma.

LEMMA 9. *The resulted pattern always has at most one complete corner.*

THEOREM 3. *For any fixed k as $r\to\infty$,*

$$m^*(r,k) \le \left(\frac{k^2-1}{k^2-k+1}\right)^k \frac{r^k}{(k-1)(k-1)!} + O(r^{k-1}).$$

Proof. Let $A=\{a_1,a_2,\ldots,a_k\}$ and $\mathrm{Cay}(m,A)$ be the Cayley digraph generated by A. Suppose that the average distance of $\mathrm{Cay}(m,A)$ is at most r. It follows from Lemmas 8 and 9 that the best possible resulted pattern (with respect to the number of vertices in terms of r) is $D=D(\alpha)$ as defined in Lemma 4, where α will be determined later.

Let f be as in Lemma 6. Noting the fact that D is contained in the cube $Q_k : 0 \leq x_i \leq \alpha$, we can easily verify that (3.1) is satisfied. Hence

$$r(m, A) = \frac{1}{m} \sum_{x \in \mathbb{Z}_m} f(A, n) = (1 + 0(1)) \frac{\int_{D(\alpha)} (x_1 + \cdots x_k)\, dx_1 \cdots dx_k}{\int_{D(\alpha)} dx_1 \cdots dx_k}.$$

It then follows from Lemmas 4 and 6 that

$$r(m, A) = (1 + 0(1)) \alpha \frac{k^2 - k + 1}{k^2 - 1}.$$

Let

$$\alpha = (1 + o(1)) r \frac{k^2 - 1}{k^2 - k + 1},$$

so that $r(m, A) \leq r$. Then the theorem follows from the fact that[1]

$$\begin{aligned} m &\leq (1 + o(1)) \int_{D(\alpha)} dx_1 \cdots dx_k = \frac{(1 + o(1))\alpha^k}{(k-1)(k-1)!} \\ &= \left(\frac{k^2 - 1}{k^2 - k + 1} \right)^k \frac{r^k}{(k-1)(k-1)!} + o(r^k). \end{aligned}$$

4. Distributed loop networks with small average distance

One of the fundamental goals in constructing a communication network is to minimize the transmission delay within the network when the number of nodes and the number of links are given. Because of their simplicity and regularity, Cayley graphs have been used as the topology to construct local area networks. Such networks are often referred as *distributed loop networks.* Distributed loop networks which are optimal or nearly optimal with respect to their diameter have been studied extensively by many authors. Since the average distance of a network reflects its overall performance, the investigation of loop networks with are optimal or nearly optimal with respect to their average distance is of great interest and importance.

Let us recall that $r(m, A)$ is the average distance of the Cayley digraph $\mathrm{Cay}(m, A)$. For any positive integers m, k, define

$$r(m, k) = \min\{r(m, A) \mid A \subseteq \mathbb{Z}_m \text{ with } |A| = k\}.$$

Clearly,

$$r(m, 1) = \frac{m-1}{2}.$$

In the case where $k = 2$, it follows from results of Wong and Coppersmith [**33**] and Jia [**23**] and Fiol and others [**9**] that

$$r(m, 2) = \frac{5}{3\sqrt{3}} (1 + o(1)) \sqrt{m}$$

[1] In fact, we can show that tail term is actually $O(r^{k-1})$.

for an infinite family of integers m. This infinite class of directed double loop networks are asymptotically optimal with respect to their average distances. For $k \geq 3$, Wong and Coppersmith proved that

$$\frac{k}{k+1}\sqrt[k]{k!m} \leq r(m,k) \leq \frac{k}{2}\sqrt[k]{m} - \frac{k}{2},$$

where the lower bound holds for all m, while the upper bound holds for an infinite family of positive integers m. Jia improved the upper bound to

$$r(m,k) \leq (1+o(1))\frac{3k}{4\sqrt[3]{4}}\sqrt[k]{m/\alpha_k},$$

where α_k is defined on page 2. Theorems 2 and 3 of this paper imply that

$$\begin{aligned}(1+o(1))\frac{(k^2-k+1)\sqrt[k]{(k-1)(k-1)!}}{k^2-1}\sqrt[k]{m} &\leq r(m,k)\\ &\leq (1+o(1))\frac{7k}{10\sqrt[4]{5}}\sqrt[k]{m/\beta_k},\end{aligned}$$

where the upper bound holds for all large positive integers m, while the lower bound holds for an infinite family of positive integers m.

For any m, let $d(m,A)$ denote the diameter of Cay(m,A). Given m and k, define

$$d(m,k) = \min_{|A|=k} d(m,A).$$

We know that $d(m,k)$ is not monotonely increasing in m, for instance, $d(m,3) = 8$ when $79 \leq m \leq 104$ or $m = 111$, while $d(m,3) = 9$ for $105 \leq m \leq 110$. It is possible that $r(m,k)$ is not monotonely increasing neither. Fide $k \geq 1$, let $g(k)$ be the maximum decreasing gap of $r(m,k)$ between any two consecutive m's, namely,

$$g(k) = \sup\{r(m,k) - r(m+1,k) \mid m = 1,2,\dots\}.$$

It would be interesting to calculate $g(k)$ if it is positive.

A k-element subset A, which is optimal with respect to its diameter, i.e., $d(m,A) = d(m,k)$, may not be optimal with respect to its average distance. For instance, both $\{1,3,8\}$ and $\{1,4,9\}$ provide the optimal diameter $d(20,3) = 4$, while $r(20,\{1,3,8\}) = 2.4$ and $r(20,\{1,4,9\}) = 2.35 = r(20,3)$. Here are some more examples:

$$\begin{aligned} d(30,3) = 5 &= d(30,\{1,4,9\}) & 2.966666667 &\approx r(30,\{1,4,9\})\\ &= d(30,\{1,5,8\}) & 2.933333333 &\approx r(30,\{1,5,8\})\\ &= d(30,\{1,5,19\}); & r(30,3) = 2.9 &= r(30,\{1,5,19\}); \end{aligned}$$

5. An application to asymptotic bases in additive number theory

Let A be a subset of nonnegative integers. For every $n \geq 0$, let $f(A,n)$ be the least number of elements of A (repetitions allowed) with sum n. A set A is called an *asymptotic basis of order* h if $f(A,n) \leq h$ for all $n \geq n_0$. Asymptotic bases are one of the central problems in additive number theory. For instance, Lagrange's Theorem asserts that the set of squares is an asymptotic basis of order 4, while Goldbach conjecture states that the set of prime numbers is an asymptotic basis of order two for the set of even integers. Waring's problem is to determine the order of the asymptotic basis of kth powers of integers. However, much of the recent study focus on the structure of general, "non-classical" asymptotic bases.

Let $g(A)$ be the least positive integer h such that A is an asymptotic basis of rder h. A set A is called an exact asymptotic basis of order h if every large integer is a sum of exactly h not necessarily distinct elements of A. Erdös and Graham [**7**] raised the question of computing the exact order of an asymptotic basis of order h in terms of h. In order to state the problem, we define,,

$$G_k(h) = \max_{g(A)\leq h} \max_{\substack{|F|=k, F\subset A \\ g(A\backslash F)<\infty}} g(A\backslash F)$$

for all $h \geq 2$ and $k \geq 1$. Erdös and Graham proved that

$$\frac{1}{4}(1+o(1)h^2 \leq G_1(h) \leq \frac{5}{4}(1+o(1))h^2,$$

which has been improved by Grekos [**11**] and Nash [**26**] to

$$\frac{1}{3}h^2+)(h) \leq G_1(h) \leq \frac{1}{2}h^2 + O(h).$$

The general case has been instigated by many authors, including Nathanson [**28, 27**], Nash [**26**], Jia [**19, 20, 21, 22**], Chen and Gu [**6**], and Su [**31**], etc.. The "average version" of this problem can defined as follows:

Let A be a subset of nonnegative integers such that

$$f(A,n) < \infty \quad \text{for all} \quad n \geq n_0.$$

If

$$r(A) = \limsup_{n\to\infty} \frac{1}{n} \sum_{\mu=n_0}^{n+n_0} f(A,\mu) < \infty,$$

we call $r(A)$ the *average order of* A. Note that a set A which has average order r may not be an asymptotic basis of any order. We are interested in the growth of the average order $r(A\backslash F)$ of $A\backslash F$ in terms of $r(A)$. Let $r \geq 1$ be any real number, k a positive integer. Define

$$G_k^*(r) = \max_{r(A)\leq r} \max_{\substack{|F|=k, F\subset A \\ r(A\backslash F)<\infty}} r(A\backslash F).$$

Recently, Jia [**23**] proved that

$$G_k^*(r) \geq \lambda_k \left(\frac{4\sqrt[3]{4}}{3}\right)^k \left(\frac{r}{k+1}\right)^{k+1} + O(r^k) \quad \text{for all } \ k \geq 1,$$

where

$$\lambda_k = \begin{cases} \dfrac{2\sqrt[3]{4}}{3} \approx 1.058 & \text{if } \ k \equiv 0 \pmod 3 \\ 1 & \text{if } \ k \equiv 1 \pmod 3 \\ \dfrac{81\sqrt[3]{2}}{100} \approx 1.021 & \text{if } \ k \equiv 2 \pmod 3 \end{cases}.$$

In particular, we have

$$G_1^*(r) \geq \frac{27}{50}r^2 + O(r).$$

In this paper, we prove the following

THEOREM 4. *For every integer $k \geq 1$,*

$$G_k^*(r) \geq \frac{\beta_{k+1}}{2}\left(\frac{10\sqrt[4]{5}}{7}\right)^{k+1}\left(\frac{r}{k+1}\right)^{k+1} + O(r^k),$$

where β_k is as defined in Theorem 2 on page 9.

Proof. Theorem follows immediately from Theorem 2 and the following lemma which we proved in [**23**].

LEMMA 10. *For any integers $r \geq 2$ and $k \geq 1$,*

$$G_k^*(r) \geq \frac{1}{2}m_1^*(r-1, k+1).$$

REFERENCES

1. S. B. Akers and B. Krishnamurthy, A group-theoretic model for symmetric interconnection networks, *IEEE Trans. Computers*, April 1989.
2. Beivide et al, Optimal distance networks of low degree for parallel computers, *IEEE Trans. Computer*, Vol. 40, No. 10, 1991.
3. J.-C. Bermond, F. Comellas and D. F. Hsu, Distributed loop computer networks: A survey, to appear.
4. Bermond, Illiades, and Peyrat, An optimization problem in distributed loop computer networks, *in*: 3rd Int. Conf. on Combinatorial Math., June 1987.
5. F. T. Boesch and J.-F. Wang, Reliable circulant networks with minimum transmission delay, *IEEE Trans. Circuits Syst.*. Vol. CAS-32, 1985.
6. S. Chen and W. Gu, Exact order of subsets of asymptotic bases, *J. Number Theory*, **41**(1992), 15-21.
7. P. Erdös and R. L. Graham, On bases with an exact order, *Acta Arith.*, **37**(1980), 201–207.
8. P. Erdös and D. F. Hsu, Distributed loop networks with minimum transmission delay, *Theoretical Computer Science*, to appear.
9. M. A. Fiol, J. L. A. Yebra, I. Alegre and M. Valero, A discrete optimization problem in local networks and data alignment, *IEEE Trans. Computers*, **C-36**(1987), 702–713.
10. R. L. Graham and N. J. A. Sloane, On additive bases and Harmonious graphs, *SIAM J. Alg. Discrete Math.*, **1**(1980), 382–404.
11. G. Grekos, *Quelques Aspects de la Théorie Additive des Nombres*, Thesis, Université de Bordeaux I, 1982.
12. R. K. Guy, *Unsolved Problems in Number Theory*, Unsolved Problems in Intuitive Mathematics, Vol. 1, Springer-Verlag, New York, 1981.
13. G. Hofmeister, Asympoticsche Abschatzungen für dreielementige Extremalbasen in naturlichen Zahlen, *J. Reine Angew. Math.*, **232**(1968), 77–101.
14. G. Hofmeister, Die dreielementigen Extremalbasen, *J. Reine Angew. Math.*, **339**(1983), 207–214.
15. D. F. Hsu and X.-D. Jia, Extremal problems in the construction of distributed loop networks, *SIAM J. Discrete Math.*, **7**(1994), 57–71.
16. D. F. Hsu and X.-D. Jia, A combinatorial problem in the construction of distributed loop networks, preprint.
17. D. F, Hsu and J. Shapiro, A census of tight one-optimal double loop networks, *Networks*, to appear.
18. D. F, Hsu and J. Shapiro, Bounds for the minimal number of transmission delays in double loop networks, preprint.
19. X.-D. Jia, *Some Results in Additive Number Theory*, Ph.D. Thesis, CUNY Graduate School 1990.
20. X.-D. Jia, Exact order of subsets of asymptotic bases in additive number theory, *J. Number Theory*, **28**(1988), 205–218.
21. X.-D. Jia, On the order of subsets of asymptotic bases, *J. Number Theory*, **37**(1991), 37–46.
22. X.-D. Jia, Extremal bases for finite cyclic groups, *J. Number Theory*, **41**(1992), 116–127.
23. X.-D. Jia, Extremal Cayley digraphs of finite cyclic groups, *SIAM J. Discrete Math.*, to appear.
24. Ein rekursive Konstruktionverfahren fur Abschnittsbasen, *J. Reine Angew. Math.*, **271**(1974), 214–217.
25. S. Mossige, Algorithms for computing the h-range of the Postage Stamp problem, *Math. Comput.*, **36**(1981), 575–582.
26. J. C. M. Nash, *Results on Bases in Additive Number Theory*, Thesis, Rutgers University, New Jersey, 1985.
27. J. C. M. Nash and M. B. Nathanson, Cofinite subsets of asymptotic bases for positive integers, *J. Number Theory*, **20**(1985), 363–372.
28. M. B. Nathanson, The exact order of subsets of additive bases, in: Proceedings, Number Theory Seminar, 1982, *Lecture Notes in Math.*, **1052**, Springer-Verlag, 1984, 273–277.

29. E. S. Selmer, *The Local Postage Stamp Problem*, Research Monograph, Department of Mathemacs, University of Bergen, Bergen, Norway.
30. A. Stöhr, Delöste und ungelöste Fragen über Basen der natürlichen Zahlenreihe I und II, *J. Rein Angew. Math.*, **194**(1955), 40–65, 111–140.
31. W. Su, A combinatorial problem in the construction of distributed loop networks, M. S. Thesis, Southwest Texas State University, 1993.
32. R. Windecker, Eine Abschnittsbasis dritter Ordnung, *Det Kongelige Norske Videnskabers Selskab Skr.*, **9**(1976), 1–3.
33. C. K. Wong and D. Coppersmith, A combinatorial problem related to multimodule memory organization, *J. Assoc. Computing Machinery*, (1974), 392–401.

DEPARTMENT OF MATHEMATICS, SOUTHWEST TEXAS STATE UNIVERSITY, SAN MARCOS, TEXAS 78666, USA, AND, DEPARTMENT OF MATHEMATICS AND COMPUTER SCIENCE, QUFU NORMAL UNIVERSITY, QUFU, SHANDONG PEOPLE'S REPUBLIC OF CHINA

E-mail address: xj01@acadima.swt.edu, or jia@xingde.math.swt.edu

DIMACS Series in Discrete Mathematics
and Theoretical Computer Science
Volume **21**, 1995

Shuffled Tree Based Fault-tolerant Hierarchical Interconnection Networks

Omar H. Karam and Dharma P. Agrawal

ABSTRACT. We introduce a class of Fully Shuffled Tree (FST) based Fault-tolerant Hierarchical Interconnection Networks (HIN's), with an r-ary shuffled-tree is used to link disjoint clusters such as ring, hypercube, and full connection. We derive relations for the performance parameters and compare them with existing networks. We also compute the execution complexity of Descend/Ascend algorithm. A very low cost Incomplete Shuffled Tree (IST) configuration is also discussed in detail.

1. Introduction

In a typical network for massively parallel systems [1], it has been observed that the message transfer is usually limited to just few adjacent nodes and that too within a predefined distance. This localized communication feature has encouraged the use of hierarchical interconnection networks (HINs) [3, 4, 5, 7, 8] to utilize communication feature in reducing the number of links of a network, while maintaining other desirable features. HINs primarily employ multiple levels of explicitly defined topologies. The lowest level starts with a number of basic clusters of nodes, and each higher level leads to a fewer number of disjoint clusters by connecting together several of the lower level clusters. At each level the cluster sizes, the connection schemes, and the corresponding network parameters can be specified, thereby providing required reduction in the number of links. There are several important characteristics that aid in evaluating different network configurations [2]. These are: the degree of a node, which is defined as the number of incoming/outgoing links per node, the diameter of the network which is the maximum value of all shortest paths between any two nodes, and the average distance among all node pairs. Since a network that has a low diameter usually requires a higher node degree, its cost is defined as $(diameter \times nodedegree)$.

1991 *Mathematics Subject Classification.* Primary 68M10, 68R10; Secondary 94C15, 54H99.
This work has been supported in part by the National Science Foundation under grant MIP-9403191

Several HIN's have been suggested in the literature such as the Hierarchical Cubic Network [5], the Extended Hypercube [7], the dB-Cube [3, 4]], and fat trees [8]]. The basic module for the Extended Hypercube (EH) consists of a k-cube of 2^k processing elements (PEs) and an additional node for handling communication - the network controller (NC). The NC is connected to every PE in the cluster. In general an EH(k,L) has L levels, $(L-1)$ to 0, and consists of one NC at level $L-1$ and a k-cube of NC's at level $L-2$. The NCs at level $L-2$ form a k-cube, and each is connected to 2^k NCs at level L-3, and so on.

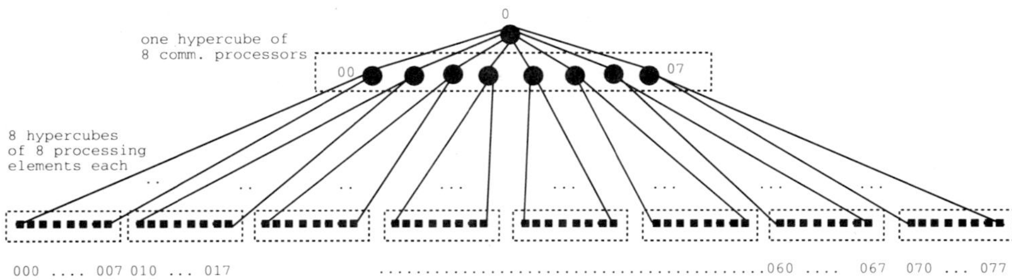

Figure 1: An Extended Hypercube [EH(3,3)] of 64 Processing Elements.

At level 1 the NCs are connected to k PEs each. Figure 1 shows an EH(3,3). The number of PE's in a network is $(r^k)^{L-1}$. The diameter of the EH is $D = 2(L-2)+k$. The degree of a PE is $deg_n = k+1$. The degree of an NC is $deg_{nc} = (2^k+k+1)$.

The Hierarchical Cubic Network, HCN(n,n), is a compound network of 2^n clusters. Each cluster is an n-dimensional hypercube (G_1) of 2^n nodes. G_2 is a fully connected scheme connecting the 2^n clusters such that node x in cluster y is connected to node y in cluster x, and node x in cluster x is connected to node $\overline{x}$ in cluster $\overline{x}$. Hence all nodes are of degree (n+1). The resultant network diameter is $n+\lceil\frac{n}{2}\rceil+1$. The network is highly fault tolerant as the scheme G_2 is fully connected. Expanding a complete HCN(n,n) must be in powers of four by doubling the size of each cluster and the number of clusters to form a complete HCN(2n,2n). But it is also possible to have incomplete HCN's.

A dBCube is obtained by replacing each node of a de Bruijn graph of by a hypercube of size 2^c. One of the important features of a be Bruijn graph is that it can be defined for any arbitrary number of nodes, including the prime numbers; while its diameter wth a given connectivity is known to be the minimum among all known graphs in the literature. A radix r N-node $N \leq r^n$ general de Bruijn graph with $2r$ connections per node is defined as follows: The address of a vetrex is represented by n digits as $(x_{n-1}x_{n-2}\cdots x_1x_0)$, with $0 \leq x_i < r$, for $i = 0, 1, \ldots, n-1$ and the node S and D are connected if

$$S = (rD + \delta) \mod N, \tag{1}$$

or

$$D = (rS + \delta) \mod N, \tag{2}$$

where $\delta = 0, 1, \ldots, r-1$ and $r \leq N$.

One of the basic problems with the basic binaary tree is that the traffic increases as we moves toward the root node. To alleviate this problem, a fat tree is defined

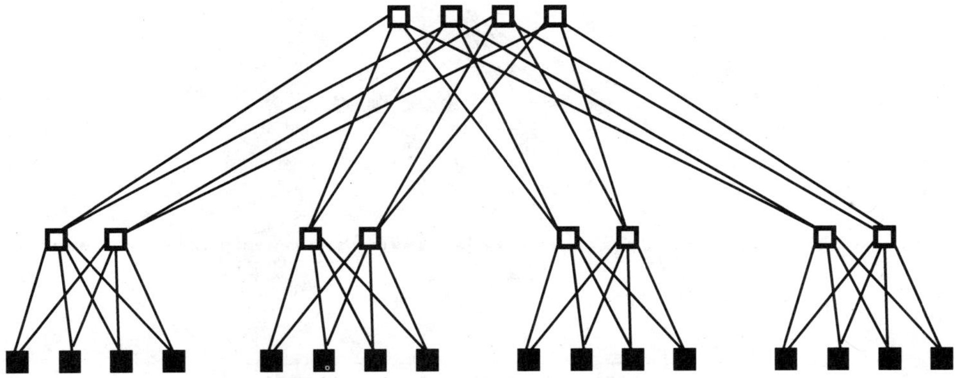

Figure 2: A 4-ary Fat Tree of 16 Processing Elements.

such that we have increasing channel capacities as we ascend from its leaves to the root. The CM-5 multiprocessor employs a 4-ary fat tree (Figure 2) as its data network. The leaves of the tree are processing elements of degree 2 while the nodes of the tree are communication switches each connected to 4 children and 2 parents (or 4). In this paper, we introduce a new class of networks called fully shuffled tree (FST) that has a limited connectivity as the fat-tree, while providing multiple disjoint paths and hence enhanced fault-tolerance.

2. The Fully Shuffled Tree, FST(r,L)

A Fully Shuffled Tree FST(r,L) is defined as a hierarchical network with L levels, $(0 \cdot (L-1))$, and r^L nodes at any level L. A node's address is given by $a_0a_1a_2...a_{L-1}$, where $a_i \in (0, 1, \cdot, r-1)$. At level 1, there is one node. Its address, a_0 is 0. At level 2, there are r nodes. Their addresses are 00 to 0(r-1) and each is connected to node 0. At level 3 there are are r^2 nodes [000, 001, 002, 003, ..., 00(r-1), ..., 0r0, 0(r-1)1, ..., 0(r-1)(r-1)]. A node $a_0a_1a_2$ in level 3 has a parent node in level 2 whose address is given with a_0a_1. Every node is connected to its parent node. Hence every parent node a_0a_1 has r children with the addresses $(a_0a_10).....(a_0a_1(r-1))$.
At any level l, the parent of a node $a_0a_1a_2...a_{l-2}a_{l-1}$ is node $a_0a_1...a_{l-2}$ in level $(l-1)$. Every parent is connected to all its children. Also at a level l the nodes are divided into r^{l-1} groups. Every group contains the nodes whose addresses have the same $(l-1)$ most significant digits. These digits are the group's address and every group has r nodes. Various groups at the lowest level, L, are called basic clusters. The nodes within each of these basic clusters are interconnected by a basic interconnection scheme which can be any suitable network such as ring (R), hypercube (HC), or fully connected (FC). Figure 3 shows an FST(8,3) and its groups at each level. Figure 4 shows an FST/HC(4,3) without the root node. The nodes at level $(L-1)$ are processing elements (PE's) while all other nodes in the network are communication processors or network controllers (NC's). The number of PE's in the network is

$$N_p = r^{L-1}$$

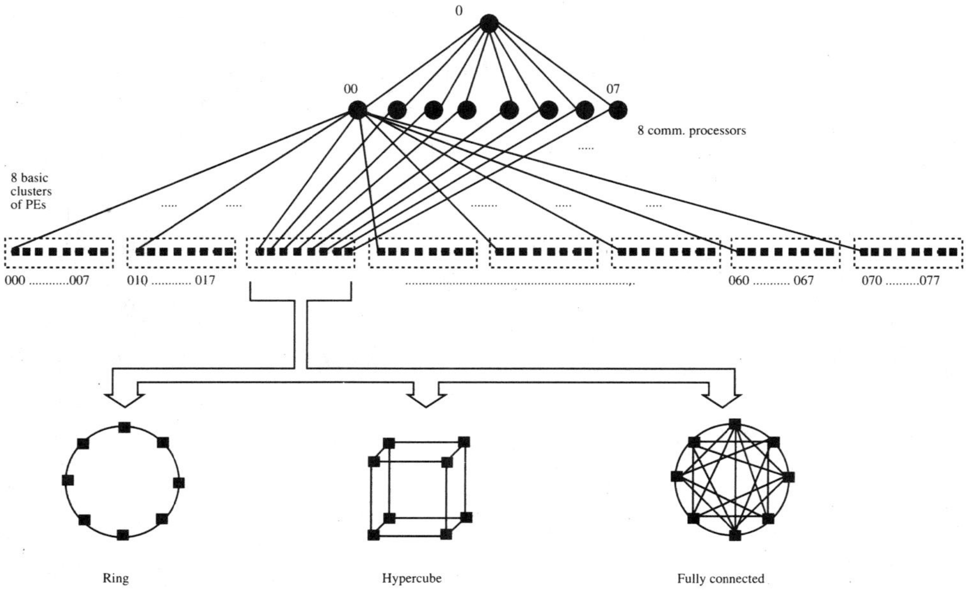

Figure 3: An FST(8,3) showing different possible configurations of a basic cluster.

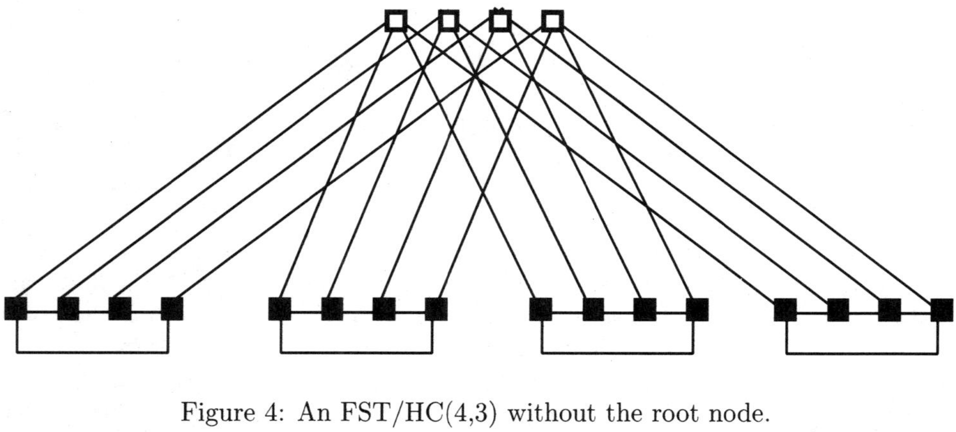

Figure 4: An FST/HC(4,3) without the root node.

and the number of network controllers is

$$N_c = \frac{r^{L-1} - 1}{r - 1}$$

2.1 Node Degree

The node degree for PE's, deg_p is $deg_b + 1$ where deg_b is the node degree for the basic network. The node degree for NC's, deg_c is $r + 1$. The average node degree is interpreted to indicate the average number of I/O ports per node and is therefore given by:

$$\begin{aligned} deg_{av} &= \frac{N_p deg_p + N_c deg_c}{N_p} \\ &= \frac{r^{L-1}(deg_b+1) + \frac{r^{L-1}-1}{r-1}(r+1)}{\frac{r^{L-1}-1}{r-1}} \end{aligned}$$

2.2 Routing

If a source node A whose address is $a_0 a_1 ... a_{L-1}$ sends a message to a destination node B $(b_0 b_1 ... b_{L-1})$ within the same basic cluster, then the routing is that of the basic network. If B is in another basic cluster and the first common ancestor between the two nodes is in level h, i.e.

$$a_0 a_1 ... a_{h-1} = b_0 b_1 ... b_{h-1}$$

then the message can follow any of the paths defined as follows where $0 \leq x_{L-1} \leq (r-1)$:

$$b_0 b_1 ... b_{h-1} a_h ... a_{L-1}$$

.

.

$$b_0 b_1 ... b_{h-1} a_{h+1} ... a_{L-2} x_{L-1}$$

.

.

$$b_0 b_1 ... b_{h-1} x_{L-1}$$

.

.

$$b_0 b_1 ... b_{h-1} b_h ... b_{L-2} x_{L-1}$$

.

.

$$b_0 b_1 ... b_{h-1} b_h ... b_{L-2} b_{L-1}$$

The distance between the two nodes A and B defined as $dist(A, B)$, can then be written as:

$$\begin{aligned} dist(A, B) = \;& 2(L - h - 1) + dist(a_{L-1}, x_{L-1}) \\ & + dist(x_{L-1}, b_{L-1}) \end{aligned}$$

and the shortest distance between them is

$$dist(A, B)_{min} = 2(L - h - 1) + dist(a_{L-1}, b_{L-1})$$

2.3 Diameter

It follows that the network diameter can be written in terms of the number of levels and the basic network diameter D_b as:

$$D = 2(L-1) + D_b$$

The product of a network's diameter and its average degree of node is referred to as the cost.

2.4 Average Distance

Defining δ_b as the sum of distances from one node to all other (r-1) nodes in the basic cluster, then the network average distance is:

$$d_{av} = \frac{1}{(r^{L-1}-1)}\left(\sum_{h=1}^{L-1}(r^{L-h}-r^{L-h-1})(2(L-h-1)+\frac{\delta_b}{r}\right)$$

The normalized average distance is the product of a network's average distance and its average degree of node. This is one measure of cost with respect to average distances.

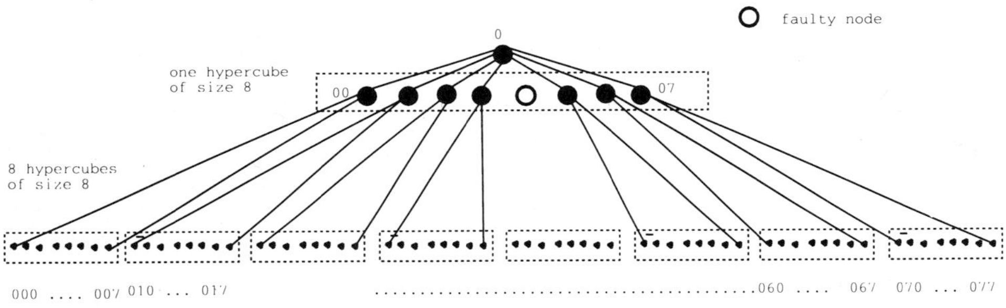

Figure 7: An EH(3,2) with one faulty node

Figure 5: An EH with one faulty network controller processor resulting in a disjoint cluster.

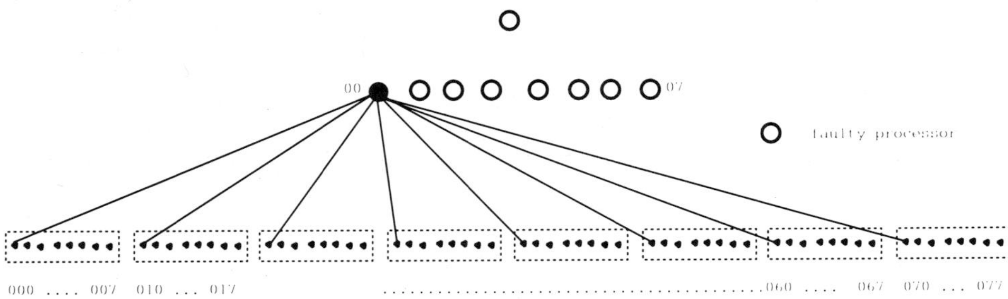

Figure 6: An FST(8,3) with r faulty network controller processors and a connected network.

2.5 Fault Tolerance

The importance of fault tolerance in interconnection networks lies in providing alternative routes in case of node or link failures and also under heavy traffic.
i) Maximum number of permissible NC failures, f_{nc}: There are r paths between a basic cluster and the r NC's at level 1. For the network to be partitioned in to disjoint groups, all these routes must be disconnected. Therefore the minimum number of permissible NC failures is $f_{nc} = r - 1$. Figure 5 shows the failure of a single node in an EH(3,3) resulting in the disconnection of a full cluster. Figure 6 shows the case where r-1 nodes of a single group in an FST(8,3) have failed without disconnecting the network.
ii) The number of disjoint paths, p, between any two nodes in the network: For intercluster communications, the number of disjoint routes between a source node in one cluster and a destination node in another cluster is $p_b + 1$ where p_b is the number of disjoint paths between two nodes within a basic cluster. For rings, $p_b = 2$, for hypercubes, $lg(r)$, and for fully connected, $r - 1$.

3. The Incomplete Shuffled Tree: IST(r,L)

The number of links in the network and the number of NC's can be reduced by using fewer PE's in a basic cluster at level (L-1) as gateway nodes and directly connecting to level (L-2). These chosen PE's would therefore act as the gateways of a basic cluster to all other clusters. The minimum possible number of gateway nodes per cluster is 1, and if more than one path to other clusters ought to be maintained, at least 2 gateways are required. The choice of these 2 nodes depends on the type of basic network used. An IST(r,L) therefore has L levels. Level (L-1) has r^{L-1} PE's divided into r^{L-2} basic clusters and the PE's addresses are the same as for an FST(r,L). Any other level, l with $l \geq 1$, has 2^{l-1} NC's. A node's address is given by $a_0 a_1 ... a_{l-1}$ where $a_0 = 0$ and a_{l-1} is the same as the address of a gateway node, and its parent is node $a_0 a_1 ... a_{l-2}$. Figure 7 shows an IST(4,4) with gateway nodes 0 and 3.

3.1 Routing

Routing is similar to the FST case except that in its basic cluster, the messages have to move to the nearest gateway node and thereafter follow the fully shuffled route to the destination cluster gateway and then to the destination node.

3.2 Network Parameters

$$N_p = r^{L-1}$$
$$N_c = \frac{2^{L-1}-1}{r-1} + 1$$

The node degrees within a cluster are different. A gateway node has one more I/O port than a non-gateway node. If deg_{p_g} is the gateway node's degree then:

$$deg_p = deg_b$$
$$deg_{p_g} = deg_b + 1$$
$$deg_c = r + 1$$
$$deg_{av} = \frac{N_p deg_p + 2N_p/r + N_c(r+1)}{N_p}$$

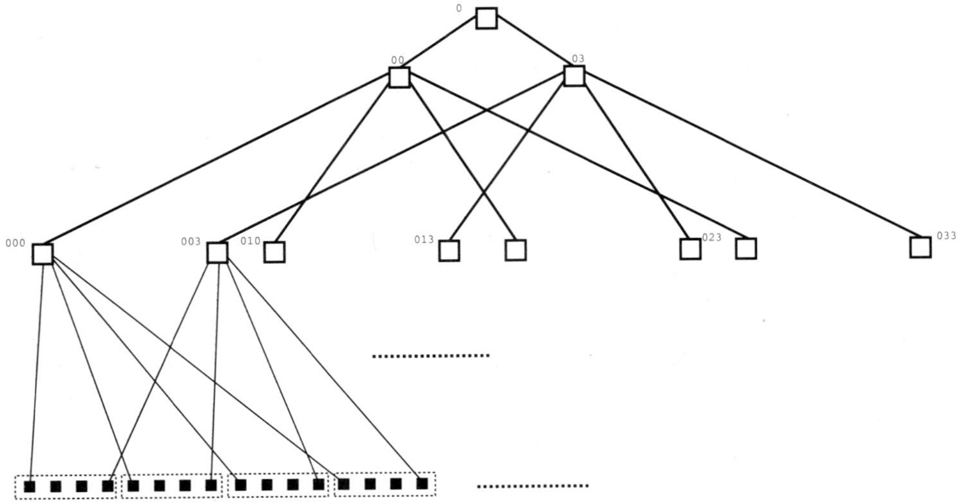

Figure 7: An Incomplete Shuffled Tree, IST(4,4)

The distance equations again depend on the distance parameters D_b and δ_g of the specific basic network used in addition to D_g, the maximum distance from any node to reach the nearest gateway node, and δ_g, the sum of distances from the gateway node to all nodes which are nearest to it. Therefore:

$$
\begin{aligned}
D &= D_g + 2(L-2) + D_b \\
d_{av} &= \frac{r^{(L-2)}}{r^{(L-1)}(r^{(L-1)}-1)} \times \\
&\quad [(2\delta_g(r^{(L-h)} - r^{(L-h-1)}))(2\delta_g(r^{(L-h)} - r^{(L-h-1)}) \\
&\quad + r(r^{L-h} - r^{(L-h-1)})2(L-h-1)) \\
&\quad + r(r^{(L-h-1)} - r^{(L-h-2)}\delta_b)]
\end{aligned}
$$

4. Network Comparisons

4.1 Comparison of Communication Patterns

4.1.1 Broadcasting

Broadcasting, or one-to-all pattern is the case where a single source node is sending the same message to all other $N-1$ nodes in the network. The number of steps necessary is equal to the network diameter.

4.1.2 All-to-one

In this pattern, there is a single destination node receiving simultaneous different messages from all other $N-1$ nodes in the network. The minimum possible number of steps is N which is the case for FST's.

4.1.3 Cluster-to-cluster

Each node within a single source cluster sends a message to the node in a destination cluster with the same basic address. In FST's the maximum number of steps for this pattern is $2(L-2)$.

4.1.4 Cluster-to-all clusters

Each node within a single source cluster sends a message to all nodes having the same basic address a_{L-1}. This can be done in $2(L-2)$ steps. Table 1 compares the communication patterns of FST's, EH's, hypercubes, and 4-ary fat trees.

Communication Pattern	FST	EH	HC	4-ary fat tree
Broadcasting	$D_b+2(L-1)$	lg(r)+2(L-1)	lg(N)	lg(N)
all nodes to one	N-1	N-1	N-1	N-1
cluster to cluster	2(L-1)	2(L-1) +r-1	lg(N)-lg(r)	lg(N)+r/4-1
cluster to all clusters	2(L-1)	2(L-1) +r-1	lg(N)-lg(r)	lg(N)+r/4-1

Table 1: A comparison of communication patterns for a cluster size=4.

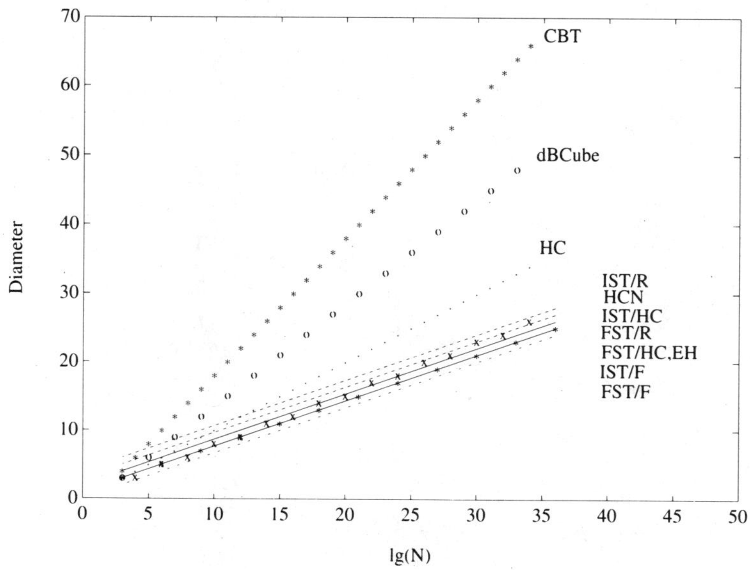

Figure 8: Comparison of network diameters, r = 8.

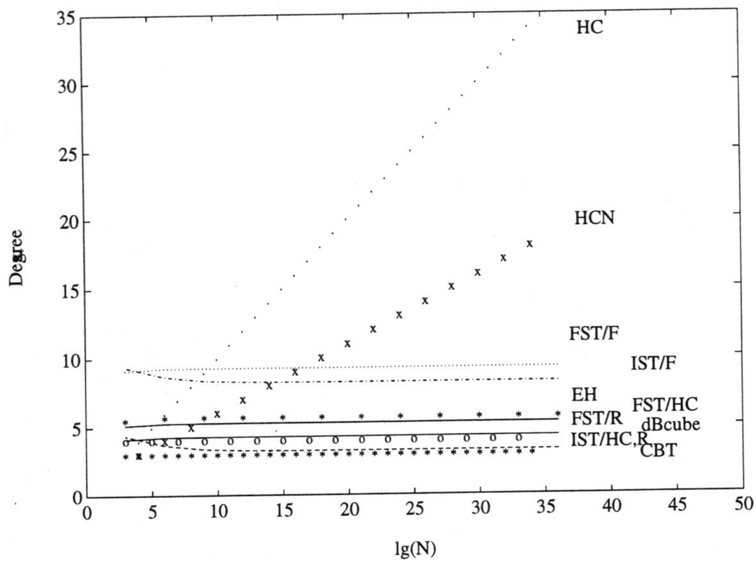

Figure 9: Comprison of network node degrees, r = 8.

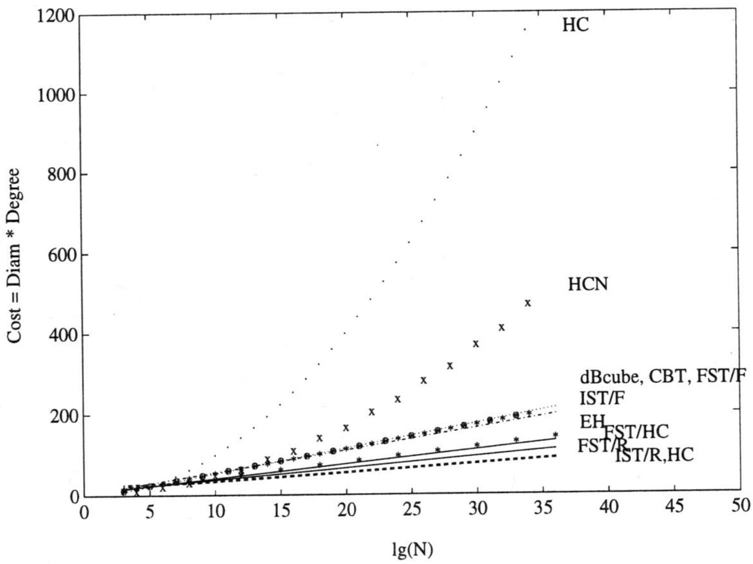

Figure 10: Comprison of network costs, r = 8.

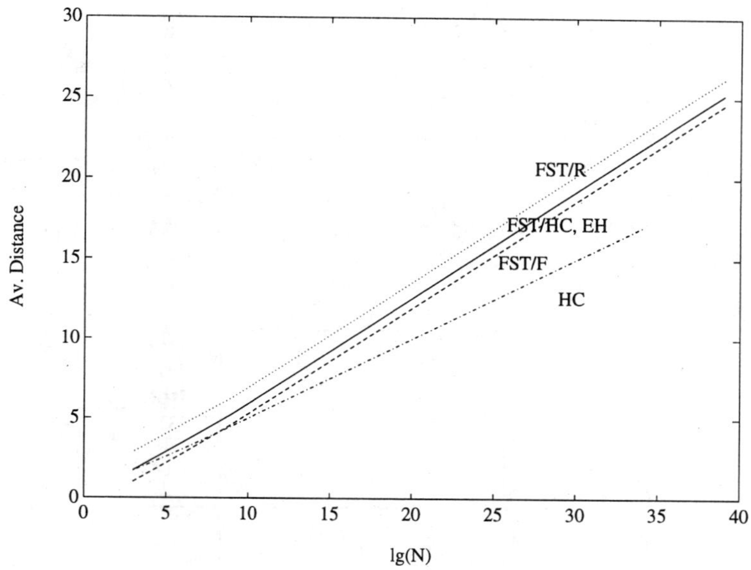

Figure 11: Comparison of Average distances, r=8.

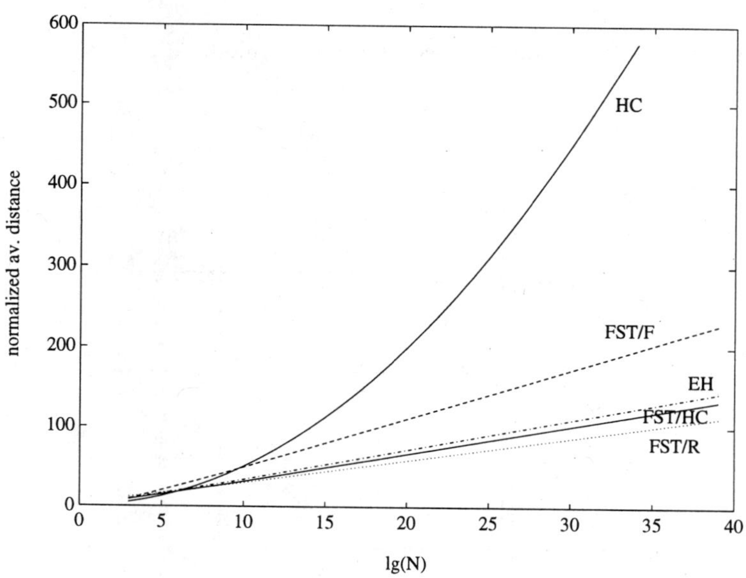

Figure 12: Comparison of normalized av. distances, r=8.

4.2 Comparison of Network Parameters

FST(4,L) and IST(4,L) need to be compared with 4-ary fat trees with respect to number of communication elements, degree of node for communication elements, network diameter, and number of disjoint intra-cluster and are given in Tables 2-5. In Figures 8 to 10 FSTs and ISTs with a basic cluster size 8, are compared to complete binary tree (CBT), hypercube, dBCube, and Extended Hypercube (EH). In Figures 11 and 12 the average distances and normalized average distances of FSTs are compared to those of HCs.

N	FST	IST	r-ary fat tree
16	5	2	12
64	21	11	68
256	84	42	304
1024	341	171	1344

Table 2: A comparison of required communication processors, r=4

N	FST	IST	r-ary fat tree
16	4	4	6
64	5	5	6
256	5	5	6
1024	5	5	6

Table 3: Degree of a node for communication processors, r=4

N	FST		IST		r-ary fat tree
	R,HC	F	R,HC	F	
16	4	3	5	4	4
64	6	5	7	6	6
256	8	7	9	8	8
1024	10	9	11	10	10

Table 4: A comparison of network diameters, r=4

	FST		IST		r-ary fat tree
	R,HC	F	R,HC	F	
number disjoint inter-cluster paths.	4	4	2	2	2

Table 5: A comparison of number of disjoint paths

5. Complexity of Parallel Algorithms Execution

An important class of parallel algorithms is the Ascend/Descend class [7, 9]. Although simple, many fundamental algorithms such as merging, FFT, permutations, sorting, convolution, and matrix operations are either instances of this scheme or simple combinations of such instances. They are either direct, cascaded, or composite A/D algorithms performing in $O(lgN), O(lgN), andO(log_2^2N)$ respectively. If the input data $t_0, t_1, ..., t_{n-1}$ are stored, respectively, in storage locations $T[0], T[1], ..., T[n-1]$, and $n = 2^k$, then a DESCEND class algorithm performs a sequence of basic operations on pairs of data that are successively $2^{k-1}, 2^{k-2}, ..., 2^0$ locations apart. Its dual class ASCEND performs basic operations on pairs of data that are successively $2^0, 2^1, ..., 2^{k-1}$ locations apart. This basic operation can be represented by OPER(m,j;U,V), which modifies the two data items present in storage locations U and V. The computation performed affects only the contents of U, V and may depend upon parameters m and j, which are integers, $0 \leq m \leq n, 0 \leq j \leq k$. DESCEND class algorithms can be written as:

proc DESCEND
for $j \leftarrow k - 1$ *step* -1 *until* $j = 0$
do for each $m : 0 \leq m < n$
pardo if $bit_j(m) = 0$ then
$OPER(m, j; T[m], T[m + 2^j])$
fi
odpar
od
corp DESCEND

The A/D algorithms run in $O(log_2(n))$ as long as the pardo can be executed in parallel for all m. They are either direct, cascaded, or compsite D/A algorithms performing in $O(lg(N)), O(lg(N)), O(log_2^2(N))$ respectively. For FST based networks, D/A algorithms can be run while maintaining two important characteristics:
1) Intercluster communication involves communication nodes only and no PE's.
2) For any basic cluster configuration (ring, hypercube, fully connected, ... etc.), the order of execution is $O(log_2(N))$.
These 2 hypotheses can be proven as follows:
A source PE with address A needs to communicate with the destination PE B with address $(S + 2^j)modN, lg(r) < j \leq log_2(N) - 1)$. From the network architecture, this source node is connected this destination node, which is in a different cluster, through only communication nodes (1). It follows that a message from A will traverse $2 * (L - h - 1)$ links to reach B, where h is the first common level between source and destination. The maximum value of $(L - h - 1)$ is $(L - 2)$, therefore an A/D algorithm is executed in $O(lgr + 2(L - 1)lg(N/r)) = O(lgN)$.

6. Conclusion

We have introduced Fully Shuffled Trees (FST's) interconnection network, and shown that it provides a low node degree, a low diameter, a low cost, and a high fault tolerance. FST's can have basic clusters of any known architecture with virtually the same performance. We have also shown that the order of execution for the Ascend/Descend class of algorithms is $O(log_2(N))$ where N is the number of processing elements in the network.

References

[1] D.P. Agrawal, *Advanced Computer Architecture*, IEEE Computer Society Press, 1986, 392 pages.

[2] D. P. Agrawal, V. K. Janakiram, and G. C. Pathak, *Evaluating the Performance of Multicomputer Configurations,* Computer, Vol. 19, No. 5, May 1986, pp. 23-37.

[3] C. Chen, D. P. Agrawal, and J. R. Burke, *dBCube: A New Class of Hierarchical Multiprocessor Interconnection Networks with Area Efficient Layout,* IEEE Transactions on Parallel and Distributed Processing, Vol. 4, No. 12, December 1993, pp. 1332-1343.

[4] C. Chen, D.P. Agrawal, and J.R. Burke, *Design and Analysis of a Class of Highly Scalable Hierarchical Networks: PdBCube,* Journal of Parallel and Distributed Computing, Special issue on Scalable Algorithms and Architectures, 1994, Vol. 22, No. 3, Sept. 1994, pp. 555-564.

[5] K. Ghose and K. R. Desai, *The Design and Evaluation of the Hierarchical Cubic Network,* Proceedings of the 1990 International Conference on Parallel Processing, pp. 355-362.

[6] K. Hwang, *Advanced Computer Architectures: Parallelism, Scalability, Programmability,* McGraw-Hill, 1993.

[7] J. M. Kumar and L. M. Patnaik, *Extended Hypercube: A Hierarchical Interconneection Network of Hypercubes,* IEEE Transactions on Parallel and Distributed Systems, Vol. 3, No. 1, January 1992, pp. 45-57.

[8] C. E. Leiserson, *Fat-Trees: Universal Networks for Hardware-Efficient Supercomputing,* IEEE Transactions on Computers, Vol. C-34, No. 10, October 1985, pp. 892-901.

[9] F. P. Preparata, and J. Vueillmin, *The Cube Connected Cycles: A Versatile Network for Parallel Computation,* Communications of the ACM, Vol. 9, No. 5, May 1981, pp. 300-309.

DEPARTMENT OF ELECTRICAL AND COMPUTER ENGINEERING, BOX 7911, NORTH CAROLINA STATE UNIVERSITY, RALEIGH, NC 27695-7911, E-Mail: dpa@ncsu.edu, TEL: 919-515-3984, FAX: 919-515-5523.

DIMACS Series in Discrete Mathematics
and Theoretical Computer Science
Volume **21**, 1995

Restricted Connectivity and Restricted Fault Diameter of Some Interconnection Networks

LI QIAO AND ZHANG YI

ABSTRACT. The restricted connectivity and the restricted fault diameter are two reliability measures for interconnection networks, in which we assume that all the neighbors of a vertex do not fail at the same time. By computing these two parameters for some well-known network families, we reaffirm that "These good network families are really good".

1. Introduction

When the interconnection networks are modeled by a graph, some graph parameters such as diameter, connectivity and fault diameter [4,5] are critical to the performance evaluation of the network topology. Here we consider two not so well-known parameters. One is the restricted connectivity which was first introduced by Esfahanian and Hakimi [3] based on the assumption that all the neighbors of any given vertex do not fail at the same time. The restricted connectivity naturally leads to another parameter called restricted fault diameter. The last wo parameters in conjunction with other well-known parameters may provide a more accurate measurement of network reliability properties.

In [2] Esfahanian showed that the restricted connectivity of the ℓ-dimensional binary hypercube is $2\ell - 2$ and the corresponding restricted fault diameter is at most $\ell + 6$. In this paper, similar results for other network families, such as the cube connected cycles, the undirected binary de Bruijn and Kautz networks, are derived.

1991 *Mathematics Subject Classification.* Primary 05C40, 90B12.
This work was supported by NSF of China

2. Notations and Definitions

A graph means an undirected graph without loops and parallel edges. For a given graph G we denote: $D = D(G) =$ the diameter $=$ the maximum distance over all pairs of vertices in G; $V = V(G) =$ the vertex set; $\kappa = \kappa(G) =$ the (vertex) connectivity $= \min\{|S| : S \subset V$ and $G - S$ is disconnected or a single vertex$\}$; $D_\kappa = D_\kappa(G) =$ the fault diameter $= \max\{D(G-F) : F \subset V, |F| = \kappa - 1\}$; $\Gamma(x) =$ the set of neighbors of a vertex x. Moreover, a set of vertices S is called a separating set if $G - S$ is disconnected; a set of vertices R is called restricted if R does not contain $\Gamma(x)$ for any vertex x; a restricted separating set (RSS) is a separating set which is also restricted.

Now we define the restricted connectivity $\kappa' = \kappa'(G)$ as

$$\kappa' = \begin{cases} |V| - 1, & \text{if } G \text{ has no RSS} \\ \min\{|S| : \ S \text{ is a RSS}\}, & \text{otherwise} \end{cases}$$

(Note that we always have $\kappa \le \kappa' \le |V| - 1$, and $\kappa' = |V| - 1$ iff every separating set of G contains a $\Gamma(v)$ for some vertex v iff every restricted set of G is not a separating set. For example, the complete graph K_n has $\kappa = \kappa' = n - 1$, the complete bipartite graph $K_{m,n}$ has $\kappa' = m + n - 1$, the n-cycle C_n has $\kappa' = n - 1$ if $n = 3, 4, 5$ and $\kappa' = 2$ if $n > 5$.)

Evidently, $G - F$ is connected for any restricted set of $\kappa' - 1$ vertices. Similar to the fault diameter D_κ, we define the restricted fault diameter $D'_{\kappa'}$ as

$$D'_{\kappa'} = \max\{D(G - F) : F \text{ is a restricted set of } \kappa' - 1 \text{ vertices}\}.$$

We will denote by Q_ℓ, $C^3(\ell)$, $UB(\ell)$ and $UK(\ell)$ the ℓ-dimensional binary hypercube, the ℓ-dimensional cube connected cycle, the ℓ-dimensional binary de Bruijn graph (with 2^ℓ vertices) and Kautz graph (with $2^\ell + 2^{\ell-1}$ vertices), respectively. See [5], [1] for the definitions of above graphs.

The above mentioned Esfahanian's results can be described as $\kappa'(Q_\ell) = 2\ell - 2$ and $D'_{2\ell-2}(Q_\ell) \le \ell + 6$ for $\ell \ge 3$.

3. Results

No polynomial-time algorithm for the computation of κ' for a general graph is known, nor do we know any nontrivial tight upper bound of κ'. However, κ' and nontrivial upper bounds for $D'_{\kappa'}$ can be obtained for some well-known graph families besides hypercubes. The following table gives an up-to-date record as we know, where the bottom right 3×2 values are our results.

From the table we can see that two desired nice properties, namely, $\kappa' > \kappa$

and $D'_{\kappa'} \leq D_\kappa +$ const. $\leq D+$ const, hold for the listed network families.

	$\lvert V\rvert$	D	κ	D_κ	κ'	$D'_{\kappa'}$
Q_ℓ	2^ℓ	ℓ	ℓ	$\ell+1$	$2\ell-2$	$\leq \ell+6$
$C^3(\ell)$ $(\ell \geq 4)$	$\ell 2^\ell$	$\lfloor \frac{5\ell-2}{2} \rfloor$	3	$\leq \frac{5\ell+8}{2}$[4]	4	$\leq \frac{5\ell+18}{2}$
$UB(\ell)$	2^ℓ	ℓ	2	ℓ[1]	3	$\leq \ell+8$
$UK(\ell)$	$2^\ell+2^{\ell-1}$	ℓ	3	$\leq \ell+2$[1]	4	$\leq \ell+14$

4. Proofs

All the proofs consist of two general steps. First we choose an edge pq of G such that the vertex set $\Gamma(pq) = \Gamma(p) \cup \Gamma(q) - \{p,q\}$ is a RSS, then $\kappa' \leq |\Gamma(pq)|$. Let $\xi = \min\{|\Gamma(pq)| : pq$ is an edge and $\Gamma(pq)$ is a RSS$\}$. Next we try to prove $\kappa' = \xi$ and obtain an upper bound of $D'_{\kappa'}$ simultaneously in the following way. For any restricted set F of $\xi - 1$ vertices and any pair of vertices x and y of $G - F$, if we always can find a path P in $G - F$ between x and y, then $\kappa' = \xi$ and an upper bound of the length of P is an upper bound of $D'_{\kappa'}$. Although the first step is almost trivial for many cases, it provides a definite upper bound ξ of κ'. The second step is not easy, in fact it is a conjecture for any given graph G. Luckily we succeeded in proving this conjecture for the above mentioned graph families. Now we give a full proof to the graph $UB(\ell)$. The proof to the graph $UK(\ell)$ is similar but lengthy, and the proof to the graph $C^3(\ell)$ is somewhat different but not very hard, so both proofs are omitted here.

The ℓ-dimensional binary de Bruijn digraph $B(\ell)$ is the digraph with the vertex set $V = \{(x_1, x_2, \ldots, x_\ell) : x_i = 0$ or $1, 1 \leq i \leq \ell\}$. There is an arc from the vertex $x = (x_1, x_2, \ldots, x_\ell)$ to the vertex y iff $y = (x_2, \ldots, x_\ell, \lambda)$, where $\lambda = 0$ or 1. The (undirected) de Bruijn graph $UB(\ell)$ is obtained from the digraph $B(\ell)$ by removing the orientation, the loops (at $(0,0,\ldots,0)$ and $(1,1,\ldots,1)$) and the parallel edge (between $(0,1,0,1,0,\ldots)$ and $(1,0,1,0,1,\ldots)$).

It was known that $\kappa(UB(\ell)) = 2$ and $D_2(UB(\ell)) = D(UB(\ell)) = \ell$ [1]. Now we prove the results mentioned before.

THEOREM. $\kappa'(UB(\ell)) = 3$ *and* $D'_3(UB(\ell)) \leq \ell + 8$.

PROOF. Let $B = B(\ell)$, $UB = UB(\ell)$, $(\ell \geq 3)$. First we choose two adjacent vertices $p = (0,0,\ldots,0)$ and $q = (0,\ldots,0,1)$ in UB. It's easy to verify that

$$\Gamma(p) \cup \Gamma(q) - \{p,q\} = \{(1,0,\ldots,0), (0,\ldots,0,1,0), (0,\ldots,0,1,1)\}$$

is a RSS, so $\kappa'(UB) \leq 3$. Next we prove $\kappa'(UB) = 3$ and $D'_3(UB) \leq \ell + 8$ simultaneously. Let F be any given restricted set of two vertices u and v, namely,

$$F = \{u,v\} \neq \{(\alpha,\ldots,\alpha,\overline{\alpha}), (\overline{\alpha},\alpha,\ldots,\alpha)\} \quad (\alpha = 0 \text{ or } 1, \overline{\alpha} = 1-\alpha).$$

We claim that for any pair of vertices x and y in $UB - F$, there is a path in $UB - F$ between x and y of length at most $\ell + 8$. This claim is proved by the method of detour.

First note that as shown for the case of deletion of one vertex we can always find in UB a path P of length at most $D(B) = D(UB) = \ell$ not containing u ([1, Th. 6]). If P does not contain v we are done. Thus let us suppose that $v \in P$ and let w and z be the two neighbors of v on the path P.

CASE 1. (w, v, z) or (z, v, w) is not a directed path in B. Suppose for instance that (v, w) and (v, z) are arcs in B. If w or z is $(\alpha, \alpha, \dots, \alpha)$, then (w, z) or (z, w) is an arc of B, therefore we have a path $P[x, w] \cdot [w, z] \cdot P[z, y]$ in $UB - \{u, v\}$ of length at most $\ell - 1$. (If Q is a path containing two vertices r and s. We denote by $Q[r, s]$ the subpath of Q with end vertices r and s and by $Q[r, s] \cdot Q'[s, t]$ the concatenation of the two paths $Q[r, s]$ and $Q'[s, t]$.) So we assume that $w, z \neq (\alpha, \alpha, \dots, \alpha)$ $(\alpha = 0$ or $1)$ from now on.

Denote $w = (x_1, x_2, \dots, x_\ell)$, $v = (x, x_1, \dots, x_{\ell-1})$ and $z = (x_1, \dots, x_{\ell-1}, \overline{x}_\ell)$, we consider the fourth vertex $v' = (\overline{x}, x_1, \dots, x_{\ell-1})$, from which (v', w) and (v', z) are also arcs of B. If $v' \neq u$, we can simply replace v by v' in P to obtain a path in $UB - \{u, v\}$ between x and y of length at most ℓ. If $v' = u$, consider the path $P' = P[x, w] \cdot [w, w_1, w_2, w_3, w_4, w_5, z] \cdot P[z, y]$, where $w = (x_1, x_2, \dots, x_\ell)$, $w_1 = (x_2, \dots, x_\ell, \overline{x}_{\ell-1})$, $w_2 = (\overline{x}_1, x_2, \dots, x_\ell)$, $w_3 = (\overline{x}_1, \overline{x}_1, x_2, \dots, x_{\ell-1})$, $w_4 = (\overline{x}_1, x_2, \dots, x_{\ell-1}, \overline{x}_\ell)$, $w_5 = (x_2, \dots, x_{\ell-1}, \overline{x}_\ell, \overline{x}_{\ell-1})$, $z = (x_1, x_2, \dots, x_{\ell-1}, \overline{x}_\ell)$. We show that $w_i \notin \{u, v\}$ for $1 \le i \le 5$. It's easy to verify that $w_1, w_3, w_5, \notin \{u, v\}$. For w_2 and w_4, we have

$$\begin{aligned} w_2 \in \{u, v\} &\Rightarrow (\overline{x}_1, x_2, \dots, x_\ell) = (*, x_1, \dots, x_{\ell-1}) \Rightarrow x_1 = x_2 = \cdots = x_\ell \\ &\Rightarrow w = (\alpha, \alpha, \dots, \alpha) \text{ and} \\ w_4 \in \{u, v\} &\Rightarrow (\overline{x}_1, x_2, \dots, x_{\ell-1}, \overline{x}_\ell) = (*, x_1, \dots, x_{\ell-1}) \\ &\Rightarrow x_1 = x_2 = \cdots = x_{\ell-1} = \overline{x}_\ell \Rightarrow z = (\alpha, \alpha, \dots, \alpha), \end{aligned}$$

both contradict to the assumption $w, z \neq (\alpha, \alpha, \dots, \alpha)$. The path P' is in $UB - \{u, v\}$ of length at most $\ell + 4$.

CASE 2. (w, v, z) or (z, v, w) is a directed path in B. Suppose for instance that (w, v, z) is a dipath. Denote $w = (x_1, x_2, \dots, x_\ell)$, $v = (x_2, \dots, x_\ell, 0)$ and $z = (x_3, \dots, x_\ell, 0, \lambda)$. Let a, b, f, d, c be vertices as shown in Fig. 1, where

$$\begin{aligned} a &= (x_2, \dots, x_\ell, 1), \\ b &= (x_3, \dots, x_\ell, 1, 1), \\ f &= (\overline{x}_2, x_3, \dots, x_\ell, 1), \\ d &= (\overline{x}_2, \overline{x}_2, x_3, \dots, x_\ell) \end{aligned}$$

and

$$c = (\overline{x}_2, x_3, \dots, x_\ell, 0).$$

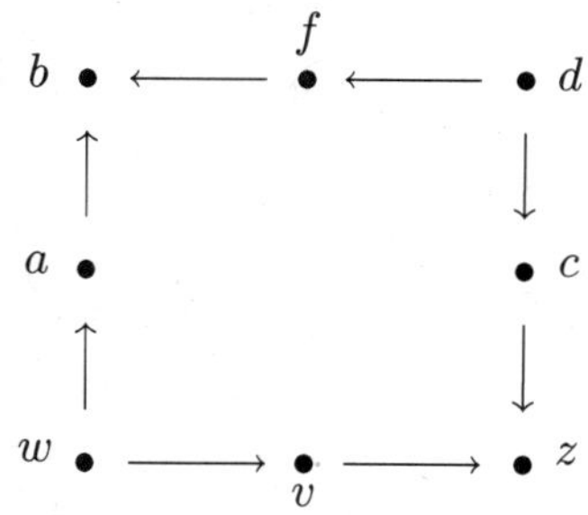

Fig. 1

It's easy to verify that all a, b, f, d, c are different from v. If they are also different for u, then we have a path

$$P[x,w] \cdot [w,a,b,f,d,c,z] \cdot P[z,y]$$

between x and y in $UB - \{u,v\}$ of length at most $\ell+4$. Otherwise, we consider the following five subcases.

2.1. $a = (x_2, \dots, x_\ell, 1) = u$ $(v = (x_2, \dots, x_\ell, 0))$.

Consider a path $Q = [w, w_1, w_2, \dots, w_q, z]$ in UB as follows:

$$\begin{aligned}
&w = (x_1, x_2, \dots, x_\ell), w_1 = (\overline{x}_2, x_1, \dots, x_{\ell-1}), w_2 = (x_1, \dots, x_{\ell-1}, \overline{x}_\ell),\\
&w_3 = (x_2, \dots, x_{\ell-1}, \overline{x}_\ell, \overline{x}_\ell), w_4 = (x_3, \dots, x_{\ell-1}, \overline{x}_\ell, \overline{x}_\ell, \overline{x}_\ell),\\
&w_5 = (\overline{x}_2, x_3, \dots, x_{\ell-1}, \overline{x}_\ell, \overline{x}_\ell), w_6 = (\overline{x}_2, \overline{x}_2, x_3, \dots, x_{\ell-1}, \overline{x}_\ell),\\
&w_7 = (\overline{x}_2, \overline{x}_2, \overline{x}_2, x_3, \dots, x_{\ell-1}), w_8 = (\overline{x}_2, \overline{x}_2, x_3, \dots, x_{\ell-1}, x_\ell),\\
&w_9 = (\overline{x}_2, x_3, \dots, x_{\ell-1}, x_\ell, 0), z = (x_3, \dots, x_\ell, 0, \lambda).
\end{aligned}$$

Obviously, all w_i except w_2 are different from u and v. If $w_2 = u$ or v, then $(x_1, \dots, x_{\ell-1}, \overline{x}_\ell) = (x_2, \dots, x_\ell, *)$, hence $x_1 = x_2 = \cdots = x_\ell$, a contradictions with the assumption $w = (x_1, x_2, \dots, x_\ell) \neq (\alpha, \alpha, \dots, \alpha)$. So we have a path $P[x,w] \cdot Q \cdot P[z,y]$ in $UB - \{u,v\}$ between x and y of length at most $\ell+8$.

2.2. $c = (\overline{x}_2, x_3, \dots, x_\ell, 0) = u$ $(v = (x_2, \dots, x_\ell, 0))$.

Consider a path $Q = [w, w_1, w_2, \dots, w_9, z]$ in UB as follows:

$$\begin{aligned}
&w = (x_1, x_2, \dots, x_\ell), w_1 = (x_2, \dots, x_\ell, 1), w_2 = (x_3, \dots, x_\ell, 1, 1),\\
&w_3 = (x_4, \dots, x_\ell, 1, 1, 1), w_4 = (\overline{x}_3, x_4, \dots, x_\ell, 1, 1),\\
&w_5 = (\overline{x}_3, \overline{x}_3, x_4, \dots, x_\ell, 1), w_6 = (\overline{x}_3, \overline{x}_3, \overline{x}_3, x_4, \dots, x_\ell),\\
&w_7 = (\overline{x}_3, \overline{x}_3, x_4, \dots, x_\ell, 0), w_8 = (\overline{x}_3, x_4, \dots, x_\ell, 0, \lambda)\\
&w_9 = (x_4, \dots, x_\ell, 0, \lambda, 1), z = (x_3, \dots, x_\ell, 0, \lambda).
\end{aligned}$$

Obviously, all w_i except w_8 are different from u and v. If $w_8 = u$ or v, then $(\overline{x}_3, x_4, \dots, x_\ell, 0, \lambda) = (*, x_3, \dots, x_\ell, 0)$, hence $x_3 = x_4 = \cdots = x_\ell = 0 = \lambda$, a contradiction with the assumption that $z \neq (\alpha, \alpha, \dots, \alpha)$. So we have a path $P[x,w] \cdot Q \cdot P[z,y]$ in $UB - \{u,v\}$ between x and y of length at most $\ell+8$.

2.3. $f = (\overline{x}_2, x_3, \dots, x_\ell, 1) = u$ $(v = (x_2, \dots, x_\ell, 0))$.

Let $[w, w_1, w_2, \ldots, w_q, z]$ be a path in UB where

$$w = (x_1, x_2, \ldots, x_\ell), w_1 = (x_2, x_1, x_2, \ldots, x_{\ell-1}), w_2 = (x_1, x_2, \ldots, x_{\ell-1}, \overline{x}_\ell),$$
$$w_3 = (x_2, \ldots, x_{\ell-1}, \overline{x}_\ell, \overline{x}_\ell), w_4 = (x_3, \ldots, x_{\ell-1}, \overline{x}_\ell, \overline{x}_\ell, \overline{x}_\ell),$$
$$w_5 = (\overline{x}_2, x_3, \ldots, x_{\ell-1}, \overline{x}_\ell, \overline{x}_\ell), w_6 = (\overline{x}_3, \overline{x}_2, x_3, \ldots, x_{\ell-1}, \overline{x}_\ell),$$
$$w_7 = (\overline{x}_3, \overline{x}_3, \overline{x}_2, x_3, \ldots, x_{\ell-1}), w_8 = (\overline{x}_3, \overline{x}_2, x_3, \ldots, x_{\ell-1}, x_\ell),$$
$$w_9 = (\overline{x}_2, x_3, \ldots, x_\ell, 0), z = (x_3, \ldots, x_\ell, 0, \lambda).$$

It's easy to verify that w_3, w_4, w_5, w_7, w_9 are different from u and v. Now we consider the other w_i's.

2.3.1. $w_1 \neq u$. If $w_1 = v$, i.e., $(x_2, x_1, x_2, \ldots, x_{\ell-1}) = (x_2, x_3, \ldots, x_\ell, 0)$, then $x_1 = x_3 = x_5 = \cdots$, $x_2 = x_4 = x_6 = \cdots$, $x_{\ell-1} = 0$. There are two cases.

1) If ℓ is even, then

$$w = (0, 1, 0, 1, \ldots, 0, 1),$$
$$z = (0, 1, 0, 1, \ldots, 0, 0),$$
$$u = (0, 0, 1, 0, \ldots, 0, 1, 1),$$
$$v = (1, 0, 1, 0, \ldots, 1, 0).$$

We can replace v by $v' = (0, 0, 1, 0, \ldots, 1, 0)$ in P to obtain a path between x and y in $UB - \{u, v\}$ of length at most ℓ.

2) If ℓ is odd, then $w = (1, 0, 1, 0, \ldots, 1, 0, 1)$, $z = (1, 0, 1, 0, \ldots, 1, 0, 0)$, $u = (1, 1, 0, \ldots, 0, 1, 1)$, $v = (0, 1, 0, 1, \ldots, 0, 1, 0)$. We can replace v by $v' = (1, 1, 0, 1, \ldots, 0, 1, 0)$ in P to obtain a path between x and y in $UB - \{u, v\}$ of length at most ℓ.

2.3.2. If $w_2 = u$, i.e., $(x_1, \ldots, x_{\ell-1}, \overline{x}_\ell) = (\overline{x}_2, x_3, \ldots, x_\ell, 1)$, then $x_1 = 1$, $x_2 = \cdots = x_\ell = 0$, hence $w = (1, 0, \ldots, 0)$, $z = (0, \ldots, 0, \lambda)$ are adjacent in UB. So $P[x, w] \cdot [w, z] \cdot P[z, y]$ is a path between x and y in $UB - \{u, v\}$ of length less than ℓ.

If $w_2 = v$, i.e., $(x_1, \ldots, x_{\ell-1}, \overline{x}_\ell) = (x_2, \ldots, x_\ell, 0)$, then $x_1 = x_2 = \cdots = x_\ell = 1$, hence $u = (0, 1, \ldots, 1)$, $v = (1, \ldots, 1, 0)$, a contradiction with the assumption that $\{u, v\}$ is a restricted set.

2.3.3. $w_6 \neq u$. If $w_6 = v$, i.e., $(\overline{x}_3, \overline{x}_2, x_3, \ldots, x_{\ell-1}, \overline{x}_\ell) = (x_2, x_3, \ldots, x_\ell, 0)$, then $\overline{x}_2 = x_3 = \cdots = x_\ell = 1$, hence $w = (x_1, 0, 1, \ldots, 1)$, $z = (1, \ldots, 1, 0, \lambda)$, $u = (1, \ldots, 1)$, $v = (0, 1, \ldots, 1, 0)$. Let $v' = (0, 1, 1, \ldots, 1)$, $v'' = (1, \ldots, 1, 0)$, then $P[x, w] \cdot [w, v', v'', z] \cdot P[z, y]$ is a path between x and y in $UB - \{u, v\}$ of length at most $\ell + 1$.

2.3.4. $w_8 \neq u$. If $w_8 = v$, i.e., $(\overline{x}_3, \overline{x}_2, x_3, \ldots, x_\ell) = (x_2, x_3, \ldots, x_\ell, 0)$, then $\overline{x}_2 = x_3 = \cdots = x_\ell = 0$, hence $u = (0, \ldots, 0, 1)$, $v = (1, 0, \ldots, 0)$, a contradiction with the assumption that $\{u, v\}$ is a restricted set of UB.

2.4. $b = u$ ($a, b, c \neq u$ and a, f, c are different.)

2.4.1. If $b = d$, i.e., $(x_3, \dots, x_\ell, 1, 1) = (\overline{x}_2, \overline{x}_2, x_3, \dots, x_\ell)$, then $x_2 = 0$, $x_3 = x_4 = \dots = x_\ell = 1$, hence $a = (0, 1, \dots, 1)$, $f = (1, 1, \dots, 1)$, $c = (1, 1, \dots, 1, 0)$. So $P[x, w] \cdot [w, a, f, c, z] \cdot P[z, y]$ is a path in $UB = \{u, v\}$ between x and y of length at most $\ell + 2$.

2.4.2. If $b \neq d$. By the argument of case 1 we have a path $Q[a, f]$ between a and f in $UB-u-v$ of length at most 6. So $P[x, w] \cdot [w, a] \cdot Q[a, f] \cdot [f, d, c, z] \cdot P[z, y]$ is a path between x and y in $UB - u - v$ of length at most $\ell + 8$.

2.5. $d = u$. It's similar to the case 2.4.

The proof of the claim, therefore the theorem is completed.

ACKNOWLEDGMENTS. The first author wishes to thank the NSF of China and the DIMACS Workshop organizers for financial assistance of attending the Workshop. Both authors would like to thank Professor R. A. Brualdi, Ms. Diane Reppert and Mr. Daode Huang for their kindful help in typing our manuscripts.

1. J. B. Bond and C. Peyrat, *Diameter vulnerability of some large interconnection networks*, Congr. Numer. **66** (1988), 267–282.
2. A. H. Esfahanian, *Generalized measures of fault tolerance with application to N-cube networks*, IEEE Trans. Comput. **38** (1989), 1586–1591.
3. A. H. Esfahanian and S. L. Hakimi, *On computing a conditional edge*, J. Inform Processing Lett. **27** (1988), 195–199.
4. M. S. Krishnamoorthy and B. Krishnamurthy, *Fault diameter of interconnection networks*, Comput. Math. Applic. **13** (1987), 557–582.
5. F. T. Leighton, *Introduction to Parallel Algorithms and Architectures: Arrays, Trees, Hypercubes*, Morgan Kaufmann, San Mateo, CA, 1992.
6. D. K. Pradhan and F. J. Meyer, *Communication structures in fault-tolerant distributed systems*, Networks **23** (1993), 379–389.

DEPARTMENT OF MATHEMATICS, UNIVERSITY OF SCIENCE AND TECHNOLOGY OF CHINA, HEFEI 230026, CHINA

DEPARTMENT OF MATHEMATICS, UNIVERSITY OF SCIENCE AND TECHNOLOGY OF CHINA, HEFEI 230026, CHINA

DIMACS Series in Discrete Mathematics
and Theoretical Computer Science
Volume **21**, 1995

SORTING AND SELECTION ON INTERCONNECTION NETWORKS

SANGUTHEVAR RAJASEKARAN

ABSTRACT. In this paper we identify techniques that have been employed in the design of sorting and selection algorithms for various interconnection networks. We consider both randomized and deterministic techniques. Interconnection Networks of interest include the mesh, the mesh with fixed and reconfigurable buses, the hypercube family, and the star graph. For the sake of comparisons, we also list PRAM algorithms.

1 Introduction

The problem of sorting a given sequence of n keys is to rearrange this sequence in nondecreasing order. Given a sequence X of n keys, and an integer i $(1 \leq i \leq n)$, the problem of selection is to find the ith smallest key in the sequence. Such a key will be denoted as select(i, X).

These two important comparison problems have been studied extensively by computer scientists. Both sorting and selection have asymptotically optimal sequential algorithms. There are several sorting algorithms that run in time $O(n \log n)$ in the worst case (see e.g., [1]). The problem of selection has an elegant linear time sequential algorithm [8].

1991 Mathematics Subject Classification: Primary: 68P10, 68Q22; Secondary: 68Q25, 62D05.

Keywords and phrases: Sorting, Selection, Parallel Algorithms.

This research was supported in part by an NSF Research Initiation Award CCR-92-09260.

Optimal algorithms for these two problems also exist for certain parallel models like the EREW PRAM, the comparison tree model, etc. This paper deals with the solution of sorting and selection on interconnection networks. We show how the idea of sampling (both randomized and deterministic) has been repeatedly used to obtain optimal or near optimal algorithms for these problems.

1.1 Models Definition

The Mesh. A mesh connected computer (referred to as the Mesh from hereon) is a $\sqrt{p} \times \sqrt{p}$ square grid where there is a processor at each grid point. Each processor is connected to its four or less neighbors through bidirectional links. It is assumed that in one unit of time a processor can perform a local computation and/or communicate with all its neighbors.

A Mesh with Buses. Two variants of the Mesh have attracted the attention of many a researcher lately: 1) the mesh connected computer with fixed buses (denoted as M_f), and 2) the Mesh with reconfigurable buses (denoted as M_r).

In M_f we assume that each row and each column has been augmented with a broadcast bus. Only one message can be broadcast along any bus at any time, and this message can be read by all the processors connected to this bus in the same time unit.

In the model M_r, processors are connected to a reconfigurable broadcast bus. At any given time, the broadcast bus can be partitioned (i.e., reconfigured) dynamically into subbuses with the help of locally controllable switches. Each processor has 4 I/O ports. There are many variations of M_r found in the literature. In PARBUS model, any combination of 4 port connections is permitted for each processor [22]. Each subbus connects a collection of successive processors. One of the processors in this collection can choose to broadcast a message which is assumed to be readable in one unit of time by all the other processors in this collection.

For instance, in a $\sqrt{p} \times \sqrt{p}$ mesh, the different columns (or different rows) can form subbuses. Even within a column (or row) there could be many subbuses, and so on. It is up to the algorithm designer to decide what configuration of the bus should be used at each time unit. The model assumed in this paper is essentially the same as PARBUS.

Both M_r and M_f are becoming popular models of computing because of the absence of diameter consideration and because of the commercial implementations [5, 40, 62, 35]. Even as theoretical models, both M_r and M_f are very interesting. For instance, n keys can be sorted in $O(1)$ time on an $n \times n$ mesh M_r, whereas we know that $\Omega(\frac{\log n}{\log \log n})$ time is needed even on the CRCW PRAM given only a polynomial number of processors.

The Hypercube. A hypercube of dimension ℓ consists of $p = 2^\ell$ nodes (or vertices) and $\ell 2^{\ell-1}$ edges. Thus each node in the hypercube can be named with an ℓ-bit binary number. If x is any node in V, then there is a bidirectional link from x to a node y if and only if x and y (considered as binary numbers) differ in exactly one bit position (i.e., the hamming distance between x and y is 1.) Therefore, there are exactly ℓ edges going out of (and coming into) any vertex.

If a hypercube processor can handle only one edge at any time step, this version of the hypercube will be called the *sequential model.* Handling (or processing) an edge here means either sending or receiving a key along that edge. A hypercube model where each processor can process all its incoming and outgoing edges in a unit step is called the *parallel model.*

Star Graphs Let $s_1 s_2 \ldots s_n$ be a permutation of n symbols, *e.g.*, $1 \ldots n$. For $1 < j \leq n$, we define $SWAP_j(s_1 s_2 \ldots s_n) = s_j s_2 \ldots s_{j-1} s_1 s_{j+1} \ldots s_n$. An n-star graph is a graph $S_n = (V, E)$ with $\mid V \mid = n!$ nodes, where V $= \{s_1 s_2 \ldots s_n \mid s_1 s_2 \ldots s_n$ is a permutation of n different symbols$\}$, and E $= \{(u, v) \mid u, v \in V$ and $v = SWAP_j(u)$ for some $j, 1 < j \leq n\}$.

A Notation. Throughout this paper we let n denote the input size and p denote the number of processors available.

2 Randomized and Deterministic Sampling

Sampling has served as an effective tool in the design of sorting and selection algorithms over a variety of models of computing. The use of sampling in comparison problems is at least as old as that of Frazer and McKellar's paper [16]. The following strategy has been proposed for sequential sorting in [16]: 1) Randomly sample $o(n)$ keys from the input and sort them using any nonoptimal algorithm; 2) Partition the input into independent subsequences using the sample keys as splitter elements; and 3) Sort each subsequence separately. Clearly, this approach can be thought of as a generalization of Hoare's quicksort algorithm [20] where a single key is employed to partition the input. This elegant approach of Frazer and McKellar has been adapted to design sorting algorithms on various interconnection networks as will be shown in this paper.

Random sampling has also played an important role in the development of efficient algorithms for selection. For example, Floyd and Rivest [15] have introduced the following scheme for sequential selection: 1) Sample $o(n)$ keys from the input and pick two elements from the sample (call them ℓ_1 and ℓ_2) such that the element to be selected has a value in the range $[\ell_1, \ell_2]$ and the number of input keys in the range $[\ell_1, \ell_2]$ is 'small'; 2) Eliminate all the input keys that fall outside the range $[\ell_1, \ell_2]$; and 3) Perform an appropriate selection from out of the remaining keys (the remaining keys can even be sorted to perform this selection). The sequential run time of this algorithm can be shown to be $n + \min\{i, n-i\} + o(n)$ with high probability, if we want to select the ith smallest key.

Recently various selection and sorting algorithms (both deterministic and randomized) have been implemented on different parallel machines. These experimental results indicate that randomized algorithms perform better in practice than their deterministic counterparts (see e.g., [7], [19], [61], [51]).

The idea of sampling has been employed in deterministic algorithms as well. Classical examples are the selection algorithms of 1) Blum, Floyd, Pratt, Rivest, and Tarjan [8]; and 2) Munro and Paterson [41]. A synopsis of deterministic sampling is: 1) To group the numbers into groups of ℓ numbers each (for an appropriate ℓ); 2) Sort each group independently; 3) Collect every qth element from each group (for some q). This collection serves as a 'sample' for the original input. For example, the median of this sample can be shown to be an approximate median for the input. Deterministic sampling has also found numerous applications as will be demonstrated in this paper.

2.1 Sampling Lemmas

Random Sampling. Let X be a sequence of n numbers from a linear order and let $S = \{k_1, k_2, \ldots, k_s\}$ be a random sample from X. Also let $k'_1, k'_2, \ldots, k'_s$ be the sorted order of this sample. If r_i is the rank of k'_i in X, the following lemma provides a high probability confidence interval for r_i. (The rank of any element k in X is the number of elements $\leq k$ in X.)

Lemma 2.1 *For every* α, *Prob.* $\left(|r_i - i\frac{n}{s}| > \sqrt{3\alpha}\frac{n}{\sqrt{s}}\sqrt{\log n}\right) < n^{-\alpha}$.

A proof of the above Lemma can be found in [54]. This lemma can be used to analyze many of the selection and sorting algorithms based on random sampling.

A Notation. We say a randomized algorithm has a resource bound of $\widetilde{O}(f(n))$ if there exists a constant c such that the amount of resource used is no more than $c\alpha f(n)$ **on any input** of size n with probability $\geq (1 - n^{-\alpha})$ (for any $\alpha > 0$). In an analogous manner, we could also define the functions $\widetilde{o}(.)$, $\widetilde{\Omega}(.)$, etc.

Deterministic Sampling. Consider a sequence X of n numbers. A deterministic sample S from X is obtained as follows: Partition the numbers into groups with ℓ elements in each group. Pick every qth element from each group. There will be roughly $\frac{\ell}{q}$ elements picked from each group and $\frac{n}{q}$ elements in all. Let k_i' be the element of S whose rank (in S) is i. Let r_i be the rank of k_i' in X. Then, the following Lemma is easy to prove:

Lemma 2.2 *r_i is in the range $[iq,\ iq + \frac{n}{\ell}q]$.*

2.2 Organization of this Paper

The rest of this paper is organized as follows. In Section 3, we survey known sorting algorithms on the network models of our interest. In Section 4, we provide a summary of known selection algorithms and in Section 5 we provide a list of open problems.

3 Sorting on Interconnection Networks

3.1 A Generic Algorithm

Algorithm I

Step 1. Randomly sample n^ϵ (for some constant $\epsilon < 1$) keys.

Step 2. Sort this sample (using any nonoptimal algorithm).

Step 3. Partition the input using the sorted sample as splitter keys.

Step 4. Sort each part separately in parallel.

3.2 The PRAM

One of the classical results in parallel sorting is Batcher's algorithm [3]. This algorithm is based on the idea of bitonic sorting and was proposed for the hypercube and hence can be run on any of the PRAMs as well. Batcher's algorithm runs in $O(\log^2 n)$ time using n processors. Followed by this, a very nearly optimal algorithm was given by Preparata [47]. Preparata's algorithm used $n \log n$ processors and took $O(\log n)$ time. Finding a logarithmic time optimal parallel algorithm for sorting remained an open problem for a long time in spite of numerous attempts. Finally in 1981, Reischuk was able to design a randomized logarithmic time optimal algorithm for the CREW PRAM [58]. At around the same time Ajtai, Komlós, and Szemerédi announced their sorting network of depth $O(\log n)$ [2]. However the size of the circuit was $O(n \log n)$ and also the underlying constant in the time bound was enormous. Leighton subsequently was able to reduce the circuit size to $O(n)$ using the technique of columnsort [33]. Though several attempts have been made to improve the constant in the time bound, the algorithm of [2] remains vastly as a result of theoretical interest.

In 1987 Cole presented an optimal logarithmic time EREW PRAM algorithm for sorting, the constant in the time bound being reasonably small [10]. In the same paper, a sub-logarithmic time algorithm for sorting on the CRCW PRAM is also given. This algorithm uses $n(\log n)^\epsilon$ processors, the run time being $O(\frac{\log n}{\log\log\log n})$. Here ϵ is any constant > 0. The lower bound result of Beam and Hastad states that any CRCW PRAM sorting algorithm will have to take $\Omega(\frac{\log n}{\log\log n})$ time in the worst case as long as the processor bound is only a polynomial in the input size n [4]. Rajasekaran and Reif [53] were able to obtain a randomized algorithm for sorting on the CRCW PRAM that runs in time $O(\frac{\log n}{\log\log n})$, the processor bound being $n(\log n)^\epsilon$, for any fixed $\epsilon > 0$. This algorithm is also processor-optimal, i.e., to achieve the same time bound the processor bound can not be decreased any further.

Table 1 summarizes these algorithms.

3.3 The Mesh

The first asymptotically optimal sorting algorithm for the mesh was given by Thompson and Kung [63]. Their algorithm can sort n numbers on a $\sqrt{n} \times \sqrt{n}$ Mesh in $O(\sqrt{n})$ time. Since the diameter of an n-node Mesh is $2\sqrt{n} - 2$, [63]'s algorithm is clearly optimal. Thompsun and Kung's algorithm is based on the idea of odd-even merging. Since a Mesh has a large diameter, it is

AUTHOR(S)	YEAR	MODEL	P	T
Preparata [47]	1978	CREW	$n \log n$	$O(\log n)$
Reischuk [58]	1981	CREW	n	$\tilde{O}(\log n)$
Cole [10]	1987	EREW	n	$O(\log n)$
Cole [10]	1987	CRCW	$n(\log n)^\epsilon$	$O(\frac{\log n}{\log \log \log n})$
Rajasekaran & Reif [53]	1988	CRCW	$n(\log n)^\epsilon$	$\tilde{O}(\frac{\log n}{\log \log n})$

Table 1: Sorting on the PRAM

imperative to have not only asymptotically optimal algorithms but also they should have small underlying constants in their time bounds. Often times, the challenge in designing Mesh algorithms lies in reducing the constants in time bounds.

Subsequent to Thompson and Kung's algorithm, Schnorr and Shamir gave a $3\sqrt{n} + o(\sqrt{n})$ time algorithm [59]. They also proved a lower bound of $3\sqrt{n} - o(\sqrt{n})$ for sorting. However, both the upper bound and the lower bound were derived under the assumption of no queueing. Ma, Sen, and Scherson [37] gave a near optimal algorithm for a related model. Kaklamanis, Krizanc, Narayanan, and Tsantilas presented a very interesting algorithm for sorting with a run time of $2.5\sqrt{n} + o(\sqrt{n})$ [25]. This algorithm was randomized and used queues of size $O(1)$. The underlying idea here is the same as that of Algorithm I. The same authors latter improved this time bound to $2\sqrt{n} + \tilde{o}(\sqrt{n})$ [24].

The idea of using $O(1)$ sized queues has been successfully employed to design better deterministic sorting algorithms as well. Kunde has presented a $2.5\sqrt{n} + o(\sqrt{n})$ step algorithm [31]; Nissam and Sahni have given a $(2 + \epsilon)\sqrt{n} + o(\sqrt{n})$ time algorithm (for any fixed $\epsilon > 0$) [44]; Also Kaufmann, Sibeyn, and Torsten have offered a $2\sqrt{n} + o(\sqrt{n})$ time algorithm [26]. The third algorithm closely resembles the one given by [25] and Algorithm I.

The problem of $k - k$ sorting is to sort a Mesh where k elements are input at each node. The bisection lower bound for this problem is $\frac{k\sqrt{n}}{2}$. For example, if we have to interchange data from one half of the mesh with data from the other half, $\frac{k\sqrt{n}}{2}$ routing steps will be needed. A very nearly optimal randomized algorithm for $k - k$ sorting is given in [49]. Recently, Kunde [32] has matched this result with a deterministic algorithm.

Table 2 summarizes sorting results for the Mesh.

AUTHOR(S)	YEAR	PROBLEM	TIME
Thompson & Kung [63]	1977	$1-1$ Sort	$7\sqrt{n}+o(\sqrt{n})$
Schnorr & Shamir [59]	1986	$1-1$ Sort	$3\sqrt{n}+o(\sqrt{n})$
Ma, Sen, & Scherson [37]	1986	$1-1$ Sort	$4\sqrt{n}+o(\sqrt{n})$
KKNT [25]	1991	$1-1$ Sort	$2.5\sqrt{n}+\tilde{o}(\sqrt{n})$
Kunde [31]	1991	$1-1$ Sort	$2.5\sqrt{n}+o(\sqrt{n})$
KKNT [25]	1992	$1-1$ Sort	$2\sqrt{n}+\tilde{o}(\sqrt{n})$
Nigam & Sahni [44]	1993	$1-1$ Sort	$(2+\epsilon)\sqrt{n}+o(\sqrt{n})$
KST [26]	1993	$1-1$ Sort	$2\sqrt{n}+o(\sqrt{n})$
Kunde [31]	1991	$k-k$ Sort	$k\sqrt{n}+o(k\sqrt{n})$
Rajasekaran [49]	1991	$k-k$ Sort	$\frac{k\sqrt{n}}{2}+2\sqrt{n}+\tilde{o}(k\sqrt{n})$

Table 2: Mesh Algorithms for Sorting

3.4 Meshes with Buses

For a Mesh with fixed buses, it is easy to design a logarithmic time algorithm for sorting n numbers using a polynomial (in n) number of processors (see e.g., [50]). However, if the Mesh is of size $\sqrt{n}\times\sqrt{n}$, then the bisection lower bound for sorting will be $\Omega(\sqrt{n})$. The same lower bound holds for a Mesh with a reconfigurable bus system also. In general, we can obtain impressive speedups on M_r and M_f if the number of processors used is much more than the input size.

When the input size n is the same as that of the network size, sorting can be done using a randomized algorithm on M_r in time that is only $\tilde{o}(\sqrt{n})$ more than the time needed for packet routing under the same settings as has been proven in [52]. This randomized algorithm is also similar to Algorithm I. In [29], Krizanc, Rajasekaran, and Shende show that on M_f also, sorting can be done in time that is nearly the same as the time needed for packet routing. The best known algorithm for packet routing on M_r takes time $\frac{17}{18}\sqrt{n}+\tilde{o}(\sqrt{n})$ [9]. For M_f, the best known packet routing time is $0.79\sqrt{n}+\tilde{o}(\sqrt{n})$ [60]. Therefore, sorting can be done on M_r in time $\frac{17}{18}\sqrt{n}+\tilde{o}(\sqrt{n})$ and on M_f in time $0.79\sqrt{n}+\tilde{o}(\sqrt{n})$.

An interesting feature of M_r is that sorting can be done on it in time $O(1)$ using a quadratic number of processors. In contrast, sorting can not be done in $O(1)$ time even on the CRCW PRAM, given only a polynomial

AUTHOR(S)	YEAR	MODEL	P	T
Batcher [3]	1968	Network	n	$\frac{1}{2}\log^2 n$
Nassimi & Sahni [43]	1981	Hypercube	$n^{1+\epsilon}$	$O(\log n)$
Reif & Valiant [57]	1983	CCC	n	$\widetilde{O}(\log n)$
Cypher & Plaxton [12]	1990	Hypercube	n	$O(\log n \log\log n)$

Table 3: Sorting on the Hypercube

number of processors [4]. A constant time algorithm using n^3 processors appears in [64]. The processor bound was improved to n^2 in independent works [23], [36], [42], [45].

3.5 The Hypercube

Batcher's algorithm runs in $O(\log^2 n)$ time on an n-node hypercube [3]. This algorithm uses the technique of bitonic sorting. Odd-even merge sorting can also be employed on the hypercube to obtain the same time bound. Nassimi and Sahni [43] gave an elegant $O(\log n)$ time algorithm for sorting which uses $n^{1+\epsilon}$ processors (for any fixed $\epsilon > 0$). This algorithm, known as *sparse enumeration sort*, has found numerous applications in the design of other sorting algorithms on various interconnection networks. A variant of Algorithm I was employed by Reif and Valiant to derive an optimal randomized algorithm for sorting on the CCC [57]. The best known deterministic algorithm for sorting on the hypercube (or any variant) is due to Cypher and Plaxton and it takes $O(\log n \log\log n)$ time [12]. This algorithm makes use of the technique of deterministic sampling and the underlying constant in the time bound is rather large. An excellent description of this algorithm can be found in [34]. See Table 3.

3.6 The Star Graph

Menn and Somani [39] employed an algorithm similar to that of Schnorr and Shamir [59] to show that sorting can be done on a star graph with $n!$ nodes in $O(n^3 \log n)$ time. Rajasekaran and Wei [56] have offered a randomized algorithm with a time bound of $\widetilde{O}(n^3)$. This algorithm is based on a randomized selection algorithm that they derive. A summary of this algorithm follows:

AUTHOR(S)	YEAR	P	T
Menn and Somani [39]	1990	$n!$	$O(n^3 \log n)$
Rajasekaran and Wei [56]	1992	$n!$	$\widetilde{O}(n^3)$

Table 4: Sorting on the Star Graph

There are n phases in the algorithm. A star graph with $n!$ nodes is denoted as S_n. In the first phase they perform a selection of n uniformly distributed keys and as a consequence route each key to the correct sub-star graph S_{n-1} it belongs to. In the second phase, sorting is local to each S_{n-1}. At the end of second phase each key will be in its correct S_{n-2}. In general, at the end of the ℓth phase, each key will be in its right $S_{n-\ell}$ (for $1 \leq \ell \leq n-1$). Selection in each phase takes $\widetilde{O}(n^2)$ time. Making use of these selected keys, every input key figures out the $S_{n-\ell}$ it belongs to in $O(n^2)$ time. The keys are routed to the correct $S_{n-\ell}$'s in $\widetilde{O}(n)$ time. Thus each phase takes $\widetilde{O}(n^2)$ time, accounting for a total of $\widetilde{O}(n^3)$ time.

The above approach differs from Algorithm I. However, random sampling is used in the selection algorithm of [56]. For a summary of these results see Table 4.

3.7 The de Bruijn Network

A directed de Bruijn network $DB(d, n)$ has $N = d^n$ nodes. A node v can be labeled as $d_n d_{n-1} \ldots d_1$ where each d_i is a d-ary digit. Node $v = d_n d_{n-1} \ldots d_1$ is connected to the nodes labeled $d_{n-1} \ldots d_2 d_1 l$, denoted as $SH(v, l)$, where l is an arbitrary d-ary digit. Note that a rotation of n digits of v, denoted as $SH(v, d_n)$, represents an adjacent node of v. If v is also connected to the nodes labeled $l d_n d_{n-1} \ldots d_2$, then the graph is referred to as undirected de Bruijn network. Hsu and Wei [21] have recently presented an $O(dn^2 \log d)$ time algorithm for sorting. If $d = 2$, their algorithm runs in time $2 \log^2 N$.

4 Selection Algorithms

The sequential selection algorithm of Blum, et. al. works as follows: 1) Partition the input of n numbers into groups with 5 elements in each group; 2) Find the median of each group; 3) Recursively compute the median M of the group medians; 4) Partition the input into two using M as the splitter key. Part I has all the input keys $\leq M$ and Part II has the remaining keys. Identify the part that has the key to be selected and recursively perform an appropriate selection in this part.

One can easily show that the above algorithm runs in time $O(n)$. This is a good example of how deterministic sampling can be employed. A variant of the above has been used in all the deterministic parallel algorithms for selection.

Likewise, random sampling has been effectively applied to derive optimal or near optimal selections algorithms on various networks. A summary of such an algorithm is given below. To begin with all the input keys are *alive.* We are interested in selecting the ith smallest key.

Algorithm II

Step 1. Sample a set S of $o(n)$ keys at random from the collection X of alive keys.

Step 2. Sort the set S.

Step 3. Identify two keys l_1 and l_2 in S whose ranks in S are $i\frac{s}{n}-\delta$ and $i\frac{s}{n}+\delta$ respectively, δ being a 'small' integer.
(* The rank of l_1 in X is $< i$, and the rank of l_2 in X is $> i$, with high probability. *)

Step 4. Eliminate all the keys in X which are either $< l_1$ or $> l_2$.

Step 5. Repeat Steps 1 through 4 until the number of alive keys is 'small'.

Step 6. Finally, concentrate and sort the alive keys.

Step 7. Perform an appropriate selection on the alive keys.

Next we enumerate known parallel selection algorithms on various models and show how the above two themes have been used repeatedly.

AUTHOR(S)	P	T	PROBLEM
Meggido [38]	n	$\widetilde{O}(1)$	Maximal
Reischuk [58]	n	$\widetilde{O}(1)$	Maximal
Rajasekaran and Sen [55]	n	$\widetilde{O}(1)$	Maximal
Folklore	$\frac{n}{\log n}$	$\widetilde{O}(\log n)$	General
Cole [11]	$\frac{n}{\log n \log^* n}$	$O(\log n \log^* n)$	General
Hagerup and Raman [17]	$\frac{n}{\log n}$	$O(\log n)$	General

Table 5: PRAM Selection

4.1 The PRAM

A straight forward implementation of Algorithm II on any of the PRAMs will yield an optimal randomized $\widetilde{O}(\log n)$ time parallel algorithm for selection. On the CRCW PRAM, a similar algorithm can be used to solve the problem of finding the maximum of n given numbers in $\widetilde{O}(1)$ time using n processors [55]. This result was proven in [38, 58] for the parallel comparison tree model. Cole used the idea of deterministic sampling to design an $O(\log n \log^* n)$ time $\frac{n}{\log n \log^* n}$ processor EREW PRAM algorithm [11]. The time bound of this algorithm has recently been improved to $O(\log n)$ using deterministic sampling as well as algorithms for approximate prefix computation [17].

4.2 The Mesh

The problem of selection where the number of processors is equal to the input size has been studied well by many researchers. The best known algorithm is due to Krizanc and Narayanan [27]. This randomized algorithm has a run time of $1.22\sqrt{n} + \tilde{o}(\sqrt{n})$ and is similar to Algorithm II. They also presented a deterministic algorithm for the case $n \neq p$ [28]. This algorithm had a run time of $O(\min\{p \log \frac{n}{p}, \max\{\frac{n}{p^{2/3}}, \sqrt{p}\}\})$. Rajasekaran, Chen, and Yooseph [51] have recently presented both deterministic and randomized algorithms for selection when $n \neq p$. Their deterministic algorithm has a run time of $O(\frac{n}{p} \log \log p + \sqrt{p} \log n)$ and the randomized algorithm resembles Algorithm II and runs in time $\widetilde{O}((\frac{n}{p} + \sqrt{p}) \log \log p)$. A new deterministic selection scheme has been proposed in [51]. The idea is to employ the sequential algorithm of Blum et. al. [8] with some crucial modifications.

A summary of the selection scheme of [51] is given below since it can be applied to any interconnection network to obtain good performance. To begin with, each one of the p processors has $\frac{n}{p}$ keys.

Algorithm III

$N := n$

Step 0. *if* $\log(n/p)$ is $\leq \log\log p$ *then*
  sort the elements at each node
else
  partition the keys at each node into $\log p$ equal parts such that keys in one part will be $\leq$ keys in parts to the right.

repeat
  Step 1. In parallel find the median of keys at each node. Let M_q be the median and N_q be the number of remaining keys at node q, $1 \leq q \leq p$.
  Step 2. Find the weighted median of $M_1, M_2, \ldots, M_p$ where key M_q has a weight of N_q, $1 \leq q \leq p$. Let M be the weighted median.
  Step 3. Count the rank r_M of M from out of all the remaining keys.
  Step 4. *if* $i \leq r_M$ *then*
    eliminate all the remaining keys that are $> M$
  else
    eliminate all the remaining keys that are $\leq M$.
  Step 5. Compute E, the number of keys eliminated.
  if $i > r_M$ *then* $i := i - E$; $N := N - E$.

until $N \leq c$, c being a constant.
Output the ith smallest key from out of the remaining keys.

An implementation of the above algorithm on the Mesh yields the time bound of $O(\frac{n}{p}\log\log p + \sqrt{p}\log n)$. Table 6 summarizes known results.

4.3 Meshes with Buses

Here we consider the problem of selection when $n = p$. On a Mesh with reconfigurable buses, a lower bound of $\Omega(\log\log n)$ applies for deterministic selection, since selection even on the parallel comparison tree model has the same lower bound. ElGindy and Wegrowicz [14] applied an algorithm

AUTHOR(S)	YEAR	RUN TIME
Krizanc et. al. [27]	n	$1.22\sqrt{n} + \tilde{o}(\sqrt{n})$
Krizanc et. al. [28]	p	$O(\min\{p \log \frac{n}{p}, \max\{\frac{n}{p^{2/3}}, \sqrt{p}\}\})$
Rajasekaran et. al. [51]	p	$O(\frac{n}{p} \log\log p + \sqrt{p} \log n)$
Rajasekaran et. al. [51]	p	$\tilde{O}((\frac{n}{p} + \sqrt{p}) \log\log p)$

Table 6: Selection on the Mesh

AUTHOR(S)	YEAR	RUN TIME	DET./RAND.
ElGindy et. al. [14]	1991	$O(\log^2 n)$	DET.
Doctor et. al. [13]	1992	$\tilde{O}(\log^2 n)$	RAND.
Hao et. al. [18]	1992	$O(\log n)$	DET.
Rajasekaran [50]	1992	$\tilde{O}(\log^* n \ \log\log n)$	RAND.

Table 7: Mesh with Reconfigurable Buses: Selection

similar to that of [41] and showed that selection can be done on a p-node M_r in $O(\log^2 p)$ time. Followed by this, Doctor and Krizanc [13] presented a very simple randomized algorithm (similar to Algorithm II) that achieves the same time bound with high probability. This time bound was improved to $O(\log p)$ by Hao, McKenzie, and Stout [18]. Using an algorithm similar to that of Algorithm II and some other crucial properties of M_r, Rajasekaran [50] gave an $O(\log\log p \log^* p)$ expected time randomized algorithm. See Table 7.

On the other hand, $\Omega(p^{1/6})$ is a lower bound for selection on M_f [30]. A very nearly optimal algorithm has been given in [30]. An optimal randomized algorithm can be found in [50].

4.4 The Hypercube

A plethora of algorithms have been proposed for selection on the hypercube (both for the case $p = n$ and the case $p \neq n$). For the case $p = n$, an optimal $\tilde{O}(\log n)$ time randomized algorithm has been given in [57] and [48]. The algorithm in [57] is for sorting and hence can be applied for selection as well. On the other hand, the algorithm given in [48] is very simple.

AUTHOR(S)	YEAR	RUN TIME	DET./RAND.
Kumar et. al. [30]	1987	$O(n^{1/6}(\log n)^{2/3})$	DET.
Olariu et. al.	1992	$O(n^{1/6}(\log n)^{1/3})$	DET.
Rajasekaran [50]	1992	$\widetilde{O}(n^{1/6})$	RAND.

Table 8: Selection on a Mesh with Fixed Buses

Model	Run Time	Lower Bound	Ref.
Sequential	$O(\frac{n}{p}\log\log p + \log^2 p \log(\frac{n}{p}))$	$\frac{n}{p}\log\log p + \log p$	[46]
Sequential	$\widetilde{O}(\frac{n}{p}\log\log p + \log p \log\log p)$	$\frac{n}{p}\log\log p + \log p$	[48]
Sequential	$O(\log^* n \log n)$	$\log n$	[6]
Sequential	$\widetilde{O}(\log n)$	$\log n$	[57, 48]
Weak Parallel	$O(\frac{n}{p} + \log p \log\log p)$	$\frac{n}{p} + \log p$	[46]
Weak Parallel	$\widetilde{O}(\frac{n}{p} + \log p)$	$\frac{n}{p} + \log p$	[48]

Table 9: Selection on the Hypercube

[48]'s algorithm has been implemented on CM-2 and empirical results are promising [51]. The best known deterministic algorithm is due to Berthomé, et. al. [6] and has a run time of $O(\log n \log^* n)$.

For the case of $p \neq n$, refer to Table 9 for a summary of the best known algorithms. All the algorithms in this Table use the technique of sampling (either deterministic or randomized). A slightly better deterministic algorithm can be obtained using Algorithm III as has been shown in [51]. The run time will be $O(\frac{n}{p}\log\log p + \log^2 p \log\log p)$. If a better sorting algorithm is discovered for the hypercube, this time bound will improve further.

4.5 The Star Graph

The only known selection algorithm on the star graph is due to Rajasekaran and Wei [56]. This randomized algorithm runs in time $\widetilde{O}(n^2)$ on an $n!$-node star graph. Within the same asymptotic time bound, this algorithm can perform n different selections. A sorting algorithm with a run time of $\widetilde{O}(n^3)$ follows from this algorithm.

5 Conclusions and Open Problems

In this paper we have surveyed known parallel algorithms for sorting and selection on various interconnection networks. We have also identified some very commonly used techniques for the design of such algorithms. Here is a partial list of open problems in this exciting area of algorithmic research: 1) Is there a simple $2\sqrt{n}-2$ step sorting algorithm for the Mesh with small queues? 2) What is the best run time achievable for selection on the Mesh (when $p = n$)? 3) Is there an $O(\log\log n)$ time algorithm for selection on M_r (either randomized or deterministic)? 4) Can one design a simple deterministic $O(\log n)$ time sorting algorithm for the hypercube? 5) Can the best known sorting times for the star graph and the de Bruijn graph be improved?

References

[1] A.V. Aho, J.E. Hopcroft, and J.D. Ullman, *The Design and Analysis of Computer Algorithms*, Addison-Wesley Publishing Company, 1974.

[2] M. Ajtai, J. Komlós and E. Szemerédi, An $O(n \log n)$ Sorting Network, Proc. 15th ACM Symposium on Theory of Computing, 1983, pp. 1-9.

[3] K.E. Batcher, Sorting Networks and their Applications, Proc. 1968 Spring Joint Computer Conference, vol. 32, AFIPS Press, 1968, pp. 307-314.

[4] P. Beame and J. Hastad, Optimal Bounds for Decision Problems on the CRCW PRAM, Proc. 19th ACM Symposium on Theory Of Computing, 1987, pp. 83-93.

[5] Y. Ben-Asher, D. Peleg, R. Ramaswami, and A. Schuster, The Power of Reconfiguration, Journal of Parallel and Distributed Computing, 1991, pp. 139-153.

[6] P. Berthomé, A. Ferreira, B.M. Maggs, S. Perennes, and C.G. Plaxton, Sorting-Based Selection Algorithms for Hypercubic Networks, Proc. International Parallel Processing Symposium, 1993, pp. 89-95.

[7] G.E. Blelloch, C.E. Leiserson, B.M. Maggs, C.G. Plaxton, S.J. Smith, M. Zagha, A Comparison of Sorting Algorithms for the Connection Ma-

chine CM-2, Proc. 3rd Annual ACM Symposium on Parallel Algorithms and Architectures, 1991.

[8] M. Blum, R. Floyd, V.R. Pratt, R. Rivest, and R. Tarjan, Time Bounds for Selection, Journal of Computer and System Science, 7(4), 1972, pp. 448-461.

[9] J.C. Cogolludo and S. Rajasekaran, Permutation Routing on Reconfigurable Meshes, Proc. Fourth Annual International Symposium on Algorithms and Computation, Springer-Verlag Lecture Notes in Computer Science 762, 1993, pp. 157-166.

[10] R. Cole, Parallel Merge Sort, SIAM Journal on Computing, vol. 17, no. 4, 1988, pp. 770-785.

[11] R. Cole, An Optimally Efficient Selection Algorithm, Information Processing Letters 26, Jan. 1988, pp. 295-299.

[12] R.E. Cypher and C.G. Plaxton, Deterministic Sorting in Nearly Logarithmic Time on the Hypercube and Related Computers, Proc. ACM Symposium on Theory of Computing, 1990, pp. 193-203.

[13] D.P. Doctor and D. Krizanc, Three Algorithms for Selection on the Reconfigurable Mesh, Technical Report TR-219, School of Computer Science, Carleton University, February 1993.

[14] H. ElGindy and P. Wegrowicz, Selection on the Reconfigurable Mesh, Proc. International Conference on Parallel Processing, 1991, Vol. III, pp. 26-33.

[15] R.W. Floyd, and R.L. Rivest, Expected Time Bounds for Selection, Communications of the ACM, Vol. 18, No.3, 1975, pp. 165-172.

[16] W.D. Frazer and A.C. McKellar, Samplesort: A Sampling Approach to Minimal Storage Tree Sorting, Journal of the ACM, vol.17, no.3, 1970, pp. 496-507.

[17] T. Hagerup and R. Raman, An Optimal Parallel Algorithm for Selection, Proc. 5th Annual ACM Symposium on Parallel Algorithms and Architectures, 1993.

[18] E. Hao, P.D. McKenzie and Q.F. Stout, Selection on the Reconfigurable Mesh, Proc. Frontiers of Massively Parallel Computation, 1992, pp. 38-45.

[19] W.L. Hightower, J.F. Prins, J.H. Reif, Implementation of Randomized Sorting on Large Parallel Machines, Proc. 4th Annual ACM Symposium on Parallel Algorithms and Architectures, 1992, pp. 158-167.

[20] C.A.R. Hoare, Quicksort, Computer Journal 5, 1962, pp. 10-15.

[21] D.F. Hsu, D.S.L. Wei, Permutation Routing and Sorting on Directed de Bruijn Networks, Technical Report, University of Aizu, Japan, 1994.

[22] J. Jang, H. Park, and V.K. Prasanna, A Fast Algorithm for Computing Histograms on a Reconfigurable Mesh, Proc. Frontiers of Massively Parallel Computing, 1992, pp. 244-251.

[23] J. Jang and V.K. Prasanna, An Optimal Sorting Algorithm on Reconfigurable Mesh, Proc. International Parallel Processing Symposium, 1992, pp. 130-137.

[24] C. Kaklamanis and D. Krizanc, Optimal Sorting on Mesh-Connected Processor Arrays, Proc. 4th Annual ACM Symposium on Parallel Algorithms and Architectures, 1992, pp. 50-59.

[25] C. Kaklamanis, D. Krizanc, L. Narayanan, and Th. Tsantilas, Randomized Sorting and Selection on Mesh Connected Processor Arrays, Proc. 3rd Annual ACM Symposium on Parallel Algorithms and Architectures, 1991.

[26] M. Kaufmann, S. Torsten, and J. Sibeyn, Derandomizing Algorithms for Routing and Sorting on Meshes, Proc. 5th Annual ACM-SIAM Symposium on Discrete Algorithms, 1994, pp. 669-679.

[27] D. Krizanc, and L. Narayanan, Optimal Algorithms for Selection on a Mesh-Connected Processor Array, Proc. Symposium on Parallel and Distributed Processing, 1992.

[28] D. Krizanc and L. Narayanan, Multi-packet Selection on a Mesh-Connected Processor Array, Proc. International Parallel Processing Symposium, 1992.

[29] D. Krizanc, S. Rajasekaran, and S. Shende, A Comparison of Meshes with Static Buses and Half-Duplex, to appear in Parallel Processing Letters, 1995.

[30] V.K.P. Kumar and C.S. Raghavendra, Array Processor with Multiple Broadcasting, Journal of Parallel and Distributed Computing 4, 1987, pp. 173-190.

[31] M. Kunde, Concentrated Regular Data Streams on Grids: Sorting and Routing Near to the Bisection Bound, Proc. IEEE Symposium on Foundations of Computer Science, 1991, pp. 141-150.

[32] M. Kunde, Block Gossiping on Grids and Tori: Sorting and Routing Match the Bisection Bound Deterministically, Proc. European Symposium on Algorithms, 1993, pp. 272-283.

[33] T. Leighton, Tight Bounds on the Complexity of Parallel Sorting, IEEE Transactions on Computers, vol. C34 (4), April 1985, pp. 344-354.

[34] T. Leighton, *Introduction to Parallel Algorithms and Architectures: Arrays–Trees–Hypercube*, Morgan-Kaufmann Publishers, 1992.

[35] H. Li and Q. Stout, *Reconfigurable Massively Parallel Computers*, Prentice-Hall Publishers, 1991.

[36] R. Lin, S. Olariu, J. Schwing, and J. Zhang, A VLSI-optimal Constant Time Sorting on Reconfigurable Mesh, Proc. European Workshop on Parallel Computing, 1992.

[37] Y. Ma, S. Sen, and D. Scherson, The Distance Bound for Sorting on Mesh Connected Processor Arrays is Tight, Proc. IEEE Symposium on Foundations of Computer Science, 1986, pp. 255-263.

[38] N. Meggido, Parallel Algorithms for Finding the Maximum and the Median Almost Surely in Constant Time, Preliminary Report, CS Department, Carnegie-Mellon University, Pittsburg, PA, Oct. 1982.

[39] A. Menn and A.K. Somani, An Efficient Sorting Algorithm for the Star Graph Interconnection Network, Proc. International Conference on Parallel Processing, 1990, vol. 3, pp. 1-8.

[40] O. Menzilcioglu, H.T. Kung, and S.W. Song, Comprehensive Evaluation of a Two-Dimensional Configurable Array, Proc. 19th Symposium on Fault Tolerant Computing, 1989, pp. 93-100.

[41] J.I. Munro and M.S. Paterson, Selection and Sorting with Limited Storage, Theoretical Computer Science 12, 1980, pp. 315-323.

[42] K. Nakano, D. Peleg, and A. Schuster, Constant-time Sorting on a Reconfigurable Mesh, Manuscript, 1992.

[43] D. Nassimi and S. Sahni, Parallel Permutation and Sorting Algorithms and a New Generalized Connection Network, Journal of the ACM, 29(3), 1982, pp. 642-667.

[44] M. Nigam and S. Sahni, Sorting n^2 Numbers on $n \times n$ Meshes, Proc. International Parallel Processing Symposium, 1993, pp. 73-78.

[45] M. Nigam and S. Sahni, Sorting n Numbers on $n \times n$ Reconfigurable Meshes with Buses, Proc. International Parallel Processing Symposium, 1993, pp. 174-181.

[46] C.G. Plaxton, Efficient Computation on Sparse Interconnection Networks, Ph. D. Thesis, Department of Computer Science, Stanford University, 1989.

[47] F. Preparata, New Parallel Sorting Schemes, IEEE Transactions on Computers, vol. C27, no. 7, 1978, pp. 669–673.

[48] S. Rajasekaran, Randomized Parallel Selection, Proc. Symposium on Foundations of Software Technology and Theoretical Computer Science, 1990, pp. 215-224.

[49] S. Rajasekaran, $k-k$ Routing, $k-k$ Sorting, and Cut Through Routing on the Mesh, to appear in Journal of Algorithms, 1995.

[50] S. Rajasekaran, Mesh Connected Computers with Fixed and Reconfigurable Buses: Packet Routing, Sorting, and Selection, Proc. First Annual European Symposium on Algorithms, Springer-Verlag Lecture Notes in Computer Science 726, 1993, pp. 309-320.

[51] S. Rajasekaran, W. Chen, and S. Yooseph, Unifying Themes for Parallel Selection, Proc. Fourth International Symposium on Algorithms and Computation (ISAAC), Springer-Verlag Lecture Notes in Computer Science 834, 1994, pp. 92-100.

[52] S. Rajasekaran and T. McKendall, Permutation Routing and Sorting on the Reconfigurable Mesh, Technical Report MS-CIS-92-36, Department of Computer and Information Science, University of Pennsylvania, May 1992.

[53] S. Rajasekaran and J.H. Reif, Optimal and Sub-Logarithmic Time Randomized Parallel Sorting Algorithms, SIAM Journal on Computing, vol. 18, no. 3, 1989, pp. 594-607.

[54] S. Rajasekaran and J.H. Reif, Derivation of Randomized Sorting and Selection Algorithms, in *Parallel Algorithm Derivation and Program Transformation*, Edited by R. Paige, J.H. Reif, and R. Wachter, Kluwer Academic Publishers, 1993, pp. 187-205.

[55] S. Rajasekaran, and S. Sen, Random Sampling Techniques and Parallel Algorithms Design, in *Synthesis of Parallel Algorithms*, Editor: J.H. Reif, Morgan-Kaufman Publishers, 1993, pp. 411-451.

[56] S. Rajasekaran and D.S.L. Wei, Selection, Routing, and Sorting on the Star Graph, Proc. International Parallel Processing Symposium, pp. 661-665, April 1993.

[57] J.H. Reif and L.G. Valiant, A Logarithmic Time Sort for Linear Size Networks, Journal of the ACM, 34(1), Jan. 1987, pp. 60-76.

[58] R. Reischuk, Probabilistic Parallel Algorithms for Sorting and Selection, SIAM Journal of Computing, Vol. 14, No. 2, 1985, pp. 396-409.

[59] C. Schnorr and A. Shamir, An Optimal Sorting Algorithm for Mesh-Connected Computers, Proc. ACM Symposium on Theory of Computing, 1986, pp. 255-263.

[60] J.F. Sibeyn, M. Kaufmann, and R. Raman, Randomized Routing on Meshes with Buses, Proc. European Symposium on Algorithms, 1993, pp. 333-344.

[61] T.M. Stricker, Supporting the Hypercube Programming Model on Mesh Architectures (A Fast Sorter for iWarp Tori), Proc. 4th Annual ACM Symposium on Parallel Algorithms and Architectures, 1992, pp. 148-157.

[62] X. Thibault, D. Comte, and P. Siron, A Reconfigurable Optical Interconnection Network for Highly Parallel Architecture, Proc. Symposium on the Frontiers of Massively Parallel Computation, 1989.

[63] C.D. Thompson and H.T. Kung, Sorting on a Mesh Connected Parallel Computer, Communications of the ACM, vol. 20, no.4, 1977, pp. 263-271.

[64] B.F. Wang, G.H. Chen, and F.C. Lin, Constant Time Sorting on a Processor Array with a Reconfigurable Bus System, Information Processing Letters, 34(4), April 1990.

APPENDIX A: Chernoff Bounds

Lemma A.1 *If X is binomial with parameters (n, p), and $m > np$ is an integer, then*

$$Probability(X \geq m) \leq \left(\frac{np}{m}\right)^m e^{m-np}. \quad (1)$$

Also,

$$Probability(X \leq \lfloor(1-\epsilon)pn\rfloor) \leq exp(-\epsilon^2 np/2) \quad (2)$$

and

$$Probability(X \geq \lceil(1+\epsilon)np\rceil) \leq exp(-\epsilon^2 np/3) \quad (3)$$

for all $0 < \epsilon < 1$.

DEPARTMENT OF COMPUTER AND INFORMATION SCIENCE, UNIVERSITY OF FLORIDA, GAINESVILLE, FL 32611
E-mail address: raj@cis.ufl.edu

DIMACS Series in Discrete Mathematics
and Theoretical Computer Science
Volume **21**, 1995

Towards a simple construction method for Hamiltonian decomposition of the hypercube

S. W. SONG

April 21, 1994

ABSTRACT. We consider the problem of Hamiltonian decomposition on the hypercube. It is known that there exist $\lfloor n/2 \rfloor$ edge-disjoint Hamiltonian cycles on a binary n-cube. However, there are still no simple algorithms to construct such cycles. We present some promising results that may lead to a very simple method to obtain the Hamiltonian decomposition. The binary n-cube is equivalent to the Cartesian product of cycles of length four ($C_4 \times C_4 \ldots \times C_4$). Case $n = 4$ is trivial. For the case $n = 6$, we first partition the set of edges of the $C_4 \times C_4 \times C_4$ into 12 disjoint cycles of length 16. We then present an operator to merge the cycles to produce the desired Hamiltonian cycles. In general the edge set of $n/2$ products $C_4 \times C_4 \ldots \times C_4$, can be partitioned into $n2^n/32$ disjoint cycles of length 16. It remains to formalize the merge operator in the general case.

1. Introduction

The problem of finding edge-disjoint cycles on a hypercube can be important in fault-tolerant distributed computing [**1**]. When the cube size is large, the probability of faulty communication links also increases. Thus for applications that require the usage of the nodes of the hypercube connected in a cycle, it is important to have knowledge of the existence of alternate cycles. Obviously, the mere knowledge of their existence is not sufficient. We should be able to construct the edge-disjoint cycles by using simple algorithms.

Several results concerning the existence of disjoint Hamiltonian cycles on graphs, in particular, hypercubes, are known. Alspach, Bermond and Sotteau [**2**] present a nice compilation of some of these results. The interested readers should

1991 *Mathematics Subject Classification.* Primary 65Y10, 05C38; Secondary 65Y05.

Supported by FAPESP (Fundação de Amparo à Pesquisa do Estado de São Paulo) - Grant 93/0603-1 and CNPq (Conselho Nacional de Desenvolvimento Científico e Tecnológico) - Grant 306063/88-3 and PROTEM-CC.

also refer to [**3, 4, 5, 7**] for details. Several other properties of the hypercube are described in [**6, 8**].

It is known [**2**] that on a hypercube of dimension n, there exist $\lfloor n/2 \rfloor$ disjoint Hamiltonian cycles. Unfortunately the proof of this result, based mainly on [**4**], does not lead to any simple algorithm to construct the disjoint Hamiltonian cycles. The problem of finding such algorithms remains an open problem to this date.

In this paper we present an interesting approach and some preliminary results that may lead to a very simple method to obtain the Hamiltonian decomposition of the binary n-cube. Observe first that the binary n-cube is equivalent to a 4-ary $n/2$-cube, or the Cartesian product of cycles of length four: $C_4 \times C_4 \ldots \times C_4$. First we consider the case $n = 4$. M. F. Foregger presents a construction of the two Hamiltonian cycles on $C_4 \times C_4$. We then consider the case of $n = 6$. We show how to partition the set of edges of the $C_4 \times C_4 \times C_4$ into 12 disjoint cycles of length 16. We then present an operator that merges the 16 cycles to produce the desired Hamiltonian cycles. An amazing result is that, in general, the edge set of $n/2$ products $C_4 \times C_4 \ldots \times C_4$, can be partitioned into $n2^n/32$ disjoint cycles of length 16. It remains to formalize the merge operator in the general case. However, though we considered only a small example, the same ideas can be applied to hypercubes in general. The details for this are still being worked out and will be published later.

§2 discusses Hamiltonian decomposition of the hypercube and presents a two-step approach to construct the cycles. §3 shows how to partition the edge set of the hypercube into cycles of length 16. §4 shows a merge operation that joins cycles of length 16 into cycles of length 64, thus obtaining the desired Hamiltonian cycles. Finally §5 discusses the generalization of the merge operation.

2. Hamiltonian decomposition

The Hamiltonian decomposition of a graph is the partitioning of the set of edges of a graph into Hamiltonian cycles. Obviously, for such a partition to exist, the graph should be regular of even degree. In [**2**] a more general definition of Hamiltonian decomposition is presented, that includes the case when the degree is odd. We are interested in the Hamiltonian decomposition of the binary n-cube. For the sake of simplicity, we will consider n even and throughout the paper consider its equivalent 4-ary $n/2$-cube, i.e. the Cartesian product of $n/2$ cycles of length four: $C_4 \times C_4 \ldots \times C_4$. We state without proof the following result (see [**2**] for details).

THEOREM 2.1. *The binary n-cube, with n even, or equivalently the product of $n/2$ cycles, $C_4 \times C_4 \ldots \times C_4$, can be partitioned into $n/2$ Hamiltonian cycles.*

We propose an approach that may lead to a very simple algorithm to construct the edge-disjoint Hamiltonian cycles. The following two major steps are used.

(i) Partition the edge set into cycles of length 16.
(ii) Merge the resulting cycles into larger cycles to obtain the desired Hamiltonian cycles.

3. Construction of cycles of length 16

In what follows, we consider the binary n-cube, with n even. We use m to denote $n/2$.

Consider first $n = 4$, we want to obtain two edge-disjoint cycles of length 16 on the $C_4 \times C_4$. Foregger [**5**] gives a construction of the desired cycles, shown in Figures 1 and 2. (The $C_4 \times C_4$ graph is shown in dashed lines, without the toroidal edges, for simplicity. Cycles are shown in full thick lines.) To distinguish the two cycles, consider one cycle colored with color 0 and the other color 1.

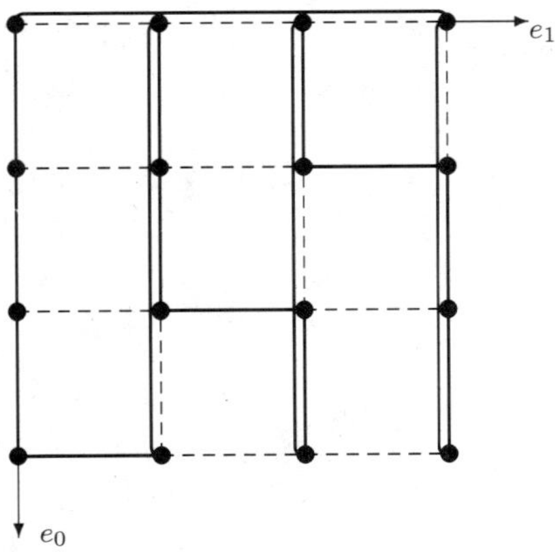

FIGURE 1. Cycle (color 0) of length 16 on $C_4 \times C_4$

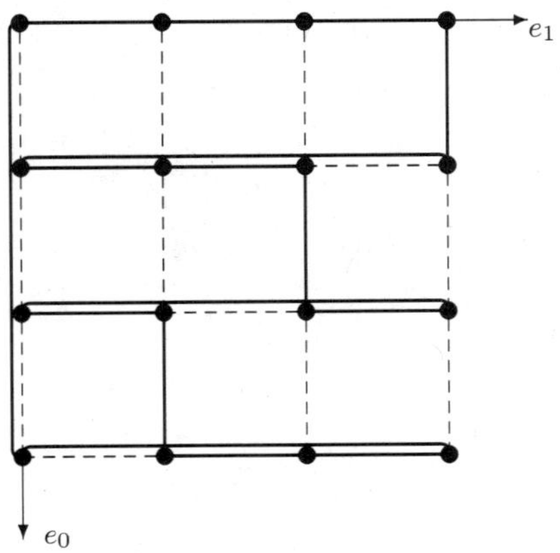

FIGURE 2. Cycles (color 1) of length 16 on $C_4 \times C_4$

In the figures, we denote the m directions by $e_0, e_1, \dots, e_{m-1}$ (in Figures 1 and 2 we have $m = 2$). Notice the symmetry of the two cycles. In Figure 1, starting from node $(0,0)$, the cycle of color 0 uses three edges along direction e_0, followed by one edge along direction e_1, and so on.

DEFINITION 3.1. *We say the a cycle is e_i-**dominant** if it alternately consists of three edges along direction e_i followed by one edge along direction $e_{(i+1 \mod m)}$.*

A e_i-dominant cycle is therefore entirely contained on a face determined by e_i and $e_{(i+1 \mod m)}$.

The cycle of color 0 of Figure 1 is e_0-dominant. Likewise the cycle of color 1 of Figure 2 is e_1-dominant. (Notice here $m = 2$.)

We now show the partitioning of the set of edges of the $C_4 \times C_4 \times C_4$ into 12 disjoint cycles of length 16. Here we divide the 12 cycles into

- 4 cycles of color 0,
- 4 cycles of color 1, and
- 4 cycles of color 2.

Cycles of length 16 of the same color will be merged in the next section to form a Hamiltonian cycle.

Figure 3 shows the 4 cycles of color 0. The cycles of color 0 are e_0-dominant. The same cycles are again shown in Figures 4, 5, 6 and 7, corresponding to the faces $e_2 = 0, 1, 2, 3$, respectively. Notice that when we move from one face to the next, the cycle is shifted to the left.

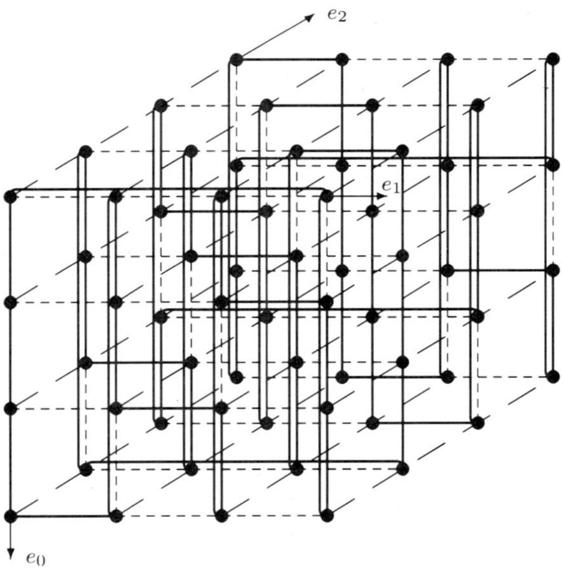

FIGURE 3. Cycles of color 0 (e_0-dominant)

Likewise, Figures 8 shows the 4 cycles of color 1. The cycles of color 1 are e_1-dominant. Notice again that when we move from one top face ($e_0 = 0, 1, 2, 3$) to the next, the cycle is shifted. Also notice that the cycles on such faces are the same, with the proper orientation, as those of Figures 4, 5, 6 and 7.

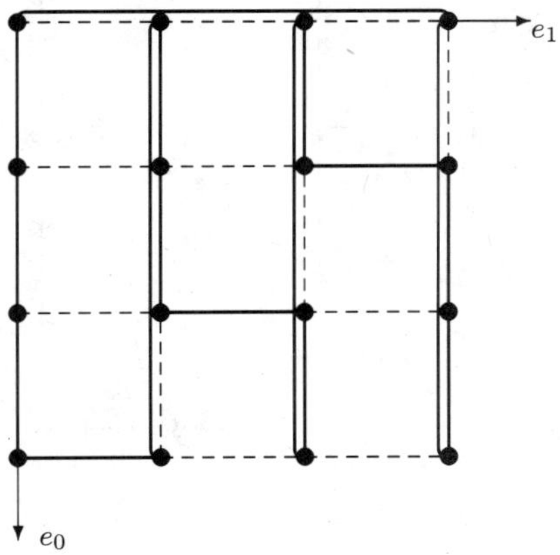

FIGURE 4. Cycle of color 0 on face $e_2 = 0$

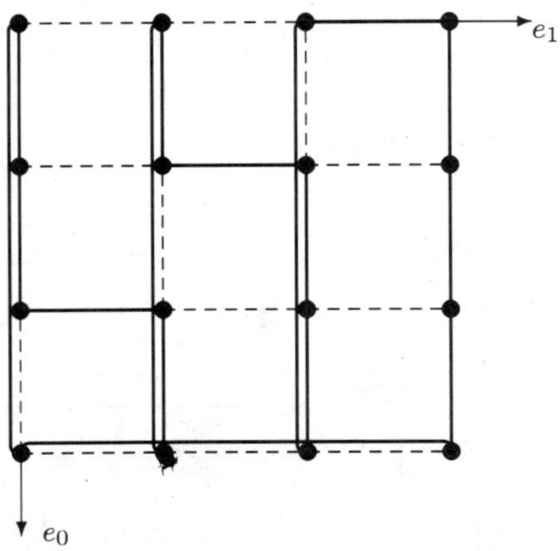

FIGURE 5. Cycle of color 0 on face $e_2 = 1$

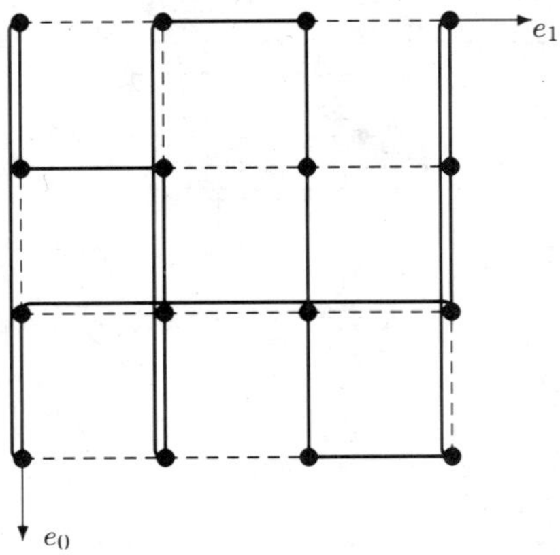

FIGURE 6. Cycle of color 0 on face $e_2 = 2$

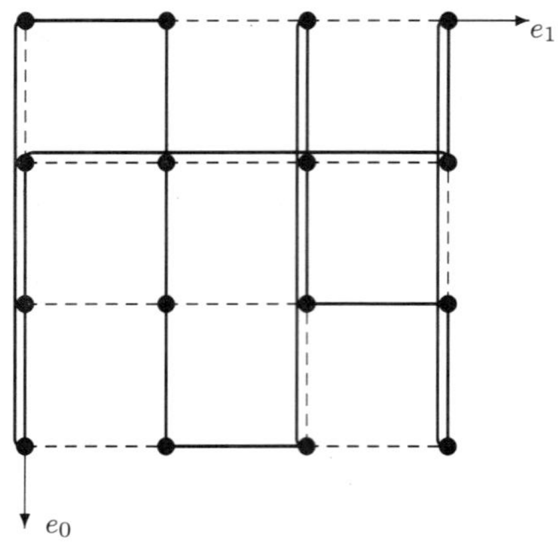

FIGURE 7. Cycle of color 0 on face $e_2 = 3$

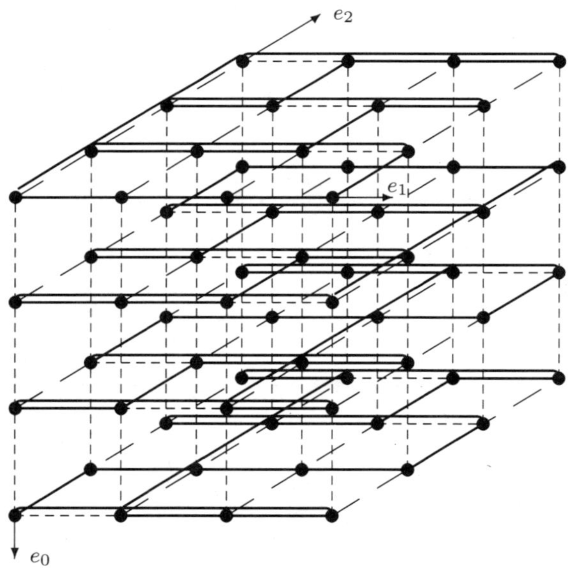

FIGURE 8. Cycles of color 1 (e_1-dominant)

Finally, Figures 9 show the 4 cycles of color 2. Again we notice that the cycles of color 2 are e_2-dominant and they are shifted when moving from one face to the next. Also, as before, these cycles are the same as those in Figures 4, 5, 6 and 7.

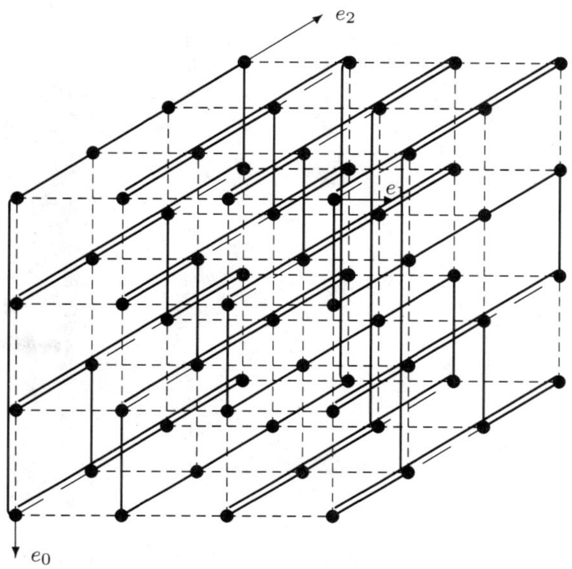

FIGURE 9. Cycles of color 2 (e_2-dominant)

In general, the edge set of $n/2$ products $C_4 \times C_4 \ldots \times C_4$, can be partitioned into $n2^n/32$ disjoint cycles of length 16. We use colors 0 to $m-1$ to color these cycles. There will be $2^n/16$ cycles of each color. Cycles of color i, $i = 0, \ldots, m-1$, are e_i-dominant.

4. Cycles merging

We show how the 12 cycles of length 16 of the previous section can be merged into the three desired Hamiltonian cycles. Cycles of the same color will be merged to form a Hamiltonian cycle.

Consider a node and edges incident with it. An edge-permutation operator is one that permutes the colors of these edges. (See Figure 10 for an example.)

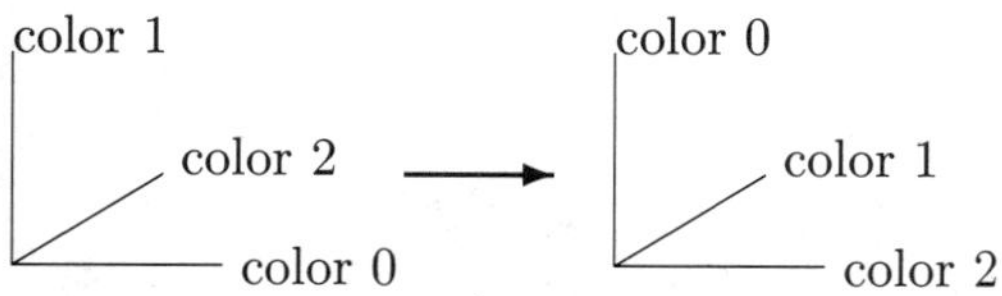

FIGURE 10. Example of an edge permutation operator

We use a set of edge-permutation operators to merge cycles of a given color to form larger cycles of the same color.

DEFINITION 4.1. *A set of edge-permutation operators constitutes a cycle merge operator (resp. cycle split operator) if it transforms a partition of the edge set of r cycles to a partition of s cycles, with $s < r$ (resp. $s > r$).*

Consider the partition of the edge set of $C_4 \times C_4 \times C_4$ into cycles of length 16 as seen in the previous section. Figure 11 shows a cycle merge operator that merges the cycles of a given color into a larger cycle of the same color. Figure 12 shows the locations (indicated by large circles) where reference point A of Figure 11 can be placed to apply the operator.

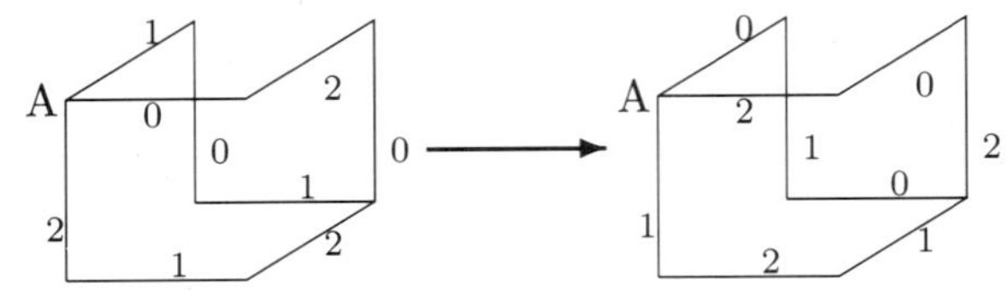

FIGURE 11. A merge operator for $C_4 \times C_4 \times C_4$

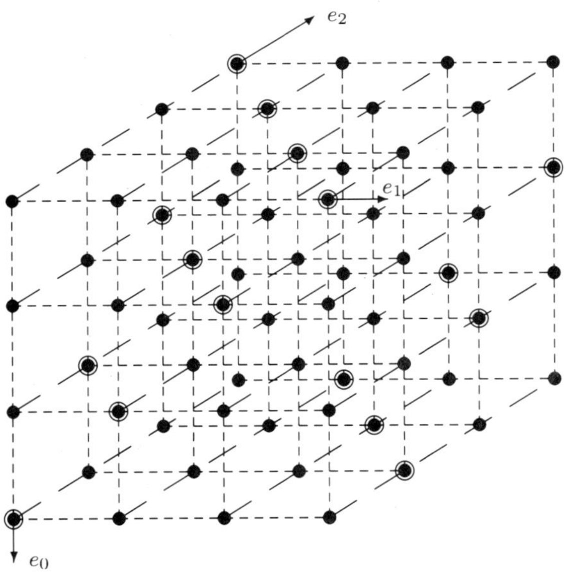

FIGURE 12. Possible locations (shown by large circles) to apply the merge operator

If the merge operator of Figure 11 is applied on one of the points of Figure 12, two cycles of each color (color 0, 1 and 2) will be merged to form a larger cycle of the same color. It will therefore transform six cycles to three.

The desired three cycles of the Hamiltonian decomposition of $C_4 \times C_4 \times C_4$ can be obtained by applying the merge operator to the points indicated in Figure 13.

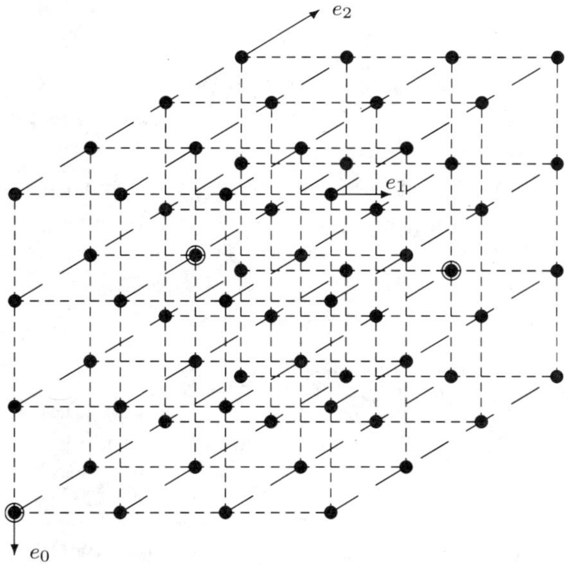

FIGURE 13. Locations to apply the merge operator for Hamiltonian decomposition

5. Generalization

To construct the Hamiltonian decomposition of $n/2$ products $C_4 \times C_4 \ldots \times C_4$, the approach is therefore as follows.

(i) Partition the edge set into cycles of length 16.
(ii) Merge the resulting cycles into larger cycles to obtain the desired Hamiltonian cycles.

The merge operator is contained in a binary $n/2$-cube. (For the case $n = 8$ the merge operator is inside a binary 4-cube, as shown in Figure 14.) It remains to formalize the merge operator in the general case.

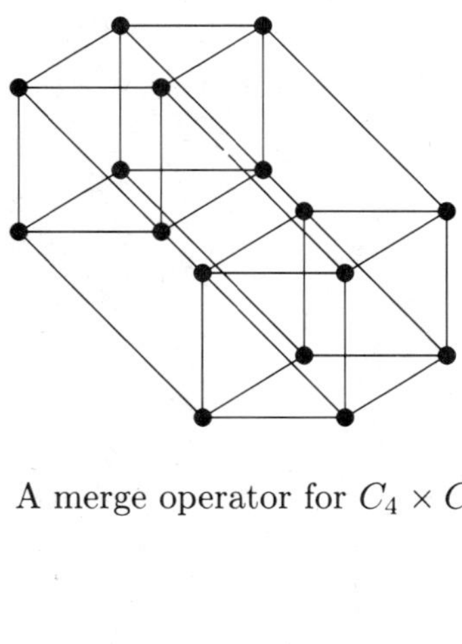

FIGURE 14. A merge operator for $C_4 \times C_4 \times C_4 \times C_4$

6. Conclusion

We have presented a simple approach to partition the edge set of a binary n-cube into Hamiltonian cycles. The starting point is the partitioning of the edge set into cycles of length 16. A merge operator constructs larger cycles until the desired cycles are obtained. Thus, this approach not only obtains the Hamiltonian decomposition, but also the partitioning into cycles of lengths 16, 32, ... , 2^n.

References

1. J. Armstrong and F. G. Gray. Fault diagnosis in a Boolean n-cube array of microprocessors. *IEEE Transactions on Computers* **38** (1981) 587 - 590.
2. B. Alspach, J.C. Bermond, D. Sotteau. Decomposition into cycles I: Hamilton decompositions. In G. H. et al. (editors) *Cycles and rays.* Kluwer Academic, 9 - 18, 1990.
3. J. Aubert et B. Schneider. Décomposition de $K_m + K_n$ en cycles Hamiltoniens. *Discrete Mathematics* **37** (1981) 19 - 27.
4. J. Aubert. Decomposition de la somme Cartesienne d'un cycle et de l'union de deus cycles Hamiltoniens en cycles Hamiltoniens. *Discrete Mathematics* **38** (1982) 7 - 16.
5. M. F. Foregger. Hamiltonian decompositions of products of cycles. *Discrete Mathematics* **24** (1978) 251 - 260.
6. P. Fraigniaud. *Communications intensives dans les architectures à mémoire distribuée et algorithmes paralléles pour la recherche de racines de polynômes.* Ph. Thesis. École Normale Supérieure de Lyon. Université Claude Bernard de Lyon I. December 1990.
7. A. Kotzig. Every Cartesian product of two circuits is decomposable into two Hamiltonian circuits. Centre de Recherche Mathématiques, Montreal (1973).
8. Y. Saad and M. H. Schultz. Topological properties of hypercubes. *IEEE Transanctions on Computers* **37** (1988) 867 - 872.

Departamento de Ciência da Computação, IME, Universidade de São Paulo, São Paulo, SP, 05508-900, Brazil

Current address: Departamento de Ciência da Computação, IME, Universidade de São Paulo, São Paulo, SP, 05508-900, Brazil

E-mail address: song@ime.usp.br

DIMACS Series in Discrete Mathematics
and Theoretical Computer Science
Volume **21**, 1995

Generalized Reduced Hypercube Interconnection Networks for Massively Parallel Computers

Sotirios G. Ziavras

ABSTRACT. The (direct binary) hypercube interconnection network has very frequently been employed in the construction of distributed-memory parallel computing systems. It offers low diameter and is so robust that it can very efficiently emulate a wide variety of other popular interconnection networks. However, the major drawback of the hypercube is the increase in the number of communication channels for each processor with an increase in the total number of processors in the system. This drawback has direct effect on the VLSI complexity of the hypercube and as a result it restricts inadvertently its adaptation in the construction of high-performance massively parallel systems. This paper investigates the properties of a new family of interconnection networks which are produced from the hypercube by a uniform reduction in the number of edges for each node. This family of networks is a generalization of the reduced hypercube (RH) family of interconnection networks introduced earlier by this author [31]. The edge reduction technique produces interconnection networks with much lower VLSI complexity than hypercubes, while the new networks maintain to high extent the very powerful hypercube properties. An extensive comparison of the proposed generalized reduced hypercube (GRH) networks with conventional hypercubes and RH's is presented. A routing algorithm is also presented, and it is concluded that GRH's are often good alternatives to hypercubes and RH's for the construction of massively parallel computers.

1. Introduction

Several interconnection networks have been proposed for the development of parallel computing systems, including rectangular meshes, trees, shuffle exchange networks, Omega networks, indirect binary n-cubes, and direct binary n-cubes. The family of direct binary n-cubes is a special case of the family of direct k-ary n-cubes with n dimensions and k nodes in each dimension [25]. In the rest of the paper, the terms *n-dimensional hypercube* and *n-cube* are used to denote the direct binary n-cube.

1991 *Mathematics Subject Classification.* Primary 68M07, 68Q22; Secondary 68R10.

The work presented in this research was supported in part by the National Science Foundation under Grants CCR-9109084 and CDA-9121475.

The n-dimensional hypercube is composed of 2^n nodes and has n edges per node. If unique n-bit binary addresses are assigned to the nodes of the hypercube, then an edge connects two nodes if and only if their binary addresses differ in a single bit. Several commercial hypercube-based systems have successfully been introduced. Such systems are the Intel iPSC, the nCUBE, the Connection Machine CM-2 [15], etc. The most important reason for the success of the hypercube has been its inherent capability of supporting the interconnection of a large number of processors with small diameter and small average internode distance. The *diameter* of an interconnection network is defined as the largest distance between two nodes, when all possible pairs of nodes are considered. The small diameter of the hypercube, which is equal to n for the n-dimensional hypercube, is the direct result of its high degree of connectivity and its regular structure. These advantages of the hypercube network result in efficient emulation of other topologies frequently used in parallel processing. Several algorithms have been proposed for embedding important topologies into the hypercube, such as rectangular meshes [5, 18], trees [10, 16, 18], pyramids [5, 21, 30], and multilevel structures [32].

The numbers of communication ports and channels per processor in the n-dimensional hypercube are equal to n, so they increase with an increase in the total number of processors in the system. This drawback affects the VLSI complexity of this network, therefore the construction of massively parallel hypercube systems with powerful processors becomes a very difficult or impossible task. Although this drawback seems to be waived in the case of incomplete hypercubes [19, 27], a large number of communication ports may be wasted and as a consequence a significant part of a system's cost may be spent for unused resources. Additionally, the VLSI complexity for part of the incomplete hypercube is similar to that of the immediately higher-dimension conventional hypercube [29].

In a similar fashion, the basic building block of hierarchical cubic networks (HCN's) [13] is the hypercube. However, the structure of HCN's is not regular. A methodology proposed by this author [29] performs slight modifications to hypercubes in order to support their incremental growth without introducing or wasting any resources for individual hypercubes ([14] presents processor allocation techniques for the resultant systems).

The cube-connected cycles [24] $CCC(n)$ interconnection network is obtained from the n-dimensional hypercube by substituting a ring with n nodes for each node in the hypercube. Each node in a ring then implements a distinct connection in one of the n dimensions. The advantage of the $CCC(n)$ is that the node connectivity is always 3, independently of the value of n. The CCC interconnection network has received special attention in the literature and several algorithms have been developed for it [4, 22].

RH's are produced by uniform "removal" at design time of several edges from conventional hypercubes [31]. Therefore, they have reduced complexity when compared to conventional hypercubes with the same numbers of nodes. This allows the construction of RH's with many more processors than in conventional hypercubes and, therefore, mitigates the architecture expansion problem. Results show that performance comparable to that of conventional hypercubes may be obtained at a much lower cost. Section 2 is devoted completely to RH's because the family of interconnection networks investigated in this paper are a generalization of RH's.

Other hypercube variations improve one or more topological properties of the hypercube or of its already discussed variations without reducing its VLSI complexity. Quite often the VLSI complexity rather increases [29]. Such variations are the twisted cube [1, 12], the enhanced incomplete hypercube [6], the crossed cube [11], the folded hypercube [2], the cubical ring connected cycles [3], the block-shuffled hypercube [17], hyper-rectangulars [20], the generalized folding cube [7], the spanning bus hypercube [28], and the generalized supercube [26].

The rest of this paper is organized as follows. Section 2 summarizes the properties of the family of reduced hypercube (RH) interconnection networks introduced in [31]. Section 3 investigates a new family of interconnection networks, namely generalized reduced hypercubes (GRH's), that are a generalization of RH's. A comparison is also carried out with RH's and conventional hypercubes. Section 4 presents a routing algorithm for GRH's. Section 5 investigates the problem of emulating conventional hypercubes on GRH's. Finally, Section 6 presents conclusions and future research objectives.

2. Reduced Hypercube (RH) Interconnection Networks

This section describes briefly the structure of reduced hypercube (RH) interconnection networks and lists their most important properties. RH's were introduced in [31] by this author. The main objective was to produce interconnection networks that carry on to a large extent the capabilities of conventional hypercubes while requiring lower cost due to lower VLSI complexity. The lower VLSI complexity of RH's permits the construction of systems with more PE's (processing elements) than in conventional hypercubes.

Although a family of generalized RH interconnection networks was proposed in [31], the properties of only a small, but very important, subset were investigated. The investigation of the properties of the most generalized family of RH's was identified as a future research goal. The most general structure of RH's is presented first. In the following discussion, $RH(k, n_1, n_2, ..., n_l)$ denotes the *reduced hypercube* interconnection network where every PE is attached to $k + l$ bidirectional communication channels and the total number of PE's is equal to 2^N, with

$$N = k + \sum_{i=1}^{l} 2^{n_i}$$

l is referred to as the number of levels in the RH. It is assumed that $k > 0$, $n_1 \leq k$, $n_{i+1} \leq 2^{n_i}$ for $1 \leq i \leq l-1$, and $n_i > 0$ for $1 \leq i \leq l$.

This RH is produced by uniformly removing some of the edges from the N-dimensional hypercube with the same number of nodes. Let the N-bit address of each node be divided into $l+1$ fields, as follows; also, l subfields are identified. More specifically, the least significant k bits of a node's address constitute the 0th field and represent the address of the node within a complete k-cube *building block*. The first subfield comprises the most significant n_1 bits of this 0th field and represents the address of a $(k - n_1)$-dimensional subcube in the k-cube building block that contains this node. In general, assuming that the subscript of the least significant bit in the address is 0, the bits with subscripts $f(j-1) = k + \sum_{i=1}^{j-1} 2^{n_i}$ through $f(j) - 1 = k + \sum_{i=1}^{j} 2^{n_i} - 1$, where $j = 1, 2, ..., l$, form the jth *field* of the address. The 0th field of the address contains the least significant k bits. In addition, the

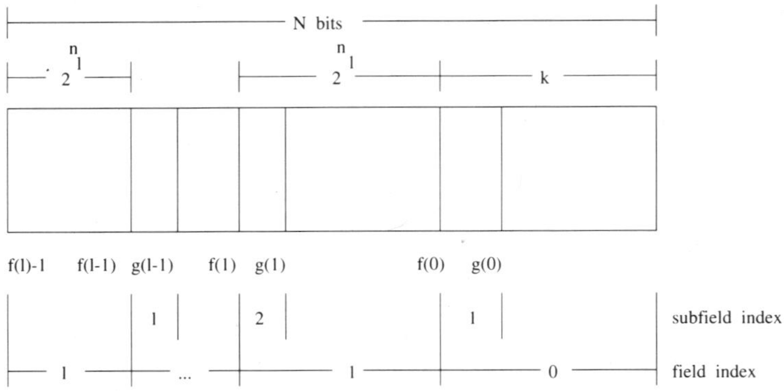

Figure 1: Fields and subfields in node addresses of the $RH(k, n_1, n_2, ..., n_l)$.

bits with subscripts $g(j-1) = f(j-1) - n_j$ through $f(j-1) - 1$ form the jth *subfield* of the address, for $j = 1, 2, ..., l$. Fig. 1 shows the location of these fields and subfields in the N-bit address.

The "removal" of edges at design time from the N-dimensional hypercube for the construction of the $RH(k, n_1, n_2, ..., n_l)$ is carried out in the following manner. Based on the aforementioned identification of fields and subfields in node addresses, from the N edges which are attached to each node in the original N-dimensional hypercube we keep the following sets that comprise a total of $k + l$ edges.

Set 1. The k edges of the N-dimensional hypercube that traverse the k lowest dimensions (i.e., dimensions 0 through $k - 1$) and connect the referenced node with k distinct nodes are kept. Therefore, a complete subcube with k dimensions that includes the referenced node is formed. This subcube was earlier referred to as a building block.

Set 2. This set is composed of l edges that were also present in the original N-dimensional hypercube. More specifically, one edge is kept for each of the remaining l fields of the node's address. For the jth field, where $1 \leq j \leq l$, the selected edge is the one that connects directly the referenced node with the node whose address differs only in the mth bit of the jth field, where m is the value in the jth subfield and the subscript of the least significant bit in a field is 0.

For the sake of simplicity, ";" is used here to separate consecutive fields in the address while a subfield lies between a "," and the first ";" to its left. For example, the neighbors attached to set 1 edges of the node with address (01101010; 101,0; 11,01) in the $RH(4, 2, 3)$ have addresses (01101010; 101,0; 11,0$\underline{0}$), (01101010; 101,0; 11,$\underline{1}$1), (01101010; 101,0; 1$\underline{0}$,01) and (01101010; 101,0; $\underline{0}$1,01). The underlined bits are those that differ from the corresponding bits in the address of the referenced node. The two neighbors that correspond to set 2 edges are (01101010; $\underline{0}$01,0; 11,01) and (01$\underline{0}$01010; 101,0; 11,01), because the values in the lower and upper subfields are 3 and 5, respectively.

Fig. 2 shows the structure of the $RH(k, 2)$ where the large squares represent the k-cube building blocks, the numbers above these squares represent the decimal values in field 1 of addresses for PE's in the corresponding building blocks, the numbers within the quadrants of large squares are subcube addresses of $(k-2)$-cubes as well as the decimal values in subfield 1 of contained-PE addresses, and each curved line represents 2^{k-2} bidirectional communication channels that interconnect pairs of PE's with the same addresses in two $(k-2)$-cubes in order to form a $(k-1)$-cube.

The $RH(k,n)$ can also be viewed as a hypercube of hypercubes. More specifically, it is obtained by replacing each node in the 2^n-dimensional hypercube with a k-dimensional hypercube, dividing the latter into 2^n $(k-n)$-dimensional subcubes, and using distinct subcubes to implement parallel connections in distinct dimensions of the original 2^n-dimensional hypercube.

Assuming a conventional hypercube with the same number of PE's (i.e., an N-cube), the ratio of the total number of communication channels in the conventional hypercube to the total number of channels in the RH network is equal to $\frac{N}{k+l}$. For example, if $k=5$, $l=2$, $n_1=2$ and $n_2=2$, then each of the two networks contains 8K PE's; the conventional hypercube contains $\frac{2^{13}\times 13}{2}=53,248$ bidirectional channels while the RH network contains $\frac{2^{13}\times(5+2)}{2}=28,672$ bidirectional channels. The reduction in the total number of bidirectional channels for this example is about 46% or 24,576 channels. Dramatic cost savings result in general from the reduction in the number of channels [31].

The investigation carried out in [31] of RH topological properties focused exclusively on RH's with $l=1$. The investigation of RH's with $l>1$ was identified as a future research objective. The main objective in the current paper is an extensive investigation of RH's with $l=2$ (such systems will be called generalized RH's, or GRH's for short), and their comparison to RH's with $l=1$ and conventional hypercubes. The remaining part of this section presents the most important properties of RH's with $l=1$ (see [31] for proofs). The investigation of RH's with $l=2$ begins in the next section.

A very significant characteristic of RH's (with $l=1$) is their capability of simultaneously emulating with dilation one several cube-connected cycles (CCC) interconnection networks of the same size. The dilation of an edge in the CCC network is the length of the path that connects the images of its two incident nodes in the RH. As emphasized in the Introduction, the CCC interconnection network has received significant attention since its inception [4, 22]. The $RH(2,2)$ is identical to the $CCC(4)$. In general, the $RH(n,n)$ contains the $CCC(2^n)$ with the same number of nodes as a proper subgraph because the ring with 2^n nodes is a proper subgraph of the n-cube building block. However, for $n<k$ the $RH(k,n)$ is viewed as 2^{k-n} $RH(n,n)$'s where all nodes with the same address in the 2^{k-n} distinct $RH(n,n)$'s are connected to form a $(k-n)$-dimensional hypercube. The nodes' addresses in the latter hypercube become the least significant $k-n$ bits of the nodes' addresses in the $RH(k,n)$. Therefore, another impressive property of the $RH(k,n)$ is its capability of emulating simultaneously in optimal manners 2^{k-n} $CCC(2^n)$'s.

Some important theorems from [31] are presented below without their proofs.

THEOREM 1. *For the emulation of the $(k+2^n)$-dimensional hypercube on the reduced hypercube $RH(k,n)$ with the same number of nodes, the dilations of the edges incident to a single node of the hypercube are: 1 for $k+1$ of them* *and* $2p+1$ *for* $\binom{n}{p}$ *of them, where* $p = 1, 2, ..., n$ *and* $\binom{n}{p}$ *represents the number of distinct* p*-combinations of* n *items. It is assumed that nodes of the hypercube are mapped onto nodes of the RH with the same address.*

For example, the dilations of the edges incident to a single node of the $RH(5,2)$ for the emulation of the 9-dimensional hypercube are 1, 3 and 5 for 6, 2 and 1 edge, respectively. Similarly, the dilations of the edges incident to a single node for the emulation of the 16-dimensional hypercube on the $RH(8,3)$ are 1, 3, 5 and 7 for 9, 3, 3 and 1 edge, respectively. The maximum dilation of edges for hypercube emulation on the $RH(k,n)$ is equal to $2n+1$. The average dilation of edges for hypercube emulation on the $RH(k,n)$ is equal to $\frac{k+(n+1)2^n}{k+2^n}$.

For example, the average dilation of edges for hypercube emulation on the $RH(5,2)$ and $RH(8,3)$ is 1.88 and 2.5, respectively. It was concluded in [31] that the average dilation of edges for hypercube emulation is relatively small in practical cases. This observation guarantees small performance degradation for low-cost implementation of hypercube algorithms on RH's. The effect of dilation is reduced from left to right for the set of four well-known packet switching techniques: store-and-forward, virtual cut-through, circuit switching, and wormhole routing [9, 23]. In advanced systems with wormhole routing the physical distance between communicating nodes may have a relatively minor effect on message latency unless it is very large. In this case, the ratio of hypercube bandwidth to RH bandwidth may prove to be an important measure. This ratio, which is simply the ratio of the numbers of edges in the two systems, is equal to $\frac{k+2^n}{k+1}$ for the $RH(k,n)$. For example, this ratio is equal to 1.5 and 1.77 for the $RH(5,2)$ and $RH(8,3)$, respectively. It is concluded that, for all practical cases (i.e., k has a relatively large value and $n = 1, 2$, or 3) RH's perform comparable to conventional hypercubes of higher cost under this performance measure as well.

THEOREM 2. *The diameter of the* $RH(k,n)$ *is equal to* $2^{n+1}+k-2$ *for* $n > 1$, *and* $k+3$ *for* $n = 1$.

For example, the diameter of the $RH(6,2)$ is 12. This is the distance of the node (0000; 00, 0000) from each of the nodes (1111; 00, 1111) and (1111; 11, 1111). In contrast, the diameter of the 10-dimensional hypercube with the same number of nodes is 10. In general, although RH's have much lower cost than conventional hypercubes with the same numbers of nodes, their diameters are comparable to those of hypercubes for practical cases [31].

The *average internode distance* in the N-dimensional hypercube is equal to $\frac{\sum_{j=1}^{N} j\binom{N}{j}}{2^N} = \frac{N}{2}$ as there exist $\binom{N}{j}$ nodes at distance j from each hypercube node. The following theorem is pertinent.

THEOREM 3. *The average internode distance in the* $RH(k,n)$ *for any routing algorithm that uses a strict order for the traversal of dimensions that interconnect building blocks is given by*

$$\frac{\sigma+k}{2} + \frac{1}{2}[\sum_{i=0}^{\sigma-2}\sum_{d=1}^{\sigma-1-i}\frac{\bar{H}(c_i, c_{i+d})}{2^d} + \sum_{d=1}^{\sigma-1}\frac{\bar{H}(c_0, c_d)}{2^d}]$$

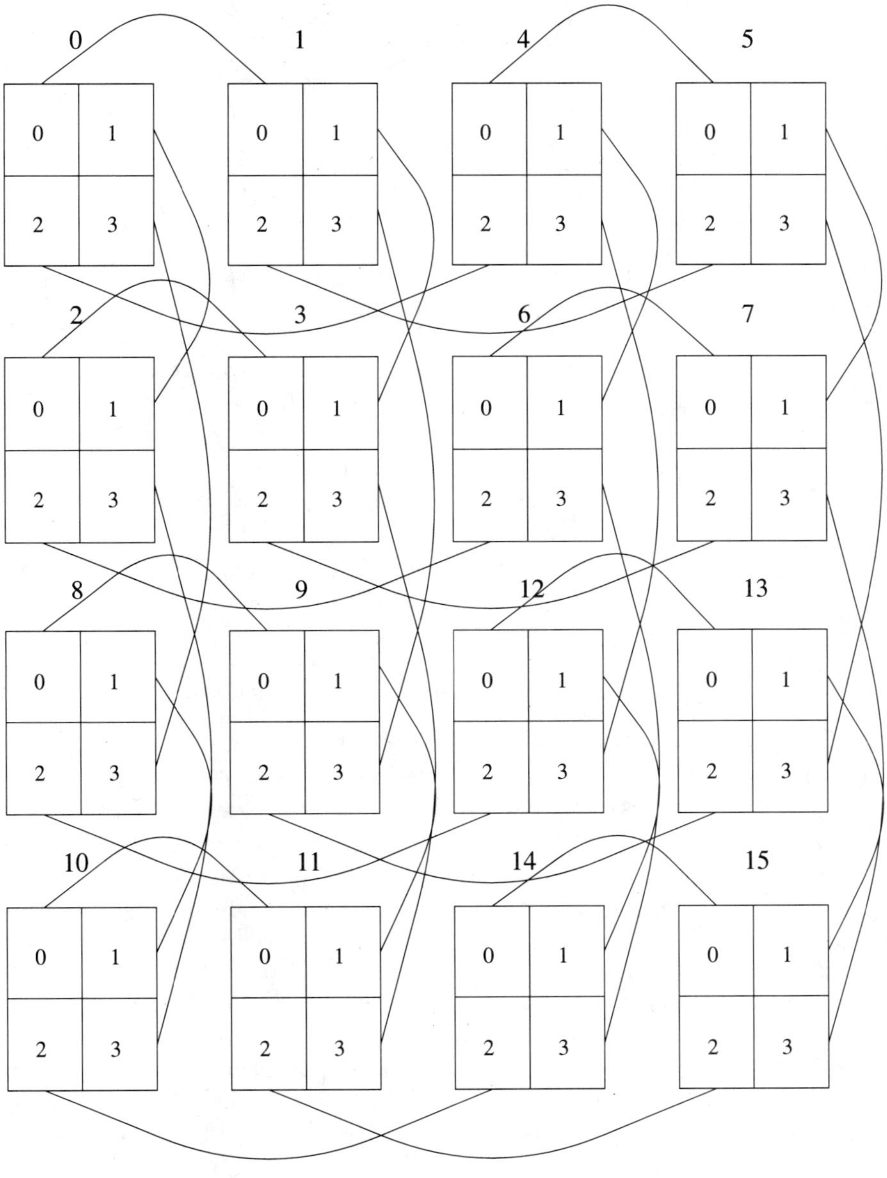

Figure 2: Structure of $RH(k, 2)$.

where

- $\sigma = 2^n$.
- c_i: *the n-bit representation of the* i^{th} *dimension in the chosen order, where* $0 \leq i \leq \sigma - 1$.
- $\bar{H}(c_i, c_{i+d})$: *the average Hamming distance between* c_i *and* c_{i+d}, *assuming all possible* 2^n *cyclic reorderings of dimensions.*

This average internode distance is proven to be very close to the average physical internode distance. For example, the average physical distance in the $RH(4,1)$ and $RH(6,2)$ is 3.5 and 6.625, respectively, and Theorem 3 gives 3.5 and 6.875, respectively, if traversal of dimensions is strictly ordered according to the binary reflected Gray code. The average internode distance in the $RH(k,n)$ is never lower than $\frac{\sigma+k}{2}$ which is the average internode distance in the respective $(2^n + k)$-dimensional hypercube with the same number of nodes. It was concluded in [31] that, for practical cases the average internode distance in RH's is comparable to that of conventional hypercubes with the same numbers of nodes.

Furthermore, optimal algorithms that embed linear arrays and multidimensional meshes into RH's appear in [33].

3. Generalized Reduced Hypercube (GRH) Interconnection Networks

An extensive investigation of RH's with $l = 2$, $n_1 = n$, and $n_2 = n$ is carried out in this paper. Such an interconnection network, which contains $2^{k+2^{n+1}}$ nodes, will be denoted by $GRH(k,n)$ and will be called *generalized reduced hypercube* throughout the paper. The $GRH(k,n)$ can also be viewed as the $RH(n,n)$ of $RH(k,n)$'s. Assume the $GRH(5,2)$ that uses 13-bit addresses for its nodes and contains 8K nodes. Each node in the $GRH(5,2)$ has seven neighbors. For example, the node (0000; 10, 00; 00, 000) has as neighbors the seven nodes (0000; 10, 00; 00, 00$\underline{1}$), (0000; 10, 00; 00, 0$\underline{1}$0), (0000; 10, 00; 00, $\underline{1}$00), (0000; 10, 00; 0$\underline{1}$, 000), (0000; 10, 00; $\underline{1}$0, 000), (0000; 10, 0$\underline{1}$; 00, 000), and (0$\underline{1}$00; 10, 00; 00, 000). In contrast, each node in the 13-dimensional hypercube with the same number of nodes has 13 neighbors, therefore, compared to the latter this GRH results in $13 \times 2^{13} - 7 \times 2^{13}$, or 49,152, or 46.15% savings in unidirectional communication channels.

In the rest of this paper, the three fields in GRH node addresses will be referred to as the lower, middle, and upper fields, from right to left, while the two subfields will be called the lower and upper subfields. The task of finding the diameter of GRH's is absolutely essential in this section. The next theorem is pertinent.

THEOREM 4. *The diameter of the* $GRH(k,n)$ *is* $5 \times 2^n + k - 6$ *for* $n > 1$, *and* $k + 6$ *for* $n = 1$.

PROOF. The proof distinguishes between $n > 1$ and $n = 1$.

a. Assume that $n > 1$. Addresses of pairs of nodes having the largest possible distance between them in the $GRH(k,n)$ differ in all $k - n$ least significant bits. This contributes $k - n$ hops to the diameter because the building block is the k-cube. Ignoring temporarily the k least significant bits in node addresses, the diameter of the $RH(2^n, n)$ is given by $2^{n+1}+2^n-2$ [31]. The lower subfield gets all possible values because all 2^n bits in the middle field must be changed, and this contributes $2^n - 1$ hops to the path (because the offset of the source node is already stored in the subfield). Using a Gray code sequence to change bits in the upper subfield, so that all bits in the upper field can be changed in the smallest possible number of steps, $2^n - 1$ distinct values are stored in the upper subfield. Due to the Gray code sequence generated for the upper subfield, the last value produced for this subfield differs in a single bit from the source's value in the upper subfield which appears first in the sequence. Since this bit can occupy any position, $n-1$ hops are required at the end in the worst case, if $n > 1$. However, for such changes to occur, a single bit in the lower subfield must also be changed each time. This contributes $(2^n-1)+(n-1)-n$ or $2^n - 2$ hops to the path; n is subtracted because a single change of each bit in the upper subfield was considered earlier. Similarly to changes performed in the upper subfield, $n-1$ hops are required in the destination building block in the worst case for changes in the lower subfield, if $n > 1$. Therefore, the diameter is $(k-n)+(2^{n+1}+2^n-2)+(2^n-1)+(2^n-2)+(n-1)$ or $5\times 2^n+k-6$.

b. Assume that $n = 1$. It was shown in [31] that the diameter of the $RH(2, 1)$ is 5. The diameter of the $GRH(1, 1)$ is equal to the diameter of the $RH(2, 1)$ plus 2; the value of 2 is added because of two additional changes required for the two one-bit subfields. Thus, the diameter of the $GRH(1, 1)$ is 7. $k - 1$ additional hops are needed for $k > 1$. Therefore, the diameter of the $GRH(k, 1)$ is $7 + k - 1$ or $k + 6$. □

For example, the diameter of the $GRH(5, 2)$ with 8K PE's is 19, while the diameter of the hypercube with the same number of PE's is 13. The numbers of unidirectional communication channels in the two systems are 7×2^{13} (i.e., 57,344) and 13×2^{13} (i.e., 106,496), respectively. The product between the diameter and the number of channels for each system is 1,089,536 and 1,384,448, respectively. The smaller the value of this product, the better the hypercube(-like) system could be because of a good balance between the diameter and the hardware cost. For this example, the $GRH(5, 2)$ is a good lower-cost alternative to the hypercube with the same number of PE's. For further comparison, the diameter and the number of unidirectional channels in the single-level $RH(5, 3)$ with the same number of PE's are 19 and 6×2^{13} (i.e., 49,152), respectively, and the value of the aforementioned product is 933,888. Although the $GRH(5, 2)$ and the $RH(5, 3)$ have the same diameter, the former has higher bandwidth while the latter has lower cost. Therefore, both structures could serve as viable alternatives to the 13-cube.

Also, the diameter of the $GRH(3, 3)$ with 512K PE's is 37, while the diameter of the hypercube with the same number of PE's is 19. The numbers of unidirectional communication channels in the two systems are 5×2^{19} (i.e., 2,621,440) and 19×2^{19} (i.e., 9,961,472), respectively. The product between the diameter and the number of channels for each system is 9.69×10^7 and 18.9×10^7, respectively. For further comparison, the diameter and the number of unidirectional channels in the single-level $RH(11, 3)$ with the same number of PE's are 25 and 12×2^{19} (i.e., 6,291,456), respectively, and the value of the aforementioned product is 15.7×10^7. Therefore, the $GRH(3, 3)$ has the lowest cost of all three systems and the lowest value for the aforementioned product. However, it has the largest diameter. These results prove that GRH's are often sound alternatives to conventional hypercubes and RH's for the construction of massively-parallel systems.

For $n_1 > 1$ and $n_2 > 1$, it can be easily shown that the diameter of the $GRH(k_2, n_2)$ is smaller than the diameter of the $RH(k_1, n_1)$ with the same number of nodes whenever $3 \times 2^{n_2} < 2^{n_1} + 4$ (it is derived from Theorems 2 and 4). The average internode distance is another important performance measure.

THEOREM 5. *The average internode distance in the $GRH(k,n)$ is upper-bounded by $\alpha \times (\alpha - 2^n + n) + 2^n + \frac{k-3n}{2}$, where α is the average internode distance in the $RH(n,n)$.*

PROOF. As mentioned earlier, the $GRH(k,n)$ can be viewed as the $RH(n,n)$ of $RH(k,n)$'s. Ignoring temporarily the least significant $k - n$ bits of the address, the $GRH(n,n)$ is viewed as the $RH(n,n)$ of $RH(n,n)$'s. For a data transfer between two $RH(n,n)$'s, any of the lower $2^n - n$ bits in the middle field need not change, and the offsets of these bits need not be created in the lower subfield in intermediate steps, except once. Therefore, an upper bound on the average distance in the $GRH(n,n)$ is $\alpha \times (\alpha - 2 \times \frac{2^n - n}{2}) + 2 \times \frac{2^n - n}{2}$ or $\alpha \times (\alpha - 2^n + n) + 2^n - n$. Finally, $\frac{k-n}{2}$ must be added to this quantity for the $GRH(k,n)$. Therefore, the average internode distance in the $GRH(k,n)$ is upper-bounded by $\alpha \times (\alpha - 2^n + n) + 2^n + \frac{k-3n}{2}$. □

For example, the average internode distance in the $GRH(6,2)$ is upper-bounded by 16.14, because this distance in the $RH(2,2)$ is 4.625. For the sake of comparison, the actual average internode distance in the 14-cube with the same number of nodes is 7. However, not only is the former an upper bound but the $GRH(6,2)$ and the 14-cube contain 8×2^{14} (i.e., 131,072) and 14×2^{14} (i.e., 229,376) unidirectional communication channels, respectively. Therefore, the $GRH(6,2)$ has much lower cost.

The next theorem computes the distance between any given pair of PE's in the $GRH(k,n)$.

THEOREM 6. *The distance between any given pair of (src, dst) PE's in the $GRH(k,n)$ is equal to $N_1 + 2 \times (N_2 - N_3) + 2 \times (N_4 - N_5) - (N_6 + N_7) - (N_8 + N_9) + N_{10} + N_{11}$, where:*

- N_1: *The total number of 1 bits in the upper field of the XOR result between src and dst.*
- N_2: *The total number of 1 bits in the chosen sequence of all different values in the upper subfields; the values in the subfields of src and dst are included.*
- N_3: *The total number of 1 bits in the results of the AND operations between all pairs of consecutive different values in the upper subfields of the chosen sequence.*
- N_4: *The total number of 1 bits in the chosen sequence of all different consecutive values in the lower subfields; the values in the subfields of src and dst are included.*

- N_5: *The total number of 1 bits in the results of the AND operations between all pairs of consecutive different values in the lower subfields of the chosen sequence.*
- N_6: *The total number of 1 bits in the upper subfield of src.*
- N_7: *The total number of 1 bits in the upper subfield of dst.*
- N_8: *The total number of 1 bits in the lower subfield of src.*
- N_9: *The total number of 1 bits in the lower subfield of dst.*
- N_{10}: *The total number of 1 bits in the least significant $k - n$ bits of the XOR result between src and dst.*
- N_{11}: *The total number of 1 bits in the lower $2^n - n$ bits of the middle field in the XOR result between src and dst.*

PROOF. N_1 bit-change operations are required in the upper field of the source address in order to match the same field of the destination address. However, a bit in the upper field can be changed only when the value in the upper subfield is equal to this bit's offset in the upper field. Ignoring temporarily the lower field of the address, $2 \times (N_2 - N_3)$ is the minimum number of steps required to produce, in the chosen order, the offsets of all bits that must be changed in the upper field of the address. Since this evaluation that uses N_2 assumes that src and dst have 0 in their upper subfields, $N_6 + N_7$ is subtracted from the calculated value. Similarly, $2 \times (N_4 - N_5)$ is the minimum number of steps required to produce, in the chosen order, the subscripts of all bits that must be changed in the middle field of the address. Since this evaluation that uses N_4 assumes that src and dst have 0 in their lower subfields, $N_8 + N_9$ is subtracted from the calculated value. N_{10} steps are required to change bits differing in the least significant $k - n$ bits of the source address because the E-cube routing algorithm for conventional hypercubes must be applied to this part of the address. Finally, N_{11} steps are required to change bits differing in the lower $2^n - n$ bits of the middle field. Hence, the result. □

For example, assume that src = (0110; 10, 10; 10, 010) and dst = (0000; 00, 00; 00, 000) in the $GRH(5, 2)$, and the intermediate nodes in the chosen path are (0110; 10, 10; 10, 0$\underline{0}$0), (0$\underline{0}$10; 10, 10; 10, 000), (0010; 1$\underline{1}$, 10; 10, 000), (0010; 11, 10; 1$\underline{1}$, 000), (0010; $\underline{0}$1, 10; 11, 000), (00$\underline{0}$0; 01, 10; 11, 000), (0000; 01, 10; $\underline{0}$1, 000), (0000; 01, $\underline{0}$0; 01, 000), (0000; 01, 00; 0$\underline{0}$, 000), (0000; 01, 00; $\underline{1}$0, 000), and (0000; 0$\underline{0}$, 00; 10, 000), in this order. Then, the chosen sequences of bit subscripts in the upper and lower subfields are (10, 11, 01, 00) and (10, 11, 01, 00, 10, 00), respectively. Thus, $N_1 = 2$, $N_2 = 4$, $N_3 = 2$, $N_4 = 5$, $N_5 = 2$, $N_6 = 1$, $N_7 = 0$, $N_8 = 1$, $N_9 = 0$, $N_{10} = 1$, and $N_{11} = 1$. Therefore, the distance between the two nodes is equal to 12.

Theorem 6 finds the distance for a particular pair of ordered sequences for values in the upper and lower subfields. The absolute shortest distance is obtained for a pair of sequences that minimize the value of $N_4 - (N_3 + N_5)$. From all parameters N_i, where $i = 1, 2, ..., 11$, the chosen routing algorithm can affect only the values of N_3, N_4, and N_5; all other parameters just depend on the source and destination addresses, independently of the chosen path. Therefore, routing algorithms for

GRH's should attempt to minimize the value of $N_4 - (N_3 + N_5)$, for the sake of efficiency. Such a routing algorithm is introduced in the next section.

4. Routing in GRH's

Routing in the conventional hypercube is performed by a distributed process that changes each time a single bit of the source address in order to match the corresponding bit of the destination address. This process produces the address of a neighbor that receives the message. The neighbor then becomes the source and this process is repeated until the message reaches the destination. The distance between any two nodes in the hypercube is equal to the Hamming distance between their binary addresses (i.e., the total number of 1 bits in the XOR result between the source and destination addresses).

To avoid deadlocks during simultaneous transmission of messages between several pairs of PE's, the E-cube routing algorithm for conventional hypercubes uses dimension ordering so that messages traverse in a predetermined increasing or decreasing order the dimensions where the source and destination addresses differ. The increasing dimension order is assumed throughout this paper. Deadlock can occur if packets are permitted to hold resources while requesting others. These resources are packet buffers in store-and-forward and virtual cut-through switching, while they are channels in circuit switching and wormhole routing [23]. For the sake of simplicity, the description immediately below of the routing algorithm for GRH's ignores potential deadlock problems. Virtual channels are introduced at the end of this section to make the routing algorithm deadlock-free.

The presentation of the routing algorithm for the $GRH(k, n)$ is verbose for the purpose of clarity. The execution steps in this routing algorithm are repeated for each PE in the path until the destination is reached. An n-bit field is appended to the destination address for the marking of bits in the upper subfield, as explained later. All bits in the upper subfield of the destination address are unmarked before the message leaves the source node.

Step 1. Apply the E-cube routing algorithm with the increasing dimension order to the least significant $k - n$ bits of the source address, if the source and destination addresses differ in any of these bits. (This step implements E-cube routing within a source subcube.) Otherwise, go to Step 2.

Step 2. Let m be the n-bit value stored in the lower subfield of the current address. If the current and destination addresses differ in the mth bit of the middle field, and either $0 \leq m < 2^n - n$ or the two addresses do not differ in the upper field, then transmit the message to the PE whose address differs from the current address only in the mth bit of the middle field (a direct connection exists for the implementation of this data transfer). Otherwise, go to Step 3.

Step 3. Let β be the n-bit value stored in the upper subfield of the current address. If the current and destination addresses differ in the βth bit of the upper field, then transmit the message to the PE whose address differs from the current address only in this bit (a direct connection exists for the implementation of this data transfer). If the two addresses do not differ in the βth bit but differ in any other bit(s) of the upper field, then go to Step 4; otherwise (i.e., the two addresses do not differ in the upper field), go to Step 5.

Step 4. If there are any marked bit(s) in the upper subfield of the destination address, then go to Step 6. Otherwise, find the bit in the upper field that differs in the current and destination addresses and whose subscript in this field differs in the smallest number of bits from the n-bit value in the upper subfield of the current address. For multiple subscripts corresponding to the smallest number of bits, choose one of them randomly. Mark the bit(s) in the upper subfield that must be changed in order to produce this subscript and go to Step 6.

Step 5. Let λ be the subscript of the most significant bit position where the current and destination addresses differ. If $\lambda < k$, then apply the E-cube routing algorithm for conventional hypercubes to the lower subfield of the current address in order to find the neighbor that then receives the message. Otherwise, go to Step 6.

Step 6. Find the bit in the middle field (including any marked bit) that differs in the current and destination addresses and whose subscript in this field differs in the smallest number of bits from the n-bit value in the lower subfield of the current address. For multiple subscripts corresponding to the smallest number of bits, choose one of them randomly. To find the neighbor that then receives the message, carry out E-cube routing within the k-cube building block to the nearest PE that can correct the aforementioned bit in the middle field.

This routing algorithm never chooses paths that contain cycles. The proof is omitted for the sake of brevity. Although this routing algorithm does not always choose a shortest path, it achieves very good performance because its main objective is to minimize or produce a small value for the term $N_4 - (N_3 + N_5)$ (see the discussion in Section 3). This is because its goal each time is to change the bit in the middle or upper field whose subscript in this field differs in the smallest number of bits from the value in the lower or upper subfield of the current address, respectively. The time complexity of this algorithm may be $O(2^n)$ because of Step 4 or Step 6. For practical cases, n is in the range from 1 to 3, therefore the complexity is always small. In addition, constant time can be achieved with a hardwired router as all operations in this algorithm are easy to implement in hardware. Any attempts to modify the routing algorithm with the objective of always minimizing the term $N_4 - (N_3 + N_5)$ will increase the time complexity quite significantly.

An example of applying this distributed routing algorithm follows. If the source and destination addresses in the $GRH(3,2)$ are (0110; 10, 10; 10,1) and (0000; 00, 00; 00,0), respectively, then the routing algorithm produces addresses in this order: (0110; 10, 10; 10,$\underline{0}$), (0$\underline{0}$10; 10, 10; 10,0), (0010; 1$\underline{1}$, 10; 10,0), (0010; 11, 10; 1$\underline{1}$,0), (0010; $\underline{0}$1, 10; 11,0), (00$\underline{0}$0; 01, 10; 11,0), (0000; 01, 10; $\underline{0}$1,0), (0000; 01, $\underline{0}$0; 01,0), (0000; 01, 00; 0$\underline{0}$,0), (0000; 01, 00; $\underline{1}$0,0), (0000; 0$\underline{0}$, 00; 10,0), and (0000; 00, 00; $\underline{0}$0,0).

The proposed routing algorithm is not deadlock-free in its current form for simultaneous data transfers between several pairs of PE's. The rest of this section deals with the problem of making this algorithm deadlock-free. Deadlock(s) can occur in GRH's because dimension ordering cannot be strictly enforced. Dimension order routing, such as E-cube routing for hypercubes, is often used in regular interconnection networks to avoid deadlocks through the elimination of cycles in channel dependency graphs [9]. A slight enhancement in the proposed routing algorithm for GRH's can make it deadlock-free. This enhancement is based on an adaptive routing algorithm for k-ary n-cubes [8]. The latter algorithm associates a dimension reversal (DR) number with each packet to avoid cycles in channel dependency graphs caused by adaptive routing. The DR number of a packet is defined as the count of the number of times the packet has been routed from a channel in a higher dimension to a channel in a lower dimension, assuming E-cube routing with the increasing dimension order for the basic non-adaptive algorithm (i.e., DR counts dimension reversals). All packets are initialized with $DR = 0$. The DR number of a packet is incremented when a dimension reversal occurs for this packet. The algorithm introduces virtual channels and divides the virtual channels of each physical channel into $r + 1$ classes, numbered 0 to r, where r represents the maximum number of dimension reversals permitted. Packets with $DR < r$ can use only virtual channels of class DR. Once a packet reaches $DR = r$, it must strictly use E-cube routing on the virtual channels of class r. Virtual channels are multiplexed very efficiently over a physical channel in a demand-driven fashion.

This technique is incorporated here in the proposed algorithm in order to achieve deadlock-free routing in the $GRH(k, n)$. $r + 1$ virtual channels are associated with each physical channel, where r is defined as the largest possible number of dimension reversals. An upper bound on r is found as follows. The largest possible number of dimension reversals due to the upper field is equal to 2^n. Similarly, the largest possible number of dimension reversals due to the lower $2^n - n$ dimensions of the middle field is $2^n - n$ while there are no dimension reversals due to the lower field. The upper subfield must be modified up to $2^n + 1$ times (2^n times to permit modification of bits in the upper field and once to match the value in the upper subfield of the destination), with the change of a single bit each time; however, these modifications do not result in dimension reversals because of subsequent modifications in the upper field. Therefore, $r = 2^{n+1} - n$.

5. Hypercube Emulation on GRH's

This section investigates the problem of emulating conventional hypercubes on GRH's. This is a very important problem to solve because of the wide variety of application algorithms that have been developed for the hypercube interconnection network. For the purpose of evaluating the degradation in performance due to missing communication channels in the $GRH(k, n)$, the dilation of edges associated with such a hypercube mapping is found. Let the function $h : G \to G'$ represent the mapping of the source graph G (i.e., the hypercube) onto the target graph G' (i.e., the GRH). It is a mapping of the vertices (i.e., nodes) of G to the vertices of G' in one-to-one fashion. The aforementioned performance measure is then defined as follows. When two neighboring nodes from G are mapped onto two distinct nodes in G', the *dilation* of the edge connecting the two nodes in G is the length of the shortest path that connects these two nodes in G'.

The dilation is an indication of the communication overhead in G' for one-hop data transfers in the source graph G. Let the source N-dimensional hypercube and the target $GRH(k,n)$ contain the same number of nodes; that is, $N = k + 2^{n+1}$. Assume that each node from the conventional hypercube is mapped to the node of the GRH with the same address (i.e., one-to-one mapping). The following theorem presents the resultant dilation of edges.

THEOREM 7. *For the emulation of the* $(k + 2^{n+1})$*-dimensional hypercube on the generalized reduced hypercube* $GRH(k,n)$ *with the same number of nodes, the dilations of all edges incident to a single node of the hypercube are as follows.* **(a)** *For the* $k + 2^n$ *least significant dimensions of the hypercube and for one dimension in the upper field the dilations of the edges are: 1 for* $k+2$ *of them, and* $2p + 1$ *for* $\binom{n}{p}$ *of them, where* $p = 1, 2, ..., n$ *and* $\binom{n}{p}$ *represents the number of distinct p-combinations of n items.* **(b)** *The dilations of edges that lie in any of the remaining* $2^n - 1$ *higher dimensions of the hypercube are given by* $2(p + q) + 1$*, where* p *is the total number of 1's in the* n*-bit number* α*, for* $\alpha = 1, 2, ..., 2^n - 1$*, and* q *is the total number of 1's in the XOR results between pairs of consecutive* n*-bit subscripts of the 1 bits in* α*, assuming that the subscripts are listed in a sequence that minimizes* q *and 0 is the first element in the sequence.*

PROOF. Without loss of generality, the proof investigates the dilations of all edges incident to the node with address 0. Therefore, the distances of the node with address 0 from the $k + 2^{n+1}$ nodes whose addresses contain a single 1 bit must be found for the GRH. If the 1 bit is one of the least significant k bits of the node's address, then the dilation is equal to 1 because the considered node belongs to the same complete k-cube building block with the node having address 0. If the least significant bit in the middle or upper field is equal to 1, then this dilation is also 1 because the lower or upper subfield, respectively, of address 0 contains the appropriate bit subscript for direct data transfer. Therefore, the dilation is 1 for $k+2$ edges. However, the dilation is larger than 1 if any of the other bits in the address is 1. For such edges, we distinguish now between two cases.

Case 1. One of the bits in the middle field is 1 (except for the least significant bit). Since any of the 2^n bits in this field may be equal to 1, to set the n-bit lower subfield to the subscripts of the 1 bits in the middle field requires p steps for $\binom{n}{p}$ of the subscripts, where $p = 1, 2, ..., n$. One more step is needed to change the corresponding bit in the middle field and the same number of steps as above (i.e., p) to set the lower subfield back to its original value (i.e., all 0's). Therefore, the dilation is equal to $2p + 1$ for $\binom{n}{p}$ edges, where $p = 1, 2, ..., n$.

Case 2. One of the bits in the upper field is 1 (except for the least significant bit). Let p be the number of 1's in the n-bit number α that represents the location of the respective dimension in the higher field of the address. Let q be the total number of 1's in the XOR results between the p subscripts of the 1 bits in α, assuming that the subscripts are listed in a sequence that minimizes q; the first value in the sequence is 0, which is the value stored in the lower subfield of node 0. To change the bit corresponding to the aforementioned dimension in the higher field, its subscript in this field must first be created in the upper subfield; $\binom{n}{p}$ such subscripts have p 1's, where $p = 1, 2, ..., n$. However, to create these p 1's in the upper subfield, their subscripts in the middle field must first be stored in the lower subfield. The chosen sequence of values for the lower subfield is the one that minimizes the value of q. q changes are carried out in the lower subfield in order to create the required subscripts of these bits for changes in the middle field. p changes are required in the middle field. Finally, one change is required in the higher field. To bring the value in the middle and lower fields back to 0, the $p + q$ previously described changes are carried out by reversing the steps described earlier. Therefore, the dilation of the corresponding edge is $2(p + q) + 1$. Hence, the result. □

For example, the dilations of all edges incident to a single node of the $GRH(5, 2)$ for the emulation of the 13-dimensional hypercube are 1, 3, 5, 7, and 9 for 7, 2, 2, 1, and 1 edge, respectively. The average dilation is 3 and the total number of unidirectional communication channels is $2^{13} \times 7$ or 57,344. In contrast, the 13-dimensional hypercube contains $2^{13} \times 13$ or 106,496 channels. Therefore, the $GRH(5, 2)$ results in savings of 46.15% in communication channels whereas the average dilation for hypercube emulation is relatively small. For further comparison, the dilations of all edges incident to a single node for emulation of the 13-dimensional hypercube on the $RH(5, 3)$ with the same number of nodes are 1, 3, 5, and 7 for 6, 3, 3, and 1 edge, respectively. The average dilation is 2.8462 and the total number of unidirectional communication channels in the $RH(5, 3)$ is $2^{13} \times 6$ or 49,152. For this emulation, the $RH(5, 3)$ is a better choice than the $GRH(5, 2)$.

Similarly, the dilations of all edges incident to a single node of the $GRH(3, 3)$ for the emulation of the 19-dimensional hypercube are 1, 3, 5, 7, 9, 11, 13, and 15 for 5, 3, 3, 3, 1, 2, 1, and 1 edge, respectively. The average dilation is 5.7368 and the total number of unidirectional channels in the $GRH(3, 3)$ is $2^{19} \times 5$ or 2,621,440. In contrast, the number of channels in the 19-dimensional hypercube is $2^{19} \times 19$ or 9,961,472. For further comparison, the dilations of all edges incident to a single node for emulation of the 19-dimensional hypercube on the $RH(11, 3)$ are 1, 3, 5, and 7 for 12, 3, 3, and 1 edge, respectively. The average dilation is 2.2632 and the total number of channels is $2^{19} \times 12$ or 6,291,456 (i.e., it has much higher cost than the $GRH(3, 3)$).

The next corollary gives the maximum dilation of edges for hypercube emulation.

COROLLARY 1. *The maximum dilation of edges for emulation of the* $(k + 2^{n+1})$*-dimensional hypercube on the* $GRH(k, n)$ *with the same number of nodes is* $2 \times (n + g) + 1$*, where* g *is the total number of 1 bits in the XOR results between all consecutive pairs of* n*-bit values in a sequence that begins with 0 and contains all numbers in the range from* $2^n - n$ *to* $2^n - 1$*; the order chosen for these numbers in the sequence minimizes the value of* g*.*

PROOF. Without loss of generality, assume the edges incident to the node with address 0. The other vertex of the hypercube edge with the largest dilation in the mapping onto the $GRH(k,n)$ is the one with address whose most significant bit is 1 while the remaining bits are 0's. To change the most significant bit in the address, the upper subfield must be all 1's. The lower subfield gets all n values for the subscripts of all bits in the upper subfield in g steps. It also requires n steps to change all the bits in the upper subfield and one step to change the most significant bit of the address. Finally, $n+g$ steps are required to bring the middle and lower fields back to 0. Therefore, the maximum dilation is $2 \times (n+g)+1$. □

For example, the maximum dilation of an edge for the mapping of the 10-cube onto the $GRH(2,2)$ is 9; the value of g is 2 and corresponds to the sequence (00, 10, 11). However, the numbers of unidirectional communication channels in the two systems are 10,240 and 4,096, respectively. Similarly, the maximum dilation of an edge for the mapping of the 19-cube onto the $GRH(3,3)$ is 15; the value of g is 4 and corresponds to the sequence (000, 110, 111, 101). However, the numbers of unidirectional communication channels in the two systems are 9,961,472 and 2,621,440, respectively.

6. Conclusions

The family of generalized reduced hypercube (GRH) interconnection networks were investigated in this paper. GRH's are a generalization of RH's (reduced hypercubes) presented in [31]. GRH's and RH's have lower VLSI complexity than conventional hypercubes with the same numbers of processors due to much smaller numbers of communication channels. RH's are viewed as hypercubes of hypercubes while GRH's are viewed as RH's of RH's. A comparative analysis has shown earlier that RH's are capable of yielding performance comparable to that of conventional hypercubes, at much lower cost. The main objective in this paper was to carry out a comparative analysis of GRH's with hypercubes and RH's. It was shown that GRH's are often viable candidates for the implementation of massively parallel systems, when compared to hypercubes and RH's. The development of processor allocation techniques for GRH's is a future research objective.

References

1. S. Abraham and K. Padmanabhan, *An Analysis of the Twisted Cube Topology*, Proc. International Conference on Parallel Processing, IEEE Computer Society, Silver Spring, MD, 1 (Aug. 1989), 116-120.

2. A.E. Amawy and S. Latifi, *Properties and Performance of Folded Hypercubes*, IEEE Trans. Parallel Distributed Systems, 2 (Jan. 1991), 31-42.

3. P. Banerjee, *The Cubical Ring Connected Cycles: A Fault-Tolerant Parallel Computation Network*, IEEE Trans. Comput., 37 (May 1988), 632-636.

4. G.E. Carlsson, J.E. Cruthirds, H.B. Sexton, and C.G. Wright, *Interconnection Networks Based on a Generalization of Cube-Connected Cycles*, IEEE Trans. Computers, 34 (Aug. 1985), 769-772.

5. T.F. Chan and Y. Saad, *Multigrid Algorithms on the Hypercube Multiprocessor*, IEEE Trans. Comput., 35 (1988), 969-977.

6. H.-L. Chen and N.-F. Tzeng, *Enhanced Incomplete Hypercubes*, Proc. International Conference on Parallel Processing, IEEE Computer Society, Silver Spring, MD, 1 (1989), 270-277.

7. S.B. Choi and A.K. Somani, *The Generalized Folding-Cube*, Proc. International Conference on Parallel Processing, IEEE Computer Society, Silver Spring, MD, 1 (Aug. 1990), 372-375.

8. W.J. Dally and H. Aoki, *Deadlock-Free Adaptive Routing in Multicomputer Networks Using Virtual Channels*, IEEE Trans. Parallel Distributed Systems, 4 (April 1993), 466-475.

9. W.J. Dally and C.L. Seitz, *The Torus Routing Chip*, Distributed Computing, 1 (1986), 187-196.

10. S.R. Deshpande and R.M. Jenevin, *Scalability of a Binary Tree on a Hypercube*, Proc. International Conference on Parallel Processing, IEEE Computer Society, Silver Spring, MD, 1 (Aug. 1986), 661-668.

11. K. Efe, *The Crossed Cube Architecture for Parallel Computation*, IEEE Trans. Parallel Distributed Systems, 3 (1992), 513-525.

12. A.-H. Esfahanian, L.M. Ni, and B.E. Sagan, *The Twisted N-Cube with Application to Multiprocessing*, IEEE Trans. Comput., 40 (Jan. 1991), 88-93.

13. K. Ghose and K.R. Desai, *The HCN: A Versatile Interconnection Network Based on Cubes*, Proc. Supercomputing '89, IEEE Computer Society and ACM SIGARCH, 1 (Nov. 1989), 426-435.

14. N.G. Haravu and S.G. Ziavras, *Processor Allocation for a Class of Hypercube-Like Supercomputers*, Proc. Supercomputing '92, IEEE Computer Society and ACM SIGARCH, Minneapolis, MN, 1 (Nov. 1992), 740-749.

15. W.D. Hillis, *The Connection Machine*, MIT Press, Cambridge, MA, 1985.

16. C.-T. Ho and S.L. Johnsson, *Spanning Balanced Trees in Boolean Cubes*, SIAM J. Sci. Stat. Comput., 10 (July 1989), 607-630.

17. W.T.-Y. Hsu, P.-C. Yew, and C.-Q. Zhu, *An Enhancement Scheme for Hypercube Interconnection Networks*, Proc. International Conference on Parallel Processing, IEEE Computer Society, Silver Spring, MD, 1 (Aug. 1987), 820-823.

18. S.L. Johnsson, *Communication Efficient Basic Linear Algebra Computation on Hypercube Architectures*, Journal Parallel Distributed Computing, 4 (April 1987), 133-172.

19. H.P. Katseff, *Incomplete Hypercubes*, IEEE Trans. Comput., 37 (May 1988), 604-608.

20. T. Kerola and A. Hartmann, *Operational Analysis on Hyper-Rectangulars*, Proc. International Conference on Parallel Processing, IEEE Computer Society, Silver Spring, MD, 1 (Aug. 1988), 78-82.

21. T.-H. Lai and W. White, *Mapping Pyramid Algorithms into Hypercubes*, Journal Parallel Distrib. Computing, 9 (1990), 42-54.

22. D.S. Meliksetian and C.Y.R. Chen, *Optimal Routing Algorithm and the Diameter of the Cube-Connected Cycles*, IEEE Trans. Parallel Distributed Systems, 4 (Oct. 1993), 1172-1178.

23. L.M. Ni and P.K. McKinley, *A Survey of Wormhole Routing Techniques in Direct Networks*, IEEE Computer, (Feb. 1993), 62-76.

24. F.P. Preparata and J. Vuillemin, *The Cube-Connected Cycles: A Versatile Network for Parallel Computation*, Comm. ACM, 24 (May 1981), 300-309.

25. C.L. Seitz, *Concurrent VLSI Architectures*, IEEE Trans. Comput., 33 (Dec. 1984), 1247-1265.

26. A. Sen, *Supercube: An Optimally Fault-Tolerant Network Architecture*, Acta Informatica, 26 (1989), 741-748.

27. N.-F. Tzeng, H.-L. Chen, and P.-J. Chuang, *Embeddings in Incomplete Hypercubes*, Proc. International Conference on Parallel Processing, IEEE Computer Society, Silver Spring, MD, 1 (Aug. 1990), 335-339.

28. L.D. Wittie, *Communication Structures for Large Networks of Microcomputers*, IEEE Trans. Comput., 30 (April 1981).

29. S.G. Ziavras, *On the Problem of Expanding Hypercube-Based Systems*, Journal Parallel Distributed Computing, 16 (Sept. 1992), 41-53.

30. S.G. Ziavras and M.A. Siddiqui, *Pyramid Mappings onto Hypercubes for Computer Vision: Connection Machine Comparative Study*, Concurrency: Practice Experience, 5 (Sept. 1993), 471-489.

31. S.G. Ziavras, *RH: A Versatile Family of Reduced Hypercube Interconnection Networks*, IEEE Trans. Parallel Distributed Systems, 5 (Nov. 1994), 1210-1220.

32. S.G. Ziavras and D.P. Shah, *High Performance Emulation of Hierarchical Structures on Hypercube Supercomputers*, Concurrency: Practice Experience, 6 (April 1994), 85-100.

33. S.G. Ziavras and M.A. Sideras, *Facilitating High-Performance Image Analysis on Reduced Hypercube (RH) Parallel Computers*, Proc. International Workshop on Parallel Image Analysis: Theory and Applications, College Park, MD (June 1994), 263-191.

Current address: Department of Electrical and Computer Engineering, New Jersey Institute of Technology, Newark, New Jersey 07102

E-mail address: ziavras@hertz.njit.edu

List of Participants

(1) **Dharma P. Agrawal**
Department of Electrical & Computer Engineering
North Carolina State University
Raleigh, NC 27695–7911
E-mail : dpa@ncsu.edu

(2) **Bill Aielo**
Bell Communications Research
435 South Street
Morristown, NJ 07960
E-mail : aiello@bellcore.com

(3) **Fred Annexstein**
Department of Computer Science
University of Cincinnati
Cincinnati, OH 45221
E-mail : annexste@san.uc.edu

(4) **Ioanna Banicescu**
Department of Computer Science
Polytechnic University
Long Island Campus
Route 110
Farmingdale, NY 11735
E-mail : ioana@rama.poly.edu

(5) **Dominique Barth**
LaBRI, Université Bordeaux I
351 Cours de la Liberation
33405 Talence cedex
FRANCE
E-mail : barth@calypso.greco-prog.fr
barth@labri.u-bordeaux.fr

(6) **Jean–Claude Bermond**
I3S CNRS

Université de Nice–Sophia Antipolis
Bat 4, 3 Rue Albert Einstein
Sophia–Antipolis, F-06560 Valbonne
FRANCE
E-mail : bermond@mimosa.unice.fr
bermond@sophia.inria.fr
bermond@diamant.unice.fr

(7) **Sandeep Bhatt**
Bell Communications Research
435 South Street, 2F-371
Morristown, NJ 07960
E-mail : bhatt@bellcore.com

(8) **Marc Breitfeld**
RUTCOR
Rutgers University
P.O. Box 5062
New Brunswick, NJ 08903-5062
E-mail : marc@rutcor.rutgers.edu

(9) **Clay P. Breshears**
107 Ayres Hall
Department of Computer Science
University of Tennessee
Knoxville, TN 37919–1301
E-mail : clay@cs.utk.edu

(10) **John Bruno**
AT&T Bell Laboratories
600 Mountain Ave.
Murray Hill, NJ 07974
E-mail : bruno@cs.ucsb.edu

(11) **William Y. C. Chen**
C–3, Mail Stop B281
Los Alamos National Laboratories
Los Alamos, NM 87545
E-mail : chen@acl.lanl.gov

(12) **Frederic T. Chong**
Artificial Intelligence Laboratory
NE43-613
Massachusetts Institute of Technology
545 Technology Square
Cambridge, MA 02139
E-mail : ftchong@ai.mit.edu

(13) **Philippe Chrétienne**
Tour 55–65 E428

Université Pierre et Marie Curie
4 place Jussieu, 75005 Paris 05
FRANCE
E-mail : chretienne@litp.ibp.fr
phc@litp.ibp.fr

(14) **Fan R. K. Chung**
Bell Communications Research
435 South Street
Morristown, NJ 07960
E-mail : frkc@flash.bellcore.com

(15) **Ed Coffman Jr.**
AT&T Bell Laboratories
600 Mountain Ave.
Murray Hill, NJ 07974
E-mail : egc@research.att.com

(16) **Gene Cooperman**
College of Computer Science, 161N
Northeastern University
Boston, MA 02115
E-mail : gene@ccs.neu.edu

(17) **Michel Cosnard**
Informatique–LIP
ENS–Lyon
46, All e d'Italie
69364 Lyon cedex 07
FRANCE
E-mail : michel.cosnard@lip.ens-lyon.fr

(18) **Robert Cypher**
IBM T. J. Watson Research Center
P.O. Box 218
Yorktown Heights, NY 10598
E-mail : cypher@watson.ibm.com

(19) **Sajal K. Das**
Department of Computer Science
University of North Texas
P.O. Box 13886
Denton, TX 76203–6886
E-mail : das@cs.unt.edu

(20) **Souripriva Das**
Department of Computer Science
Rutgers University
Piscataway, NJ 08854

(21) **Nate Dean**
Bell Communications Research
Room 2M389, 445 South Street
Morristown, NJ 07962
E-mail : nate@bellcore.com

(22) **Ralf Diekmann**
Department of Computer Science
University of Paderborn
D-33095 Paderborn
GERMANY
E-mail : diek@uni-paderborn.de

(23) **Jack Dongarra**
Computer Science Department
University of Tennessee
Knoxville, TN 37996–1301
E-mail : dongarra@cs.utk.edu

(24) **Richard Draper**
Supercomputing Research Center
17100 Science Drive
Bowie, MD 20715
E-mail : draper@super.org

(25) **Ding-Zhu Du**
Department of Computer Science
University of Minnesota
Minneapolis, MN 55455
E-mail : dzd@cs.umn.edu

(26) **José Duato**
Dept. Ingeniera Sistemas, Comp. y Automatica
Facultad de Informatica
Universidad Politecnica de Valencia
P.O.B. 22012, 46071 – Valencia
SPAIN
E-mail : jduato@aii.upv.es

(27) **Dannie Durand**
Bell Communications Research, 2D–335
445 South Street
Morristown, NJ 07960–6438
E-mail : durand@bellcore.com

(28) **Doreen L. Erickson**
Southern College of Technology
1100 South Marietta Parkway
Marietta, GA 30060–2896
E-mail : erickson@sct.du

(29) **Shimon Even**
Computer Science Department
Technion, Israel Institute of Technology
Haifa 32000
ISRAEL
E-mail : even@cs.technion.ac.il

(30) **Vance Faber**
C–3, Mail Stop B281
Los Alamos National Laboratories
Los Alamos, NM 87545
E-mail : vxf@lanl.gov

(31) **Paul Feautrier**
Université de Versailles St–Quentin
45 Avenue des Etats Unis
78035 Versailles Cedex
FRANCE
E-mail : Paul.Feautrier@masi.uvsq.fr

(32) **Afonso Ferreira**
LIP
Ecole Normale Superieure de Lyon
46 Allée dÍtalie, 69364 Lyon Cedex
FRANCE
E-mail : ferreira@lip.ens-lyon.fr

(33) **Chuck Fiduccia**
Supercomputing Research Center
17100 Science Drive
Bowie, MD 20715
E-mail : fiduccia@super.org

(34) **Larry Finkelstein**
College of Computer Science
Northeastern University
Boston, MA 02115
E-mail : laf@ccs.neu.edu

(35) **Pierre Fiorini**
ETCA/CREA
Laboratoire Systeme de Perception
16 bis av. Prieur de la Cote d'Or
94114 Arcueil cedex
FRANCE
E-mail : fiorini@etca.fr

(36) **Pierre Fraigniaud**
Laboratoire de l'Informatique du Parallelisme
ENS–Lyon

69364 Lyon cedex 07
FRANCE
E-mail : pfraign@lip.ens-lyon.fr

(37) **Li-Xin Gao**
Department of Computer Science
University of Massachusetts
Amherst, MA 01003
E-mail : gao@cs.umass.edu

(38) **Luisa Gargano**
Dipartimento di Informatica ed Applicazioni
Universita di Salerno
84081 Baronissi (SA)
ITALY
E-mail : lg@udsab.dia.unisa.it

(39) **Apostolos Gerasoulis**
Department of Computer Science
Rutgers University
New Brunswick, NJ 08903
E-mail : gerasoul@cs.rutgers.edu

(40) **Leslie Goldberg**
Department of Algorithms & Discrete Mathematics
Dept. 1423, P.O. Box 5800
Sandia National Laboratories
Albuquerque, NM 87185–1110
E-mail : lagold@cs.sandia.gov

(41) **Ronald L. Graham**
AT&T Bell Laboratories
600 Mountain Avenue
Murray Hill, NJ 07974
E-mail : rlg@research.att.com

(42) **David Greenberg**
Sandia National Laboratories
Mail Stop 1423, P.O. Box 5800
Albuquerque, NM 87185–5800
E-mail : dsgree@cs.sandia.gov

(43) **Vipul Gupta**
Department of Computer Science
Rutgers University
Piscataway, NJ 08855
E-mail : vgupta@cs.rutgers.edu
vgupta@yoko.rutgers.edu

(44) **Mounir Hamdi**
Department of Computer Science

Hong Kong University of Science & Technology
Clear Water Bay, Kowloon
HONG KONG
E-mail : hamdi@cs.ust.hk

(45) **Rabah Harbane**
Laboratoire de Recherche en Informatique
Bat. 490, Université de Paris–Sud, Orsay 91405
FRANCE
E-mail : harbane@lri.lri.fr

(46) **D. Frank Hsu**
Department of Computer & Information Science
Fordham University
Bronx, NY 10458–5198
E-mail : hsu@murray.fordham.edu

(47) **Susan Flynn Hummel**
Computer Science Department
Polytechnic University
Six MetroTech Center
Brooklyn, NY 11201
E-mail : hummel@mono.poly.edu

(48) **M. Roberto Ito**
Department of Electrical Engineering
University of British Columbia
Vancouver, B.C. V6T 1Z4
CANADA
E-mail : mito@ee.ubc.ca

(49) **Ravi Jain**
Bell Communications Research
445 South Street
Morristown, NJ 07960
E-mail : rjain@thumper.bellcore.com

(50) **Chris Jesshope**
Department of Electronics & Electrical Engineering
University of Surrey
Guildford, Surray GU2 5XH
GREAT BRITAIN
E-mail : c.jesshope@ee.surrey.ac.uk

(51) **Xing-De Jia**
Department of Mathematics
Southwest Texas State University
San Marcos, TX 78666
E-mail : jia@xingde.swt.edu
xjo1@academia.swt.edu

(52) **Jia Jiao**
Department of Computer Science
Rutgers University
BPO 23869, P.O. Box 1119
Piscataway, NJ 08855–1119
E-mail : jjiao@paul.rutgers.edu

(53) **David Johnson**
AT&T Bell Laboratories, Room 2D–150
600 Mountain Avenue
Murray Hill, NJ 07974
E-mail : dsj@research.att.com

(54) **Nabil Kahale**
DIMACS
Rutgers University
Piscataway, NJ 08855-1119
E-mail : kahale@dimacs.rutgers.edu

(55) **Luděk Kučera**
Department of Applied Mathematics
Charles University
Malostranske nam.25
118 00 Prague
CZECH REPUBLIC
E-mail : ludek@kam.ms.mff.cuni.cz
kuceral@cspguk11.bitnet

(56) **Tom Leighton**
Department of Mathematics
Massachusetts Institute of Technology
Cambridge, MA 02139
E-mail : ftl@math.mit.edu

(57) **Peiyuan Li**
CAIP Center, Busch Campus
Rutgers University
Piscataway, NJ 08855–1390
E-mail : pli@caip.rutgers.edu
pli@yinyang.rutgers.edu

(58) **Qiao Li**
Department of Mathematics
University of Science & Technology of China
Hefei, Anhui 230026
CHINA

(59) **Art Liestman**
School of Computer Science
Simon Fraser University

Burnaby, BC V5A 1S6
CANADA
E-mail : art@cs.sfu.ca

(60) **Pangfeng Liu**
DIMACS
Rutgers University
Piscataway, NJ 08855
E-mail : pangfeng@dimacs.rutgers.edu
pangfeng@bellcore.com

(61) **Zhen Liu**
INRIA Centre Sophia – Antipolis
2004 Route des Lucioles
B.P. 93, 06902 Sophia – Antipolis
06560 Valbonne Cedex
FRANCE
E-mail : zhen.liu@mirsa.inria.fr
liu@sophia.inria.fr

(62) **Yuh-Dauh Lyuu**
NEC Research Institute
4 Independence Way
Princeton, NJ 08540
E-mail : lyuu@research.nj.nec.com

(63) **Eva Ma**
Department of Computer Science
Rensselaer Polytechnic Institute
Troy, NY 12180–3590
E-mail : eva@cs.rpi.edu

(64) **Teruo Matsuzawa**
Center for Information Science
Japan Advanced Institute of Science & Technology
15 Asahidai
Tatsunokuchi, Ishikawa 923–12
JAPAN
E-mail : matuzawa@jaist.ac.jp

(65) **Graham M. Megson**
Department of Computing Science
University of Newcastle–Upon–Tyne
Claremont Tower, Claremont Rd
Newcastle–Upon–Tyne, NE1 7RU,UK
E-mail : Graham.Megson@newcastle.ac.uk

(66) **Zevi Miller**
Department of Mathematics & Statistics
Miami University

Oxford, OH 45056
E-mail : zmiller@miavx1.acs.muohio.edu

(67) **Burkhard Monien**
Fach. Informatik
University of Paderborn
Warburger Str. 100, FB17
D-33098 Paderborn
GERMANY
E-mail : bm@uni-paderborn.de

(68) **Marco Morazan**
120 West 44th Street, Apt. 1611
New York, NY 10036
E-mail : csmtm@csfaculty.ccny.cuny.edu

(69) **Felix P. Muga II**
Department of Mathematics
Ateneo de Manila University
P.O. Box 154 Manila 0917
PHILIPPINES
E-mail : fpmuga@mathsci.admu.edu.ph

(70) **Andrew Odlyzko**
AT&T Bell Laboratories, Room 2C–355
600 Mountain Avenue
Murray Hill, NJ 07974
E-mail : amo@research.att.com

(71) **Sabine R. Oehring**
Lehrstuhl Informatik I
University of Wuerzburg
Am Hubland, D-97074 Wuerzburg
GERMANY
E-mail : oehring@informatik.uni-wuerzburg.de

(72) **Petrisor Panaite**
LRI, Université Paris–Sud
Centre d'Orsay–Bât 490
91405 Orsay Cedex
FRANCE
E-mail : petr@lri.lri.fr

(73) **Aleksandar Pekec**
Department of Mathematics
Hill Center – Busch Campus
Rutgers University
New Brunswick, NJ 08903
E-mail : pekec@math.rutgers.edu

(74) **James Philbin**

NEC Research Institute
4 Independence Way
Princeton, NJ 08540
E-mail : philbin@research.nj.nec.com
(75) **Marc Picquendar**
Department of Computer Science
University of Massachusetts
Amherst, MA 01003
E-mail : picquend@cs.umass.edu
(76) **Dan Pritikin**
Department of Mathematics & Statistics
Miami University
Oxford, OH 45056
E-mail : dpritikin@miavx1.acs.muohio.edu
(77) **Donna Quammen**
Computer Science Department
George Mason University
Fairfax, VA 22030
E-mail : quammen@cs.gmu.edu
(78) **Sanguthevar Rajasekaran**
Department of Computer & Information Science
University of Pennsylvania
Philadelphia, PA 19104
E-mail : raj@central.cis.upenn.edu
(79) **Rajeev Raman**
UMIACS
A.V. Williams Building
University of Maryland
College Park, MD 20742
E-mail : raman@umiacs.umd.edu
(80) **Keith H. Randall**
NE43–336, Laboratory for Computer Science
Massachusetts Institute of Technology
545 Technology Square
Cambridge, MA 02139
E-mail : randall@theory.lcs.mit.edu
(81) **Satis Rao**
NEC Research Institute
4 Independence Way
Princeton, NJ 08540
E-mail :
(82) **Andrew Rau–Chaplin**
Room 428 CoRE Bldg., DIMACS

Rutgers University
Piscataway, NJ 08855
E-mail : arc@dimacs.rutgers.edu

(83) **Dana Richards**
National Science Foundation, Room 1145
4201 Wilson Blvd.
Arlington, VA 22230
E-mail : richards@nsf.gov

(84) **Igor Rivin**
School of Mathematics
Institute for Advanced Study
Princeton, NJ 08540
E-mail : igor@math.ias.edu

(85) **Daniel Rockmore**
Department of Mathematics & Computer Science
Sudikoff Laboratory
Dartmouth College
Hanover, NH 03755
E-mail : rockmore@gilligans.dartmouth.edu

(86) **Arnold Rosenberg**
Department of Computer Science
University of Massachusetts
Amherst, MA 01003
E-mail : rsnbrg@cs.umass.edu

(87) **Robert Rowley**
Department of Computer Science
Oregon State University
Corvallis, OR 97331
E-mail : rowley@cs.orst.edu

(88) **Larry Rudolph**
Department of Computer Science
The Hebrew University of Jerusalem
Jerusalem 91904
ISRAEL
E-mail : rudolph@cs.huji.ac.il

(89) **Isaac Saias**
MS M986 Marimac Building, C–3
Los Alāmos National Laboratories
Los Alamos, NM 87545
E-mail :
Laboratory for Computer Science
Massachusetts Institute of Technology
545 Technology Square

Cambridge, MA 02139
E-mail : saias@theory.lcs.mit.edu

(90) **Majd F. Sakr**
NEC Research Center
4 Independence Way
Princeton, NJ 08540
E-mail : sakr@research.nj.nec.com

(91) **Vittorio Scarano**
Department of Computer Science
University of Massachusetts
Amherst, MA 01003
E-mail : vittorio@cs.umass.edu

(92) **Eric Schwabe**
Electrical Engineering & Computer Science Department
Northwestern University
Evanston, IL 60208–3118
E-mail : schwabe@eecs.nwu.edu

(93) **Jacob Shapiro**
Department of Statistics & Computer Science
Baruch College
City University of New York
P.O. Box 513, 17 Lexington Ave.
New York, NY 10010
E-mail : jacbb@cunyvm.cuny.edu

(94) **Balaram Sinharov**
POB 950, MS P411, IBM
Poughkeepsie, NY 12590
E-mail : balaram@vnet.ibm.com

(95) **Ramesh Sitaraman**
Department of Computer Science
Princeton University
Princeton, NJ 08544
E-mail : rks@princeton.edu

(96) **Siang Wun Song**
Department of Computer Science
University of São Paolo
Rua do Matão 1010
05508–900 São Paolo, SP
BRAZIL
E-mail : song@ime.usp.br

(97) **Dominique Sotteau**
LRI, Bat 490
Université de Paris–Sud

91405 Orsay Cedex
FRANCE
E-mail : sotteau@lri.fr
dominique.sotteau@lri.fr

(98) **Chitoor V. Srinivasan**
Department of Computer Science
Hill Center, Busch Campus
Rutgers University
New Brunswick, NJ 08903
E-mail : srinivas@cs.rutgers.edu

(99) **Aravind Srinivasan**
School of Mathematics
Institute for Advanced Study
Princeton, NJ 08540
E-mail : aravind@math.ias.edu

(100) **Quentin F. Stout**
EECS Department
University of Michigan
Ann Arbor, MI 48109–2122
E-mail : qstout@eecs.umich.edu

(101) **Hal Sudborough**
Computer Science Program, EC3.1
The University of Texas at Dallas
Richardson, TX 75083–0688
E-mail : hal@utdallas.edu

(102) **R. Swaminathan**
Department of Computer Science
University of Cincinnati
822 Old Chemistry Building
Cincinnati, OH 45221–0008
E-mail : swamy@jupiter.csm.uc.edu

(103) **Thanasis Tsantilas**
Department of Computer Science
Columbia University
New York, NY 10027
E-mail : thanasis@cs.columbia.edu

(104) **Theodore Vassiliades**
851 S. Morgan
Department of Mathematics, Statistics & Computer Science
University of Illinois
SEO 302, M/C 249
Chicago, IL 60607–7045
E-mail : tvassil@wagner.lac.math.uic.edu

(105) **Rakesh M. Verma**
Computer Science Department
University of Houston
Houston, TX 77204
E-mail : rmverma@cs.uh.edu

(106) **Pearl Y. Wang**
Computer Science Department, MS 4A5
George Mason University
Fairfax, VA 22030–4444
E-mail : pwang@cs.gmu.edu

(107) **Douglas B. West**
Mathematics Department
University of Illinois
1409 West Green Street
Urbana, IL 61801
E-mail : west@math.uiuc.edu

(108) **Susanne Wetzel**
AT&T Bell Laboratories, Room 2D-203
P.O. Box 636
600 Mountain Avenue
Murray Hill, NJ 07974–636
E-mail : wetzel@research.att.com

(109) **Ethan Wolf**
Laboratory For Computer Science, NE43–309
Massachusetts Institute of Technology
Cambridge, MA 02139
E-mail : elwolf@math.mit.edu

(110) **Mihalis Yannakakis**
AT&T Bell Laboratories
600 Mountain Avenue
Murray Hill, NJ 07974
E-mail :

(111) **Louxin Zhang**
Department of Computer Science
University of Waterloo
Waterloo, Ontario N21 3G1
CANADA
E-mail : lzhang@neumann.uwaterloo.ca

(112) **Hong Zhu**
Department of Computer Science
Fudan University
Shanghai 200433
CHINA

E-mail : dingjia.xin@cnl.fudan.crn.cn

(113) **Sotirios Ziavras**
Department of Electrical & Computer Engineering
New Jersey Institute of Technology
University Heights
Newark, NJ 07102
E-mail : ziavras@hertz.njit.edu

(114) **Xiaoling Zuo**
Computer Science Division
University of California
517 Evans Hall
Berkeley, CA 94720